Woodlot Biodiversity
2nd Edition

Editor
Dr Steven Newmaster

Authors
Dr Steven Newmaster
Chris Earley
Dr Aron Fazekas
Carole Ann Lacroix
Troy McMullin
Brian Lacey
Thomas Henry
Jose Maloles
Dr Subramanyam Ragupathy
Kevan Berg
Peter Williams

Illustrations
Nicole Daoust & Patricia Beader

Biodiversity Institute of Ontario (BIO) Herbarium
University of Guelph

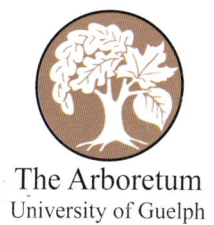

The Arboretum
University of Guelph

© 2013 Steven Newmaster

Printed in Canada

All rights reserved. No part of this publication may be reproduced, stored in a retrieval system or data base, or transmitted, in any form or by any means, without prior written permission of the publisher, or in the case of photocopying or other reprographic copying, a license from Access Copyright, The Canadian Copyright Licensing Agency, Toronto, Ontario (Access Copyright). This license excludes certain photographs licensed under a Creative Commons license (see pp. 560-561).

Publisher: Newmaster
First Edition published in 2010

Library and Archives Canada Cataloguing in Publication

Woodlot biodiversity / editor, Steven G. Newmaster; authors, Steven G. Newmaster ... [et al.]. -- 2nd ed.

Includes bibliographical references and index. ISBN 978-0-9866554-1-8

1. Woodlots--Ontario, Southern. 2. Biodiversity--Ontario, Southern.

I. Newmaster, Steven G., 1967-

SD387.W6W665 2013   634.9'909713   C2013-900999-X

Editor: Steven G. Newmaster
Cover Photo: Joan Riemer
Design: Brian Lacey, Thomas Henry & Jose Maloles
Graphics: Brian Lacey & Jose Maloles
Illustrations: Nicole Daoust & Patricia Beader
Printing: Friesens Corporation, Altona Manitoba, Canada

Contents

Foreword	5
Dedication	6
Acknowledgements	6
Woodlot Biodiversity	**9**
Woodlots in Southern Ontario	9
Woodlot Biodiversity	9
Structural Diversity	10
Spatial Variation in Woodlots	11
Good and Bad Biodiversity	12
Temporal Variation in Woodlots	14
Agricultural History and Biodiversity	14
Woodlots are Supermarkets for Wildlife	15
Woodlot Ethnoecological Diversity	16
Exploring Wootlot Diversity	**18**
Exploring Alpha Diversity	18
Tree Thinking – Exploring Variation in Species Traits	21
Exploring Beta Diversity	22
Cryptic Botany	**25**
Introduction	25
BRYOPHYTES	25
Bryoflora Diversity	25
What is a Bryophyte?	25
Where do Bryophytes Live?	26
The Importance of Microhabitats	27
Ecological Role of Bryophytes in Woodlots	28
Exploring Woodlot Bryodiversity	28
Liverworts (Hepatics)	29
Mosses (Musci)	37
FERNS AND FERN ALLIES	51
Fern Diversity	51
What is a Fern?	51
Fern Ecology	52
Club-mosses (Lycopodiaceae)	53
Horsetails and Scouring Rushes (Equisetum)	60
Ferns (Pteridophyta)	65
Important Fern Characters	66
The Key to Floral Diversity	**101**
Introduction	101
The Basic Parts of a Flower	101
Regular/Irregular Flowers	103
Counting Floral Parts	103

Sexual Orientation	103
Key to Flowering Plant Families	105
Index to Angiosperm Families	107

Woody Plants 109
Introduction 109
Winter Key to Woody Plants 110
Summer Key to Woody Plants 123
Silhouettes of Some Common and Unique Woody Plants 136
Species Profiles 140

Woodlot Herbs 371
Introduction 371
An Identification Key to Woodlot Herbs 372
Species Profiles 375

Woodlot Lichens 427
Introduction 427
What is a Lichen? 429
Lichen Structure 430
Lichen Reproduction 430
Lichen Language 431
Go Explore 431
Lichens and Allied Fungi of the Arboretum 431
The Big Six 432
An Identification Key to the Lichens of the Arboretum 433
Species Profiles 438

Woodlot Birds 475
Introduction 475
Identification 475
When Is Each Species in Our Area? 475
Anatomy 476
Predators 477
Other Birds 480

Woodlot Mammals 515
Introduction 515
Tracks 515
Species Profiles 518

Glossary 542

Bibliography 554

Photo Credits 559
Photos Licensed Under Creative Commons 560

Index 562

The Authors 578

Foreword

Alan Watson

This Field Guide to *Woodlot Biodiversity* offers an excellent tool to assess the diversity of the biota in forested areas of southern and central Ontario. The use of numerous authors with specific expertise in biodiversity of forests and the various taxa of flora and fauna assures the accuracy of the guide.

It is important to be able to identify the various plants and animals that one finds in an area in order to determine its biodiversity. What to the uninitiated may seem like a forest with a few different species, upon closer inspection, may be an area that is rich in the variety of its species. However, without a way to help in the identification of the species encountered the task of learning about the details of a woodlot is difficult. This guide, with its chapters on understanding and exploring diversity, keys to floral diversity and the identification of hundreds of plants and animals will be invaluable in learning about the biotic elements of the forest.

It is widely accepted that the more one learns about a landscape, the more the landscape is valued. Experiencing and learning about the flora and fauna of a forest, will help lead to a greater appreciation of the forested landscape. Appreciation is an important step in the process of conservation since an argument for a woodlot's existence can be based upon the importance of the biotic components that the woodlot contains. This field guide, *Woodlot Biodiversity*, will help you in one of the first steps of appreciation and conservation: identification.

Dedication

This book is dedicated to the next cohort of University of Guelph Students. It is written by University of Guelph staff, students, faculty, and alumni who gained this knowledge while studying woodlots with previous instructors and professors here on campus at the University of Guelph. There is a strong tradition of exploring biological diversity here at this University and this book is only part of the voyage – diligently follow the compass, observe and enjoy having the freedom in exploring seas of biodiversity knowledge.

Acknowledgements

We are grateful to all of the professors and instructors at the University of Guelph who taught us how to explore biological variation on the landscape and often inspired us out in the Arboretum. We would like to thank all of the students (many students from BIOL 1070, BOT 2030, BOT 2050, and BOT 3710) who helped us explore woodlot diversity at the University of Guelph Arboretum and other woodlots in southern Ontario. Several people deserve special mention as they collected data and helped to review early drafts of the book: Sean Fox, Joan Riemer, Royce Steeves, Jillian Bainard, Alan Watson, Sean Rapai, Lizbeth Elias, Lyndsay Schram, Stephanie Lyons, Natasa Nadj, Annabel Newmaster, Neil Webster, and Wayne Bell. Many thanks to Theodore Esslinger and Richard Harris for assistance with the identification of some lichen species, and to Roman Olszewski for reviewing the chapter on lichens. We are grateful to all of the Arboretum staff and volunteers who have collected biodiversity data over the years.

Many photographers have graced this book with their talents and we are grateful for their generosity in sharing the images of biodiversity in this book. Complete photo credits are found near the end of this book before the index. The artistic cover and many of the full page insets were produced by the talented Joan Riemer – thank you!

Support for this project was provided by the Gosling Foundation, William Girling Environmental Education Endowment, Reid's Heritage Homes, The Arboretum at the University of Guelph, and the Biodiversity Institute of Ontario Herbarium.

Woodlot Biodiversity

Steven Newmaster, Peter Williams, Kevan Berg & Subramanyam Ragupathy

Woodlots in Southern Ontario

Most of Ontario is forested (Fig. 1). Although most of these forests are in northern Ontario, southern Ontario still has over 2.4 million hectares of forest. These southern forests are found in relatively small (80% are less than 3 ha) fragmented woodlots within a landscape of agricultural and urban areas that are home to more than 90% of Ontario's 10 million residents. Settlement history and agricultural development have resulted in a landscape that has been widely cleared with scattered woodlots remaining on sites too poor for agriculture. Today, residents of southern Ontario own 87% of these woodlots with the remaining woodlots owned by public agencies such as counties or conservation authorities as conservation land. Most (84%) of these residents highly value the presence of woodlots in urban and rural areas and would like to see more forests in southern Ontario. In response, the province has developed forest restoration plans that have converted over 130,000 hectares of abandoned agricultural lands into forests on Crown land, and a comparable amount of new forests have been established on private lands through agreements with landowners.

Figure 1. Ecosystem Area by Land Class in Ontario

Woodlot Biodiversity

Woodlot biodiversity is considerable and provides habitat for wildlife that inhabit forest ecosystems. Biodiversity in an ecosystem is defined as the variation in organismal diversity from soil microbes to white tailed deer. The number of species (species richness) within an ecosystem is one measure of biodiversity – more species means greater biodiversity, fewer species means lower biodiversity. We often think about the number of woody plant species in a forest without considering how many other plants occupy this habitat. Bryophytes (mosses and liverworts) and lichens alone double the species diversity of woody plants in a woodlot (Fig. 2). At the University of Guelph Arboretum we have sampled woodlot diversity including over 450 species that have been included in this field guide.

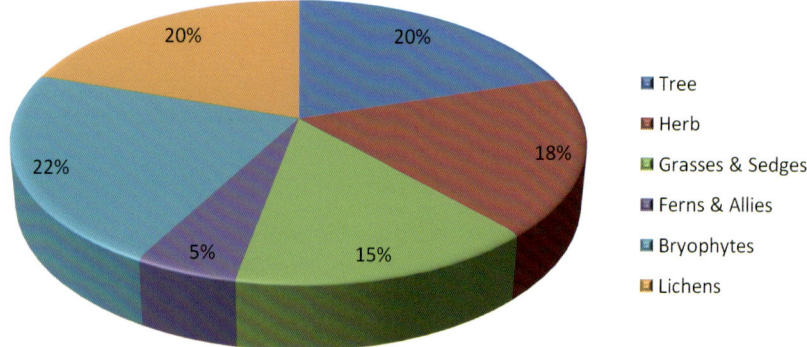

Figure 2. Woodlot biodiversity

Structural Diversity

Structural diversity can vary considerably in woodlots. The structure of a forest community refers to the vertical arrangement and spatial organization of the plants. The forest is made up of layers of vegetation including: L1 – dominant trees, L2 – subdominant trees, L3 – 2.0 to 10.0 m trees and shrubs, L4 – 0.5 to 2.0 m trees and shrubs, L5 – < 0.5 m trees and shrubs, L6 – herbs (broad-leaved herbaceous plants), L6 – grasses, sedges, and rushes, and L7 – bryophytes and lichens. The vertical layering is often dominated by one or more layers and is a consequence of variation in the growth form of plants that are in high abundance within a dominant layer. The growth form of a community is referred to as its physiognomy. A woodlot with little understory shrubs and herbs is dominated by a tree physiognomy (Fig. 3), whereas a woodlot with many shrubs is dominated by a shrub physiognomy (Fig. 4). The physiognomy of a woodlot is defined by the dominant vegetation layer and most abundant species, e.g. tall shrub physiognomy with dogwood species.

Figure 3. Tree physiognomy

Figure 4. Shrub physiognomy

Plants have traits that allow them to dominate particular layers. For example, herbs that dominate in the understory may be shade tolerant, i.e. they have the ability to photosynthesize efficiently at low light levels. The physical traits of shade tolerant plants may include broad, thin leaves and high levels of foliar nutrients. Variation in physiognomy or the number of vegetation layers in a woodlot reflects the character of the physical environment. Typically more humid environments have more layers, e.g. swamps often have thick layers of shrubs and trees. Disturbance and competition can also dramatically influence woodlot physiognomy. A windstorm can open the canopy so that it has an herb physiognomy with species that are more competitive with more light. An aggressive invasive such as common buckthorn or garlic mustard can change the community structure by eliminating native species of shrubs, herbs and bryophytes.

Spatial Variation in Woodlots

The biodiversity concept embodies spatial variation. The number of species within local woodlot is measured at a local scale (alpha diversity); scale can be decreased to microhabitats (point diversity) or increased to include multiple woodlots in a township at a broader landscape scale (gamma diversity). Regional biodiversity considers a much larger area like the Carolinian zone or southwestern Ontario. Therefore we must consider spatial biodiversity in an ecological context, which includes the number, size, and arrangement of habitats within a woodlot at a local scale, and the distribution of these ecosystems across the landscape at a regional scale (southern Ontario). There are several spatial species distributions to consider:

1) Locally and regionally common – white-tailed deer (see pg. 532) typically inhabit several different types of local woodlots, which is considered a local scale of diversity in which the organism is abundant. White-tailed deer are commonly found in woodlots throughout the province of Ontario and are therefore also abundant at regional scales.

2) Locally rare, regionally common – several small species of liverworts in the genus *Lophozia* (see pg. 33) only inhabit logs on which they are found in low abundance (locally rare). The distribution of logs in a woodlot is important for the conservation of these liverworts at local scales, but at regional scales most of these species are commonly found within any particular woodlot that has enough logs (regionally common).

3) Locally common, regionally rare – the lichen *Xanthoria parietina* (see pg. 471) is a rare woodlot species that is locally common in the Guelph Arboretum. This lichen is abundant on deciduous trees throughout the Arboretum, but is regionally rare as it is only known to occur in woodlots near Guelph.
4) Locally and regionally rare – many stubble lichens (see pg. 455) are only found on old-growth trees (locally rare) in woodlots that have not been disturbed for over 100 years and are not near large cities with considerable pollution (regionally rare).

Woodlot biodiversity must be considered at multiple scales on the landscape with respect to phenology. Phenology is the study of periodic plant and animal life cycle events (e.g., flowering, breeding, and migration) and how these are influenced by seasonal and annual variations in climate. A simple count of species records diversity (species richness) at one point in time. Individual plants do not move around woodlots, but there is a phenological response to seasonal variation in climate. Many woodlot wildflowers are spring ephemerals (e.g. trout lily, pg. 388; hepatica, pg. 393; bloodroot, pg. 406; coltsfoot, pg. 414): the aerial parts (i.e. stems, leaves, and flowers) develop early each spring and then quickly bloom, set seed, and die back to underground parts (i.e. roots, rhizomes, and bulbs) for the remainder of the year. Other plants only flower late in the season such as asters and witch-hazel (see pg. 217). Animals (insects, mammals, birds etc.) can move within and among woodlots at local and regional scales (e.g. migratory birds and animals). Consideration of an organism's range will influence the scale at which to explore woodlots diversity. Animals may only occupy woodlots at a particular time of day for feeding or a particular time of year for breeding or hibernation. A good strategy for exploring woodlot diversity will consider the life habits of the organisms you wish to study with respect to phenology and spatial scale. Several visits should be made throughout the seasons within the appropriate scale on the landscape.

Good and Bad Biodiversity
Biodiversity is often confused with the quality of an ecosystem. Local (alpha) biodiversity is an index of ecosystem complexity, but a high or low count of species may or may not be a good thing depending on the ecosystem and the organisms. Some of the most important undisturbed or fragile ecosystems have low biodiversity, usually because they occur at ecological extremes (e.g., too hot, too cold, too wet). Woodlots with good soils will generally have a high biodiversity, but a good portion of that may be undesirable. Some very productive sites may have low diversity and large trees that are young because they have grown quickly in fertile soils. It is also important to consider how ecosystems are represented in a local area or region (e.g., acreage, size, number, clustering). For example, an escarpment cliff face (mesohabitat) may cover little acreage, but adds important biodiversity regionally because there are many species that are restricted to this mesohabitat. An ecosystem that occupies a large proportion of the landscape is also important because it provides core habitat for many species. A high number of species in a woodlot can be a good thing, but sometimes high species richness reflects poor ecosystem health. For example, if 30% of the plant species in a woodlot are invasive exotics (weeds) that occupy 80% of the ground cover that woodlot may have high biodiversity but poor ecosystem health. This can be a little confounding if the canopy trees are healthy, vigorous, and diverse, but the understory is dominated by invasive alien species such as common buckthorn and garlic mustard.

Little is known about the effects of invasive and alien species in woodlots. Considerable damage can be caused by some invasive species such as the emerald ash borer. Over time, invasive alien species (IAS) can critically influence forest ecology. Nightcrawlers (worms) that were brought from Europe dominate the forest floor fauna in many woodlots. By eating almost all the leaf litter, these worms have caused huge shifts away from the natural upland forest floor with thicker humus or organic layers to forest floors that are mostly mineral soil. The exposed mineral soil makes it easier for invasive alien species like garlic mustard to become established and dominate the understory. In woodlots we do not have a comprehensive database for invasive or alien species. Some plants are invasive within ecosystems causing shifts in community structure; invasive species include dog-strangling vine, garlic mustard, Norway maple, common buckthorn, giant hogweed, Japanese knotweed, and exotic honeysuckles. Others plants are alien species that are not native to the province, but do inhabit woodlots and may compete for resources with native flora. However, many of the alien species have been here since the arrival of early settlers and they do not invade and dominate ecosystems. Approximately 28% of the Ontario flora is made up of alien species of which less than 20% are known to establish in woodlots (Fig. 5). This is much greater than the number of alien species found in native forests or forests disturbed by forest fire. It is not known if these species are causing changes in woodlot biodiversity as there is little research despite the call for action from woodlot owners. Furthermore, distribution ranges for many IAS are unknown, and there is no protocol to detect and assess the spread of IAS. The province enacted the Ontario Weed Control Act (1990) to control the spread of specific invasive plants that are detrimental to agricultural production but there is no equivalent legal instrument for the control of IAS in non-agricultural ecosystems. The Crown Forest Sustainability Act (1994) provides for healthy and productive ecosystems, not the control of particular IAS. But invasions are a natural process and ecosystems are always changing. It is important to assess biodiversity by considering whether the biodiversity is good or bad for the ecosystem and whether it is likely to improve or degrade over time.

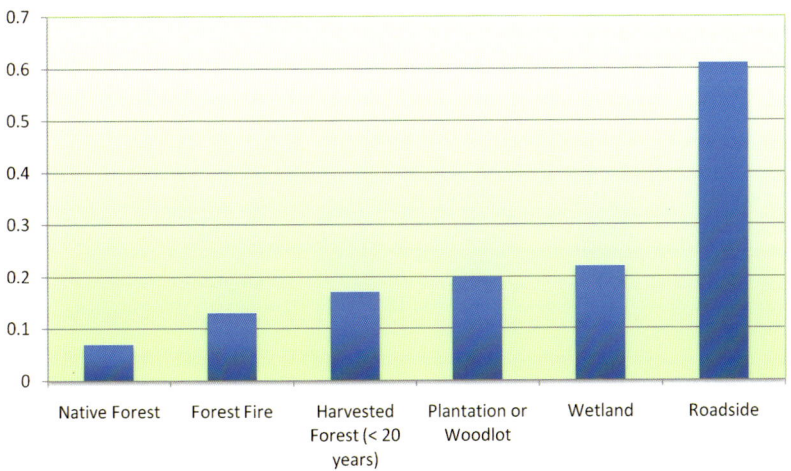

Figure 5. Percentage of Ontario's alien species known to occur in various habitats

Temporal Variation in Woodlots

Today's woodlot biodiversity is a snapshot in time. The woodlot community changes with time and this temporal variation is often referred to as succession. Succession is defined as predictable and orderly changes in the composition or structure of an ecological community through time. It may be initiated by the formation of new unoccupied habitat (e.g. bare rock following mining) or by disturbance (e.g. wind damage, fire, logging) within in a woodlot. The course of community change is influenced by environmental variables, site conditions, and interactions with the species present, and also by more stochastic factors such as availability of colonists or seeds. Communities in early succession will be dominated by fast-growing, well-dispersed species (opportunist, fugitive, or r-selected life-histories), followed by more competitive (k-selected) species that dominate the ecosystem. For example, lichens and mosses start to grow on bare rocks, providing a source of organic debris that in time will become a seed bank for grasses, goldenrods, shrubs, and eventually trees that dominate the site by shading out the early colonizers. Each stage in community assemblage changes the site, creating conditions that favour another suite of species. Although this appears to be a deterministic linear pathway it may move back several stages in community structure or completely change into another ecosystem type following a disturbance or a planned forest management intervention. For example, disturbances in a woodlot (e.g. fire, windthrow, decline in a dominant species, logging, grazing), create new conditions that favour invasion of new species of plants from available seed sources. This changes the community structure, setting a new successional pathway determined by the combination of site conditions (e.g. temperature, fertility, moisture, light availability) and suitable plant traits (e.g. light tolerance, seed dormancy) of the new community.

Agricultural History and Biodiversity

Most of the woodlots in southern Ontario have been disturbed in the past. Since the mid-1800's, European settlers have cleared and cultivated or pastured almost all of southern Ontario. Understanding historical land use is useful in interpreting the biodiversity present in most woodlots today. Since the 1900's, land degradation, changes in farm technology, the economy, government policy, and society have caused many areas to be abandoned by agriculture. These are often left to develop back into forest or have been replanted to trees. The biggest changes have been in the abandonment of degraded or marginal cultivated areas since 1900 and the reduction in livestock grazing pressure since the 1950's. The shift from heavy grazing of most natural areas to fenced pasture and feedlot systems has resulted in widespread improvement in woodland health and regeneration, and gradual movement from rough pasture to forest. Many woodlots now have areas that were likely grazed but never actually cleared for agriculture because it was too dry, too rocky, too steep, or otherwise inconvenient to farm. As farming pressure was reduced on adjacent areas, patches of various sizes returned to forest through old field succession or by tree planting. Old field succession refers to a change in community structure after a cultivated field, pasture, clearing, or roadside has been abandoned or left undisturbed. Many of these areas have good soils and available moisture, encouraging the growth of many plants (e.g. grasses, forbs, shrubs, or trees). Annual and perennial grasses and herbs often invade first and after a few years an abandoned field will become a meadow dominated by perennial grasses, goldenrods, and asters. Some early successional stages with perennial grasses, shrubs, and some trees can dominate a site for many years by creating dense root systems, chemicals (allelopathy) that affect other plants, and/or dense shade. Some of these communities will keep other plants from becoming established unless they establish from the edges of the woodlot or there is some disturbance (e.g. drought or wet weather, fire, ploughing, disease). This has led to a situation where woodlots are often structurally complex small stands of trees on fairly small properties. Patches of trees have resulted from where a corner of a field was left uncultivated, a pasture was abandoned, or where an old field was replanted with trees. Each of these small patches of trees may have limited biodiversity, but when considered together or left to expand they provide a complex and diverse ecosystem that provides habitat for many organisms.

Grazing by livestock or wildlife can influence woodlot diversity. These animals can heavily influence pioneering species during old field succession. Preferential browsing of species (e.g. maple and oak are preferred browse species, ash and cedar are not) often changes the community structure of a developing woodlot. The historical effects of livestock grazing will diminish over time, but pressure from wildlife (e.g. deer and rodents) can exert a continuing dramatic influence. This is particularly obvious in and near areas where deer populations are high, often protected from hunting. An overgrazed/browsed mature woodlot usually has low species diversity in the understory and the absence of some key overstory species that are favoured for browsing and bark chewing or sensitive to compaction and/or root damage. Grazing and browsing pressure can cause the community to shift to a suite of species that are tolerant of browsing. For example, with little browsing, a hardwood forest with poplar, white pine, maples, oaks, and other species often develops, whereas more intense browsing often results in a young forest that is dominated by white cedar with scattered cherry or ash (hardwoods that are less desired for browsing). The resulting dense cedar stand will prevent the establishment of a mixed forest until the cedar is removed by logging, fire, storm damage, or senescence. The most obvious sign of overgrazing and browsing in a woodlot is the presence of a browse line, a distinct line of foliage below which almost all the vegetation has been eaten. Grazing and browsing animals often spread invasive or alien species or create disturbance that encourages the establishment of such species.

Woodlots are Supermarkets for Wildlife

Woodlot plants provide food for wildlife. Many animals have been observed eating leaves, flowers, fruit, or even entire plants within woodlots. Although these accounts are incidental and qualitative we have included any existing records within this guide for each species of bird or mammal. At the Biodiversity Institute of Ontario (BIO) Herbarium we are developing a quantitative approach to understanding what plants are consumed by different species of animals. This began with the creation of a DNA barcoding library of plants for the Flora of Ontario. We have already barcoded over 1200 species of the 4800 known species of plants in the province. DNA barcoding is an identification tool that uses a short genetic marker in an organism's DNA to build a comprehensive DNA barcode library. This library serves as a reference to which new unknown DNA samples can be matched and given a species name. There are many useful applications of DNA barcoding that include identifying plant leaves, roots, or seeds even when flowers or fruit are not available, identifying cryptic plant products in commerce (e.g. wood or powdered leaves in herbal supplements), or identifying the diet of an animal based on stomach contents or feces. Currently we are working with students here at the University of Guelph to barcode plant fragments found within animal scat in order to understand what a particular animal ate that day. We can recreate the dietary diversity for a particular animal and relate this to the plant diversity and nutrition available in woodlots. This will help us understand the value of woodlots as supermarkets for wildlife.

Woodlot Ethnoecological Diversity

All humans have the ability to know and remember their local landscape and environment, and to communicate this understanding to others. The extent to which we do this ranges from the way we develop informal local knowledge about our backyard woodlot or our favourite hiking trail to the formal study and communication of scientific knowledge. In fields such as botany, ecology, and geography, we understand and communicate knowledge of our environment according to empirically based perceptions and classifications. This scientifically ordered system of knowing is an astounding accomplishment that helps us in industrialized societies to make reliable predictions and decisions about ecosystems and land management. However, the scientific approach also has its limitations. An alternative and perhaps complimentary approach can be found in the ways in which keepers of traditional ecological knowledge (TEK) learn about and understand the environment.

Traditional ecological knowledge is a finely tuned yet adaptive form of knowledge about the environment that is acquired through extensive observation of a species or an area. It is an ever-evolving body of knowledge, practice, and belief about the interconnected relationships of living beings and their environment that is collected over many generations by people who depend on this information for their livelihood. Typically, TEK is found within indigenous societies around the world, but not exclusively so, as there are many examples of nonindigenous groups that also hold TEK. More generally, TEK is an ecological knowledge system that accrues within any nonindustrial or rural subsistence-oriented society that exhibits historical continuity in the use of natural resources. It is shared among users of a resource in an oral tradition or through shared practical experiences, and in this way it grows and changes incrementally, tested and retested by trial and error as it is handed down through generations.

As a way of knowing, TEK is similar to science in that it is based on a cumulative body of observations that enable people to understand and make predictions about the environment. Nonetheless, there are also important differences between the two forms of knowledge. For example, science is often deeply rooted in conventional European cultural beliefs, and through its formalized and empirically based tenets, it is often overly rigid, inflexible, and insensitive to the local and fine-scale features of the environment. Traditional ecological knowledge on the other hand, is a rich and highly resolved repository of local environmental wisdom that can fill in these finer details of biodiversity. While the use of TEK is not a substitute for scientific surveys and for the process of science for understanding and confirming ecological patterns, a view of the land through the eyes of local resource user can nevertheless provide a different, yet complimentary, lens with which to observe and understand the landscape.

In recent decades, interest in TEK has grown rapidly due in part to recognition that local knowledge of the environment represents a vast and underutilized database of information with many potential applications. Research from around the world has increasingly revealed highly detailed local knowledge about species distributions, ecological interactions and processes, habitat diversity, economically important species, and sustainable management practices. It is now widely recognized that this knowledge can contribute to biodiversity conservation, to the status of rare species, and to management of protected areas. Local knowledge is important for ecologists and resource managers as a means to improve research, resource management, and environmental impact assessment, and professionals in other disciplines, including anthropologists, ethnobiologists, and the pharmaceutical industry, share an interest in TEK for scientific, social, or economic reasons.

Ethnoecology is the study of how people interact with (and understand) all aspects of the natural environment, including plants, animals, landforms, forest types, and soils, among many other things. Ethnoecology is used to encompass all studies that describe local people's interaction with the natural environment, and includes subdisciplines such as ethnozoology, ethnobotany, and ethnoentomology. Ethnobotany is the subdiscipline of ethnoecology that concerns plants. Ethnozoology and ethnobotany are studies of the complex set of relationships of plants and animals to past and present human societies, and are used by academics to learn about traditional ecological knowledge. Indigenous and folk knowledge have increasingly attracted attention in Western scientific circles. For many years there has been great concern for the increasing loss of this traditional knowledge, and much thought has been given as to how to preserve and implement this huge area of knowledge that was previously ignored by Western science. One aspect of traditional knowledge is the classification of living things; all human societies respond to the diversity of plants and animals in their local environment by grouping them into categories of greater or lesser inclusiveness. People create categories in order to make sense of this diversity and group things based on similarities and differences. Every plant will have one or more uses, either as a food resource or for medicinal or ritual purposes, and these were known by the whole local community as they have been living together with plants and the landscape for centuries. This traditional ecological knowledge of plants and animals is responsible for most of the medicine and food used in modern society. We should consider traditional knowledge of plants when searching for cures to diseases and common ailments. This knowledge is accumulated over many generations and represents a fine balance between people and their environment. It represents the successes of trials and errors that have withstood the test of time. Traditional ecological knowledge can offer important clues with respect to the exploitation of the resources by local people, and thus is useful for commercial and government planning, and for scientific researchers.

Exploring Wootlot Diversity

Steven Newmaster

Exploring Alpha Diversity

The best way to understand woodlot biodiversity is to go explore a woodlot. Inventory diversity at the scale of a woodlot or forest stand is called alpha diversity and is measured by species richness, i.e. the number of species. Wandering around looking for critters and plants by visually inspecting the natural habitat is great, but qualitative. There is no way to reproduce your interpretation of biodiversity. Field biologists have adapted this approach, calling it an 'intelligent meander' and recording date, time, and location for the visit and then a list of species by habitat. The most critical factor of this approach is to record the time spent exploring; the more time you spend the more diversity will be revealed until the variation in all species and habitats is recorded. A more quantitative approach includes the use of some bounded space such as plots or transects (e.g. Fig. 1: T1, T2, and T3) to record diversity. This ecological approach requires a systematic grid positioned over the entire woodlot (Fig. 1). Plots are randomly located on the grid; randomization reduces bias and helps to provide an accurate snapshot of diversity from the landscape. The size and shape of the plots depends on what you would like to sample; 4 m radius plots are typically used for general forest plant community sampling, whereas 10 cm square quadrats are used to sample lichen diversity on tree trunks. Recording diversity in plots results in an inventory of species (called species richness), which may include some abundance measure such as frequency of individuals or percent cover; percent cover is a measure of influence, i.e. how much space a plant is taking up within an ecosystem. In each of the woodlot diagrams you can identify the tree silhouettes (see Fig. 2) for each plot, recording the number of species and its respective frequency:

Plot	Species
1	White Birch(1), Pin Cherry(1), Honeysuckle(1)
2	Black Locust(1), Pin Cherry(2)
3	Trembling Aspen(1), Virginia Creeper(1), Black Locust(1)
4	Virginia Creeper(1), Honeysuckle(1), White Birch(1)
5	Sugar Maple(1), Alternate-leaved Dogwood(2)

Figure 1. Woodlot map with systematic grid

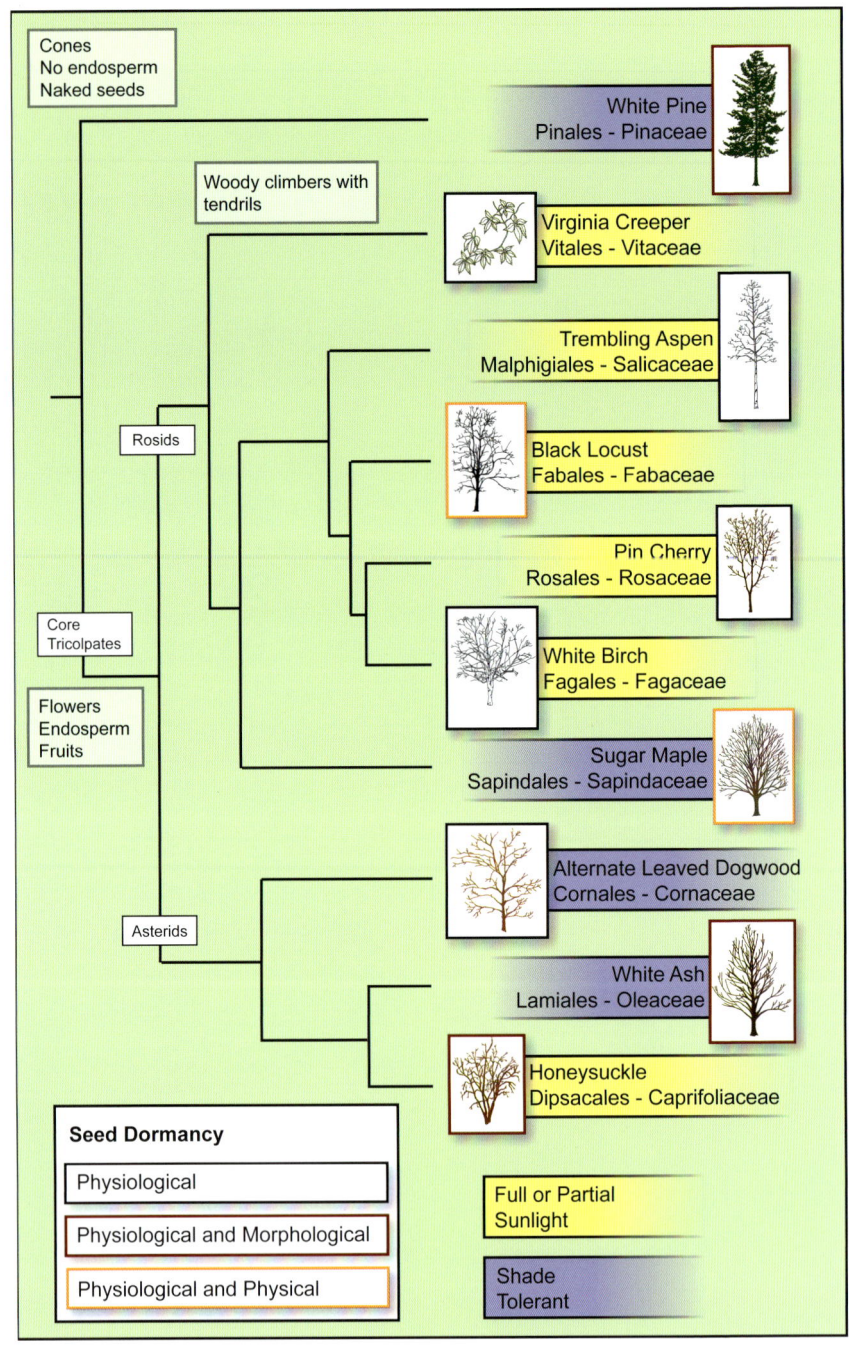

Figure 2. Classification tree of the 10 woody plant species from Fig. 1

Deciding the number of plots to sample requires a fundamental understanding of the species-area relationship in ecology. That is, as you sample more area (i.e. plots, quadrats, or transects) you will record more species until you have collected everything in the woodlot at which time adding more plots will be a waste of time as you will not record more species (see Fig. 3).

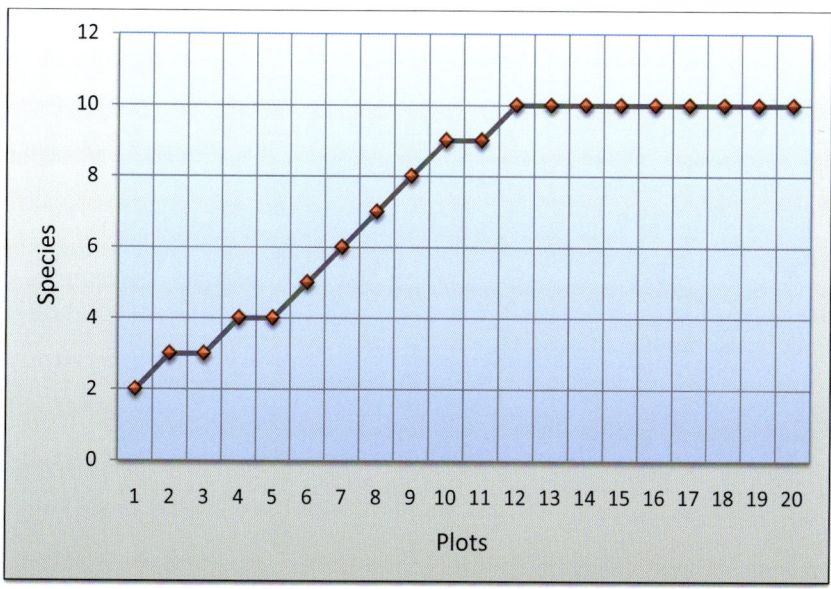

Figure 3. Species-area curve of the woodlot from Fig. 1

Tree Thinking – Exploring Variation in Species Traits

Species traits may explain difference in communities. Exploring biodiversity requires close examination of the variability in organisms. When sitting in the woodlot you will notice that some trees can grow in the shade under the forest canopy and other species make up a community on the edge of the woodlot that can tolerate exposure to full or partial sun (see Fig. 1: Woodlot Map and Fig. 2: Tree Classification). Seed-banking species store seeds in the soil that readily establish when the environmental conditions are appropriate. Some tree seeds are dispersed by wind and others by animals. Seed dormancy has resulted in divergent responses to the environment. Through this adaptation, germination is timed to avoid unfavourable weather for subsequent plant establishment, growth, and reproduction. The variation and relationship among species' traits can be used to understand alpha and beta diversity – why species are found in a particularly woodlot.

Exploring Beta Diversity

Exploring multiple woodlots raises interesting observations such as differences in the number, abundance or type species of species among these woodlots. This is an important observation because a sample of communities from two different transects within a woodlot may have the same number of species (see Fig. 2, T1 & T3 = 6 spp. and T2 = 5 spp.), but they are not the same communities; they do not have the same species. This type of diversity is not 'inventory diversity' but rather 'differentiation diversity'. At the scale of a woodlot or forest stand this is called beta diversity and compares alpha diversity among several communities. A comparison of alpha diversity among different communities via transects within or among woodlots or forest stands is called beta diversity. Beta diversity is low when two communities have similar species and respective abundance; transects 1 and 3 are similar each with the same species. Beta diversity is higher when less species are shared among sites or their abundance is quite different; transects 1 and 2 are less similar because they only share 1 species (pin cherry). Beta diversity is really a measure of community variation. Although community variation is difficult to assess for many sites, there is a tool called ordination which can be used to explore this variation quantitatively in multivariate space. This multivariate statistical technique can be illustrated for all 20 plots revealing a gradient of shade tolerance (Fig. 4).

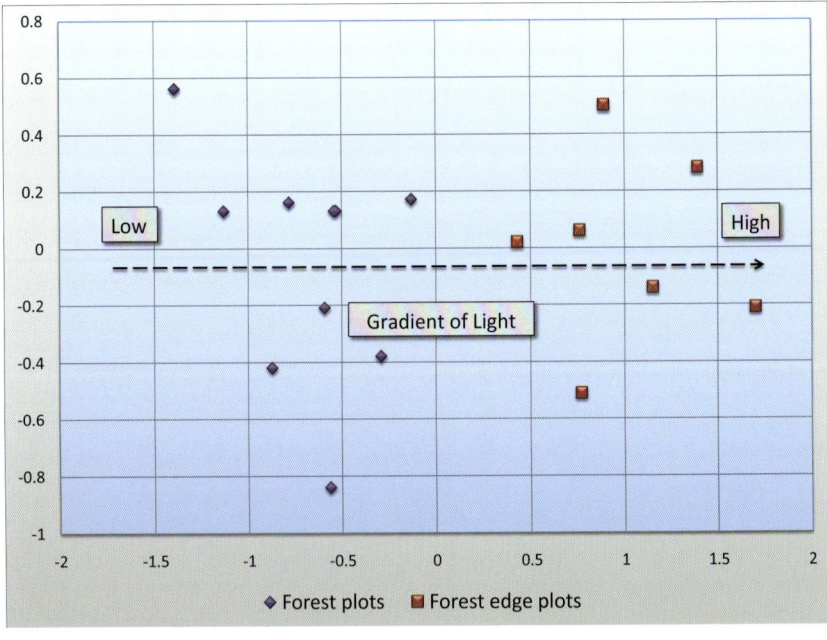

Figure 4. An ordination of the 20 plots in the woodlot from Fig. 1

CRYPTIC BOTANY

Cryptic Botany

Steven Newmaster & Thomas Henry

Introduction
The cryptic plants are somewhat mysterious with strange life cycles and many species that are hidden deep in the forest among cracks on a rock or along the shaded side of a log. As defined by botanist Carolus Linnaeus (1707–1778) the name Cryptogamae comes from the Greek word *kryptos*, meaning hidden and *gameein*, to marry, referring to the lumping of several groups of plants in one category. Cryptogamic botany is the study of spore-producing plants: bryophytes and ferns. Both bryophytes and ferns reproduce by spores (Fig. 1) and have motile sperm; in a woodlot, their sperm must swim through the forest to fertilize the female gametophyte (**archegonium**). Mosses, liverworts, and ferns produce many complex biochemicals with medicinal value that have been used by indigenous cultures for thousands of years.

Figure 1. Spores

BRYOPHYTES

Bryoflora Diversity
Bryophytes are a diverse group of plants that inhabit many places on the globe and dominate many ecosystems such as the arctic and the boreal forest. Currently, there are over 20,000 known species of bryophytes that include 12,800 mosses, 8,000 liverworts (hepatics) and 100 hornworts. In Ontario there are over 670 bryophytes (540 mosses, 130 hepatics, 3 hornworts), of which 25% are rare or uncommon. Forests ecosystems provide habitat for 375 species of bryophytes (285 mosses, 90 hepatics) and 60% of the Ontario bryoflora is located in southern part of the province. Woodlots are an important ecosystem for maintaining bryophyte diversity and conserving rare plants. Our preliminary studies of woodlot floristic diversity recorded an average woodlot inventory of over 100 species including trees, shrubs, herbs, ferns, bryophytes, and lichens (see Fig. 2, pg. 10). Bryophytes comprised 25% of this woodlot diversity and one can expect to find 20-40 species of bryophytes when exploring a woodlot.

What is a Bryophyte?
Bryophytes are unique plants that have adapted to inhabit some of the harshest conditions in nature. Unlike the rest of the plant kingdom, including ferns, they do not have roots and almost all bryophytes do not have vascular tissue. They anchor themselves to their substrate with **rhizoids**. The green leafy portion that we see is called the **gametophyte**, which is haploid (i.e. with one set of chromosomes) and is the dominant part of its life cycle. This is odd because all the other groups of plants that we can see (e.g. trees, shrubs, herbs, ferns) are diploid (i.e. with two sets of chromosomes) with haploid gametes (i.e. sperm or eggs). Bryophytes have motile sperm that must swim in a film of water from one gametophyte to another in order

CRYPTIC BOTANY

Figure 2. Sporophyte

Figure 3. Stoma with peristome

Figure 4. Capsule dispersing spores

to achieve fertilization. This is a miraculous accomplishment considering the next gametophyte may be located on a log far across the woodlot. Successful fertilization results in a diploid **sporophyte** developing on the gametophyte. This is the little **capsule** and stalk (**seta**) that is commonly seen on bryophytes, especially during the wettest seasons – spring and fall (Fig. 2). The capsule may have a hat (**calyptra**) and a lid (**operculum**) that pops off when mature revealing the mouth of the capsule (**stoma**), which is surrounded by one or two rows of teeth, called the **peristome** (Fig. 3), that aid in spore dispersal (Fig. 4). Many mosses have a ring of 16 triangular teeth in two rows that fold in to cover the stoma or fold back to open the stoma. Other mosses have a peristome that looks like a membrane with slits and functions like a salt/pepper shaker (Fig. 5). The moss sporophyte develops slowly over weeks or months. The spores mature in the upright capsule following meiosis. Conversely, the hepatic sporophyte develops while immersed within the protection of specialized leaves until the spores have matured via meiosis. The black capsules then split along the sides into 4 parts, to spread the spores with wiry structures called **elaters**. The entire development of the hepatic sporophyte can happen in hours or days. The setae of liverworts are thin, white or translucent, and quite weak. Liverwort gametophytes look different than those of mosses in that they are flattened and are either ribbon-like or have leaves in two rows, not spirally arranged leaves like most mosses. Also, liverwort leaves (Fig. 6) do not have the midrib (costa) that is common to many mosses (Fig. 7).

Where do Bryophytes Live?

Bryophytes can be found worldwide. They show a high degree of tolerance to harsh environments as they can be found in habitats such as on rock surfaces, cliffs, tree trunks, or bare ground that other plants are unable to colonize. Consequently, their distributions are much broader than those of seed plants. Bryophytes are poikilohydric, meaning they can suspend metabolic activity upon drying and are physiologically tolerant of desiccation, frost, and freezing. On a walk in a woodlot on a warm (5 °C) sunny winter day you may notice fresh bright green moss on rocks and trees; they are photosynthetically active. Most plants will freeze and die in late fall or early winter as ice crystals form in their cells puncturing the cell walls. In habitats such as tree trunks and bare rock most plants will die of dehydration. The desiccation tolerance of bryophytes is facilitated by small plant size and small cell size, which allows them to have wide geographic ranges at varying latitudes, altitudes, and levels of precipitation and temperature. The seed plants are limited by dispersal of seeds whereas bryophytes have tiny spores, often produced in great quantity, to aid in widespread dispersal throughout the globe. When sexual reproduction is not possible, asexual reproduction is rampant in bryophytes. This includes specialized branches that break off, stick onto slugs, and later fall off and grow into new plants. Some hepatics have specialized splash cups (Fig. 8) with masses of cells called gemmae that are dispersed by rain drops.

Figure 5. Hair Cap Moss peristome

The Importance of Microhabitats

The diversity of bryophytes is correlated with habitat heterogeneity at two spatial scales. Mesohabitats are localized physiographic (e.g. streams, seeps, cliffs) or physiognomic (e.g. forests) features. In a forested landscape, mesohabitats are arranged into a mosaic of dominant mesohabitats (e.g. the entire forest floor), wherein restricted mesohabitats (e.g. streams, seeps, cliffs) exist. Microhabitats (e.g. logs, rocks, stumps) are the smallest landscape units and may be unique to one type of mesohabitat. Our forest research indicates that bryophyte diversity is quite variable among different types of microhabitats. Analyses of bryophyte beta-diversity in our research indicates that the variety of microhabitats within a community has considerable influence on community structure, which tells us that specific bryophytes are restricted to or prefer particular microhabitats. Epiphytic communities,

Figure 6. Hepatic leaves

Figure 7. Moss leaves

CRYPTIC BOTANY

Figure 8. *Gemmae cup*

Figure 9. *Thallose growth form*

Figure 10. *Leafy growth form*

including bryophytes that live on trees, provide a classic example of microhabitat differentiation. Epiphytes typically exhibit both a vertical and a horizontal zonation, segragating vertically from the base to the crown along gradients of humidity, pH, and nutrient content. Perhaps the most important habitat in the woodlot is coarse woody debris, e.g. logs, big branches, etc. Each stage of decay from bark to rotten wood provides critical habitat for a specific community of bryophytes that becomes part of the successional transition from tree to organic soil.

Ecological Role of Bryophytes in Woodlots

Bryophytes have many important ecological roles in woodlots. These include buffering the underlying soil temperature (a major component in the nitrogen and phosphorus cycles), seed bed formation, early colonization of disturbed sites, creation of habitat in harsh environments, and the provision of food and habitat resources for animals and insects. In some northern forests the carpet of moss on the forest floor provides more biomass than the trees. This carpet of moss is very important in releasing nutrients into the forest soil and provides habitat for many creatures such as snails, salamanders, mites, and a host of insects. Examination of almost any moss under a microscope will also reveal another important ecological role: they provide shelter and humidity for a remarkable diversity of invertebrates, which are an integral part of the food web in many of our important forest habitats.

Exploring Woodlot Bryodiversity

Bryophyte diversity can be explored through the recognition of variability in structure. There are many different shapes and colours of bryophytes growing on a suite of different habitat types. Exploring different habitats is a great way to find lots of species. Identifying the species starts with understanding how to differentiate a liverwort (hepatic) from a moss. Although the biological difference has already been discussed the differences in appearance may seem somewhat cryptic. Described below are several common woodlot liverworts and mosses that will help you to explore their diversity.

CRYPTIC BOTANY

Liverworts (Hepatics)

Liverworts are flatter than mosses, which usually have leaves radiating around the stem. Liverworts have two growth forms; thallose (Fig. 9) and leafy (Fig. 10). The thallose liverworts are divided into two groups including:

i) Marchantiales – thallus is dichomtomously forked with loose upper tissue (in cross section), dorsal pores, and ventral scales

ii) Metzgeriales – thallus is not dichotomously forked, solid cells with no air spaces (in cross section) and lacks pores and scales

Leafy liverworts are variable in size: *Lophozia* (pg. 33) is a medium-sized (2 mm wide) leafy liverwort and *Cephalozia* is a smaller species. Leafy liverworts also have different types of leaf insertion. Leaf insertion may be **succubous,** or overlapping from the apex of the plant downward, like the shingles on a roof, each leaf overlaps the next one below (see picture of *Lophozia*). The opposite is called **incubous** leaf insertion where the leaves overlap from the base toward the stem tip (Fig. 6).

CRYPTIC BOTANY

Umbrella liverwort
Marchantia polymorpha

Family: *Marchantiaceae*

There is only one genus and species in the genus *Marchantia* in Ontario and it is distinguished by its thallose growth form with stomata (plant breathing pores), cells with slightly visible borders, circular gemmae cups, and stalked male and female receptacles. It grows on soil or wet logs near streams in woodlots. A similar species called *Lunularia cruciata* has half-circle gemmae cups.

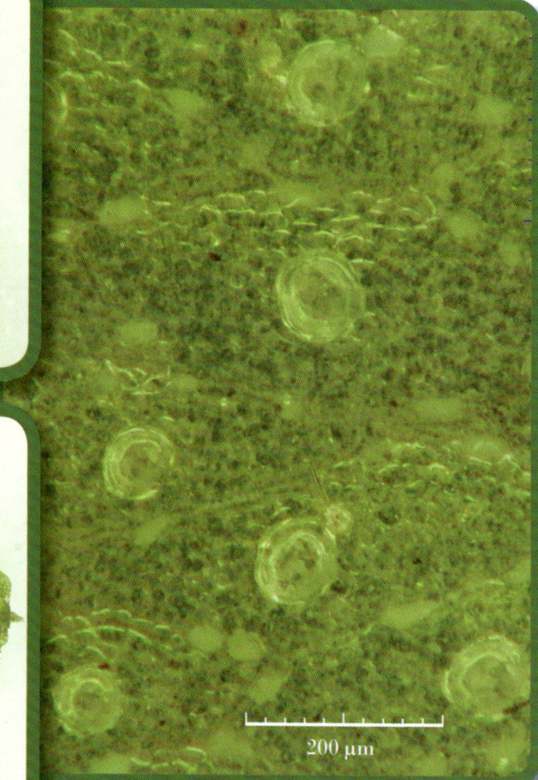

CRYPTIC BOTANY

Alligator liverwort
Conocephalum conicum

Family: *Conocephalaceae*

There is only one genus and species in the genus *Conocephalum* in Ontario and it is distinguished by its thallose growth with distinctive stomata and dark cell borders. It can be distinguished from *Marchantia polymorpha* (previous page) by noting the darker cell borders and absence of gemmae cups. The surface of this plant looks like alligator skin. The presence of sesquiterpenoids and triterpenoids give this plant a strong pungent odour when crushed.

CRYPTIC BOTANY

Riccardia latifrons

Family: *Aneuraceae*

There are three species of *Riccardia* in Ontario. All of them are thallose, with a branched thallus that differentiates the three species: 1) *Riccardia latifrons* – irregularly to once pinnate, 2) *Riccardia palmata* – palmately divided, and 3) *Riccardia multifida* – 2–3 times pinnately divided. All of these species grow on logs in wet areas of woodlots, such as swamps and streams.

CRYPTIC BOTANY

Riccia

Family: *Ricciaceae*

In Ontario there are six species of *Riccia*. In the genus *Riccia* the thallus is either ribbon-like or forms round rosettes. Some species are commonly found in woodlots growing on wood or soil and others are only found in wetlands.

Lophozia

Family: *Jungermanniaceae*

There are 16 species of *Lophozia* in Ontario. Most of these grow on logs without bark. They are small leafy liverworts with creeping stems that range from 3–8 mm long. The succubous leaves have two lobes and the tips sometimes have globular gemmae of different shapes and colours that are useful in distinguishing some of the species. Rare species in this genus are restricted to large logs in old growth forests.

CRYPTIC BOTANY

Frullania

Family: *Jubulaceae*

Most of the seven species of *Frullania* in Ontario grow on trees or rocks. The rounded leaves are complicate-bilobed, having two leaf lobes folded on one another. The lower leaf resembles an inverted cup and is often home to microscopic animals such as rotifers. This species produces sesquiterpenes that may cause skin rashes.

CRYPTIC BOTANY

Bazzania trilobata

Family: *Lepidoziaceae*

Bazzania trilobata is the only species of *Bazzania* found in Ontario, but this species ranges from the woodlots of southern Ontario to the boreal forests in the northern part of the province. This leafy liverwort is quite large (up to 3 cm in the north) and can form extensive mats (ca. 200 cm^2) on the forest floor. The main leaves are dark green and 3-toothed at the apex. It has underleaves (amphigastra) that are 4–5 toothed at the apex. The plant has a unique aromatic fragrance like that of sandalwood and lime. The stems are dichotomously forked and produce pale green root-like banches with small leaves. Unique sesquiterpene alcohols and nine known sesquiterpenoids are produced by this plant that may explain medicinal use by several indigenous cultures.

CRYPTIC BOTANY

Mosses (Musci)

There are many characters that are used to identify mosses. The sporophyte is used for classification, particularly the peristome (Fig. 2). The capsule size and shape are good characters for identification. The setae, or capsule stalks, of different species display differing colours and lengths and are sometimes twisted. The leaves of different species have characteristic cells that are long or short and may have minute protuberances called papilla. The midrib of the leaf, called a costa, may be long, short, doubled, or non-existent. The leaves may be entire (no teeth), toothed or even double-toothed. Mosses have two different growth forms called:

i) Acrocarpous – the archegonia (female sex organs) and the capsules are borne at the tips of stems or branches. Acrocarpous mosses may branch extensively; once they have fruited, branches take over the erect growth (see picture of *Atrichum*, pg. 38).

ii) Pleurocarpous – the female sex organs (archegonia) and capsules are borne on short, lateral branches, and not at the tips of branches. Pleurocarpous mosses tend to form spreading carpets rather than erect tufts (see picture of *Brachythecium*, pg. 48).

CRYPTIC BOTANY

Atrichum undulatum

Family: *Polytrichaceae*

There are six species of *Atrichum* in Ontario. All are acrocarps that look like little conifers and grow on soil. The *Polytrichaceae* are characterized by nematodontous peristomes (Fig. 5). The genus *Polytrichum* is similar to *Atrichum*, but does not have undulate leaves and the calyptra is covered in long hairs (hence the name *Polytrichum*, from Greek *poly*, meaning many, and *trichum*, hair). Two common woodlot species are *Atrichum undulatum*, which is large (20 mm) and commonly found on humus at the base of trees, and *Atrichum angustatum*, which is smaller (10 mm) and is found on mineral soil in disturbed areas such as up-turned tree roots.

2 mm

CRYPTIC BOTANY

Sidewalk moss
Bryum

Family: *Bryaceae*

This is a large genus with 27 species in Ontario. Everyone has met this moss or at least stepped on it as it commonly inhabits cracks in sidewalks. It is characterized by acrocarpous growth and a prominent costa that is often extended as a point on the end of the leaf. Many species have reddish stems and grow in dense mats or at least tufts of individuals. Most of the species grow on soil in disturbed sites, but some grow in woodlots at the bases of trees or on up-turned tree roots.

CRYPTIC BOTANY

Broom moss
Dicranum/Dicranella

Family: *Dicranaceae*

There are many species of Broom mosses of which *Dicranum* (22 species) and *Dicranella* (10 species) are the largest genera. These acrocarpous mosses look like the end of a broom with the bristles up. Many grow in woodlots on all kinds of microhabitats including tree bases, logs, rocks, stumps and bare soil. Some species are small (5 mm) and grow in dense mats that look like golf greens, such as *Dicranum montanum*, which is commonly found at the bases of trees. Other species such as *Dicranum polysetum* are large (10–15 cm), grow in loose tufts, and have wavy leaves. *Dicranum flagellare* has flagellate branches that break off, stick to slugs, and are taken to new habitats where they fall off and grow new plants. *Dicranella heteromalla* have very fine shiny leaves and yellow setae.

CRYPTIC BOTANY

Fissidens

Family: *Fissidentaceae*

There are 11 species of *Fissidens* in Ontario of which half can be found in woodlots. Three species are commonly found on the forest floor and it is very likely that if you find a terricolous (ground-dwelling) moss that it is *Fissidens*. All species are flat, have forked peristome teeth, and equitant leaves. An equitant (*equi* = horse) leaf is split into two blades that clasp the stem and leaf above, as in *Iris* leaves.

CRYPTIC BOTANY

Pin cushion moss
Leucobryum glaucum

Family: *Leucobryaceae*

There is only one genus and species in this family – *Leucobryum glaucum*, often called 'pin cushion moss'. The plants have a unique colour as they have a very large costa made up of porose cells that hold water. This is a common woodlot moss found on the forest floor or sometimes on logs. In the southern part of the province it grows in small tufts, but in northern Ontario it grows in large (10 cm high by 20 cm wide) hemispherical cushions (rarely seen in the south). The whitish-green colour give it the false appearance of a nice dry place to sit down during a hike.

CRYPTIC BOTANY

Plagiomnium

2 mm

Family: *Mniaceae*

The Mniaceae is made up of three common acrocarpous genera: *Plagiomnium* (six species, all with single-toothed leaves), *Mnium* (six species, all with double-toothed leaves) and *Rhizomnium* (five species, all with entire leaves). All of these species have a very distinctive costa and small leaf cells bordered by long cells that give the leaf a unique framed appearance. Almost all of these species are found growing on the forest floor. One of the most common mosses in eastern North American hardwood forests is *Plagiomnium cuspidatum*, which has plagiotropic branches (like strawberry stolons) that aid in spreading growth on the forest floor.

2 mm

43

CRYPTIC BOTANY

Tree moss
Climacium dendroides

Family: *Climaciaceae*

Although there are three species of *Climacium* in Ontario, only *Climacium dendroides* is common. This is an easy moss to learn because it grows upright as individuals and is dendroid (looks like a little tree) with reddish stems. This is a common woodlot moss that grows in moist areas (abundant in swamps) on soil or humus.

CRYPTIC BOTANY

Electrified cat tail moss
Rhytidiadelphus triquetrus

Family: *Hylocomiaceae*

There are two species of *Rhytidiadelphus* in Ontario, of which one is rare and the other is commonly found in woodlots growing on the forest floor. This moss grows upright like a little tree and is relatively large (8–20 cm). The leaves are plicate (longitudinal folds) with a long double costa, which gives the leaf the appearance that it has many lines. The leaves are widely spreading in many directions giving the moss a shaggy appearance and it is commonly called 'shaggy moss' or 'electrified cat tail moss'.

500 µm

CRYPTIC BOTANY

Pigtail moss
Hypnum

Family: *Hypnaceae*

Many of the 15 species of *Hypnum* in Ontario are commonly found in woodlots. These pleurocarpous species have falcate (sickle-shaped) second (turned to one side) leaves that are organized like a hair braid and are often called the 'pigtail mosses'. The long cells give the leaf a shiny appearance making this quite an attractive moss. The double costa is difficult to see as it is very short. Many species grow on logs, but others are found on tree bark and soil, or humus on the forest floor.

1 mm

CRYPTIC BOTANY

Beautiful moss
Callicladium haldanianum

Family: *Hypnaceae*

There is only one species of *Callicladium* in Ontario and it is commonly found in woodlots growing on logs and stumps. Its leaves are cupped and bright green and it grows in mats on the forest floor. This pleurocarpous moss is often misidentified as *Brachythecium* (see pg. 48), but it does not have a visible costa; you need a microscope to see its very short double costa. Although most mosses do not have common names this one does and it is aptly called 'beautiful moss'.

500 μm

2 mm

CRYPTIC BOTANY

Brachythecium

Family: *Brachytheciaceae*

There are 20 species of *Brachythecium* in Ontario and most of these are found in woodlots. The genus is characterized as a pleurocarp with lance-shaped leaves that have a long single costa and toothed border (serrate). If you hold a plant up towards the sky you can see the costa and teeth on the leaves with the aid of a 10-15x hand lens. Some species are medium-sized (1-2 cm long) and others are more robust (4-8 cm long). Most of the species grow on the forest floor, often at the base of trees. Some species grow up the trunks of saplings (5-10 cm in diameter) and others grow on well rotted logs.

1 mm

CRYPTIC BOTANY

Fern moss
Thuidium

Family: *Thuidiaceae*

There are four species of *Thuidium* in Ontario all of which are found in forests. These mosses are pleurocarpous and most of them are pinnately branched 2–3 times, much like a fern, and hence they are commonly called the 'fern mosses'. The leaves have a single costa and the cells are papillose (possessing small protuberances on the surface of the cells) giving the leaves a bumpy look. These species can be found on logs, rock, soil, and humus, and are very common in cedar swamps.

CRYPTIC BOTANY

FERNS AND FERN ALLIES

Fern Diversity

Ferns are part of a group of plants called the pteridophytes that includes true ferns, horsetails, whisk ferns, club-mosses, spike-mosses, and quillworts. Pteridophytes are seedless vascular plants, i.e. plants with xylem and phloem whose dispersal relies on spores not seeds. There are approximately 12,000–15,000 species worldwide. In Ontario there are over 150 species and hybrids (18 Ophioglossoid ferns, 82 true ferns, 22 club-mosses, 13 horsetails, 13 quillworts, and 4 spike-mosses) of which 30% are rare or uncommon. Forests ecosystems provide habitat for 80% of the pteridophyte species and over 70% of Ontario's pteridophyte flora is located in the southern part of the province. The woodlots of southern Ontario are an important ecosystem for maintaining pteridophyte diversity. Our preliminary studies of woodlot floristic diversity recorded an average woodlot inventory of over 100 species of plants including trees, shrubs, herbs, ferns, bryophytes and lichens (see Fig. 2, pg. 10). Pteridophytes comprised 8% of this woodlot diversity and one can expect to find 5–15 species of ferns and fern allies when exploring a woodlot.

What is a Fern?

Ferns and fern allies have unique life-cycles. All pteridophytes undergo an **alternation of generations**, in which a dominant sporophyte generation produces spores through meiosis, and a free-living gametophyte generation forms the gametes (eggs and sperm) by mitosis. The life cyle begins when a spore germinates to produce a short-lived, delicate gametophyte (Fig. 11), roughly the size of a loonie, that quickly matures and produces egg-forming archegonia and sperm-producing antheridia. Following a rainfall, the multi-flagellated sperm swim away from mature antheridia, attracted to chemicals that are released on the necks of the archegonia, where they fertilize the eggs.

Fig. 11. Fern gametophyte

It is amazing if you consider that in most cases the sperm produced by a gametophyte cannot successfully fertilize its own eggs and must swim to archegonia on neighbouring, genetically different gametophytes. Most pteridophytes are **homosporous** (produce spores that are all the same size) but a few groups are **heterosporous** with large megaspores and smaller microspores. The megaspores produce megagametophytes that only form eggs, and microspores produce microgametophytes that only form sperm. Heterospory evolved independently in several groups of vascular plants, diversifying the **phanerogams** (seed plants) and a few groups of pteridophytes including the water-clover ferns, *Salvinia* ferns, and all members of the spikemosses and quillworts. Once fertilized the gametophyte withers away and the diploid (2n) fern sporophytes grow. These are the **fronds** that are familiar to most people and are often used in floral arrangements. When mature, the undersides of fern leaves or specialized branches produce clusters of capsular structures called **sporangia** (Fig.

CRYPTIC BOTANY

Fig. 12. Oak fern sporangia

Fig. 13. Sensitive fern spores

12), within which meiosis forms the haploid (n) spores (Fig. 13). Specialized cells on the sporangia help to release the spores and dry wind currents cause the active release of spores from the capsules. The spores that are wind-borne often establish in shady, moist habitats and germinate to yield another multicellular gametophyte and the cycle of life continues.

Fern Ecology

Ferns are an important part of our forest ecosystem. About 300 million years ago, in an era called the Carboniferous period, huge tree-like fern allies dominated the earth in massive swamps which are a source of today's coal. Imagine a present-day club-moss (10–30 cm tall) growing up to 36 meters (120 feet) tall during the Carboniferous period! Some ferns grow equally well on various substrates while others are confined strictly to specific habitats such as rock or even cliffs. A forested cliff is often a good habitat, where ferns grow in fissures and crevices. Some ferns are only found on acidic rocks such as granites, sandstones, and quartzites, while others grow only on alkaline rocks such as calcites and dolomites. This makes various fern species very good indicators of their immediate environment, and conducting surveys of fern flora in an area provides a valuable ecological tool to measure environments and site types in forest classification. Ferns are remarkably well adapted to various disturbances, possessing the ability to accumulate toxins from their environment, providing an important role for ferns in ecosystem conservation and restoration.

CRYPTIC BOTANY

Club-mosses (Lycopodiaceae)

Club-mosses are short evergreen plants sometimes said to resemble very large mosses or small conifer seedlings. Club-mosses have simple linear leaves each with a single vein. Like other ferns and fern allies, club-mosses produce spores in sporangia. The sporangia are either produced in the axils of modified leaf-like structures or within distinct cone-like strobili (the scales of which, in reality, are simply more highly modified leaf-like structures). There are over 400 species of club-mosses in the world, many of which are tropical epiphytes. In Ontario there are 16 species and 6 hybrids in 4 genera. All of our species are typically found on the ground in cool and damp northern forests. Note that the term 'fern allies' should be used with caution when referring to club-mosses and other lycophytes (including spike-mosses and quillworts) because, even though lycophytes share similar life history characteristics with ferns and horsetails, ferns and horsetails are evolutionarily more closely related to seed plants than they are to the lycophytes.

CRYPTIC BOTANY

Genus Key to Some Common Woodlot Club-mosses of Ontario

1(a) Sporangia contained in leaf axils; upper leaf axils often bearing gemmae ....*Huperzia lucidula*, pg. 56

1(b) Sporangia produced on cone-like strobili; gemmae not present 2

2(a) Strobili with green leafy bracts; upright branches always bearing strobili ....*Lycopodiella inundata*, pg. 57

2(b) Strobili with yellow or brown scale-like bracts, green when immature but still distinctly scale-like; upright branches bearing strobili or not 3

3(a) Stems flattened; leaves in 4 ranks ....*Diphasiastrum digitatum*, pg. 55

3(b) Stems rounded; leaves in 6 or more ranks ....*Lycopodium*, pg. 58

CRYPTIC BOTANY

Southern running-pine
Diphasiastrum digitatum

Family: *Lycopodiaceae*

Vegetative Characteristics: Plants short, evergreen, with both upright and horizontal stems; stems distinctly flattened, with leaves in 4 ranks.

Reproductive Characteristics: Spores produced inside sporangia on separate cone-like structures (strobili); sporangia subtended by scale-like bracts; strobili in clusters of 3 or 4 on a long stalk.

Notes: *Diphasiastrum digitatum* (southern running-pine) is a common forest species that typically grows in somewhat dry forests. The spores are yellow and highly flammable and the dried spores have been used to create flashes for theatrical effect because they burn quickly and at a cool temperature.

CRYPTIC BOTANY

Shining fir-moss
Huperzia lucidula

Family: *Lycopodiaceae*

Vegetative Characteristics: Plants short, evergreen, with upright stems only (lacking horizontal stems); stems distinctly rounded; leaves shiny, broadest around the middle; shorter leaves produced early in the season followed by longer leaves later on; leaves in 6 ranks.

Reproductive Characteristics: Spores produced in sporangia located in leaf axils; distinctive gemmae, or vegetative reproductive structures, often borne from upper leaf axils.

Notes: *Huperzia lucidula* (shining fir-moss) is a common forest species that typically grows in moist shaded areas. Its attractive shiny green leaves and gemmae make it easily recognizable from a distance. It can sometimes be mistaken for *Lycopodium annotinum* (pg. 59) when reproductive structures are not present but look for leaves in 6 ranks (*L. annotinum* has leaves in 8 ranks).

CRYPTIC BOTANY

Northern bog club-moss
Lycopodiella inundata

Family: *Lycopodiaceae*

Vegetative Characteristics: Plants short, deciduous, with creeping vegetative stems and upright fertile stems; leaves narrow and linear, not toothed, in 8 to 10 ranks.

Reproductive Characteristics: Spores produced in sporangia on cone-like strobili; sporangia subtended by green leafy bracts; strobili solitary, distinctly wider than stems.

Notes: *Lycopodiella inundata* (northern bog club-moss) is a common species that typically grows in bogs and along lake-shores. It is the only species in the genus *Lycopodiella* in Ontario and it is easily identified by its solitary strobili with green leafy bracts on upright fertile stems and its deciduous habit.

CRYPTIC BOTANY

L. dendroideum

Club-mosses
Lycopodium spp.

Family: *Lycopodiaceae*

Vegetative Characteristics: Plants short, evergreen, with both upright and horizontal stems; plants often resemble small conifer seedlings; stems distinctly rounded with leaves in 6 or more ranks; leaves small, linear, with a single vein.

Reproductive Characteristics: Spores produced inside sporangia on separate cone-like structures (strobili); sporangia subtended by scale-like bracts; strobili produced singly or in small clusters; strobili may be sessile or stalked.

Notes: The dried spores of some species have been used to create flashes for theatrical effect because they burn quickly and at a cool temperature. The genus name *Lycopodium* is from Greek *lycos*, 'wolf', and *podes*, 'foot', a reference to how the branches of the species *L. clavatum* sometimes resemble a wolf's paw.

Key to Some Common *Lycopodium* Species

1(a) Strobili in small clusters on long-stalked stems; leaves each tipped with a fine white bristle ....*Lycopodium clavatum*

1(b) Stobili single, sessile; leaves not bristle-tipped 2

2(a) Upright stems tree-like ....*Lycopodium dendroideum*

2(b) Upright stems unbranched ....*Lycopodium annotinum*

CRYPTIC BOTANY

Lycopodium Species Descriptions

Lycopodium clavatum (common club-moss) is a common species that typically grows in dry forests and clearings. It is a distinctive species with a more sprawling habit than the other species of *Lycopodium* and its upright branches are sometimes said to resemble a wolf's paw. *Lycopodium clavatum* can be easily identified by its bristle-tipped leaves. Another species, *L. lagopus*, has been recently segregated from *L. clavatum* and can be identified by having solitary strobili rather than strobili in clusters.

L. clavatum

Lycopodium dendroideum (prickly tree club-moss; pictured on previous page) is a common species which is typical of moist or dry forests and clearings. It can be easily identified by its tree-like branching habit. There are two other species of *Lycopodium* in Ontario that very closely resemble *L. dendroideum*: *L. obscurum* (tree club-moss) and *L. hickeyi* (Hickey's tree club-moss). *Lycopodium dendroideum* can be distinguished from *L. obscurum* and *L. hickeyi* by having leaves on the lower part of the stem (below the branches) that are spreading rather than strongly appressed. *Lycopodium obscurum* can be distinguished from *L. hickeyi* by having ventral leaves that are smaller than the other leaves (in *L. hickeyi* and *L. dendroideum* all of the leaves are equal in size). *Lycopodium dendroideum* is the most common and widespread species.

L. clavatum strobili

Lycopodium annotinum (bristly club-moss) is a common species that typically grows in moist forests and has a circumpolar distribution. This species can sometimes be confused with *Huperzia lucidula* (pg. 56) when reproductive structures are not present but it can be identified by having leaves in 8 ranks instead of 6.

L. annotinum

59

CRYPTIC BOTANY

Horsetails and Scouring Rushes (*Equisetum*)

Equisetum species have slender erect stems that are hollow in the centre and have ridges around the outside. The leaves are reduced to small teeth that are fused at the base to form distinctive sheaths at each node. Some species have whorled branches at each node while other species are unbranched. Like other species of ferns and fern allies, *Equisetum* species produce spores in sporangia. The sporangia are contained in cone-like structures (strobili) at the tips of fertile stems. The strobili of *Equisetum* species are different from those of the club-mosses in that the sporangia are suspended on the underside of umbrella-like structures called sporangiophores instead of in the axils of scale-like or leafy bracts. There are 15 species of *Equisetum* worldwide and most species have very wide distributions. There are 9 species and 4 hybrids that can be found in Ontario. Horsetails grow in a range of moist places including in ditches, meadows, and woodlands, along roadsides, lakeshores and riverbanks, and in marshes, swamps, or shallow water. The species in the genus can be subdivided into two subgenera: species in the subgenus *Equisetum* have nodes with branches and are typically called horsetails, and species in the subgenus *Hippochaete* have nodes without branches and are typically called scouring rushes. The common name 'scouring rush' refers to the fact that horsetails are abrasive and high in silicates and have been used to scrub pots. The common name 'horsetail' refers to how the branched species somewhat resemble a horse's tail. The genus name *Equisetum* is from Latin *equis*, 'horse', and *seta*, 'bristle'.

Key to the Woodlot Horsetails of Southern Ontario

1(a) Vegetative stems with branches at each node (subgenus *Equisetum*) ....2

2(a) Branches rebranching; stem sheaths papery ....*Equisetum sylvaticum*, pg. 62

CRYPTIC BOTANY

2(b) Branches not rebranching; stem sheaths stiff 3

3(a) Branches ascending; stem sheath teeth solid brown; sheath teeth of branches (not stems!) attenuate ....*Equisetum arvense*, pg. 62

3(b) Branches spreading; stem sheath teeth brown with pale margins; sheath teeth of branches (not stems!) deltoid ....*Equisetum pratense*, pg. 62

1(b) Vegetative stems lacking branches at each node (subgenus *Hippochaete*) 4

4(b) Plants short (<30 cm); stems thin, solid, and wiry; sheath teeth 3 ....*Equisetum scirpoides*, pg. 63

4(b) Plants taller (up to 1.2 m); stems thick, erect, and hollow; sheaths usually with 4 or more teeth 5

5(a) Sheath teeth 3–12, persistent, brown in the centre with white margins; stem ridges 3–12 ....*Equisetum variegatum* ssp. *variegatum*, pg. 63

5(b) Sheath teeth 14–50, often deciduous; sheaths constricted at the base with dark bands above and/or below; stem ridges 14–50 ....*Equisetum hyemale* ssp. *affine*, pg. 63

CRYPTIC BOTANY

E. sylvaticum

E. arvense

E. arvense

**subgenus *Equisetum* (Horsetails)
Species Descriptions**

Equisetum sylvaticum (wood horsetail) is a common woodland species with a circumpolar distribution. Its rebranching branches and papery brown sheaths make it difficult to confuse with any other *Equisetum* species.

Equisetum arvense (field horsetail) is a very common species with a circumpolar distribution. It grows in a variety of habitats, including open woods, meadows, fields, and roadsides. *Equisetum arvense* can be easily confused with *E. pratense* and *E. palustre*. *Equisetum arvense* can be distinguished from *E. pratense* by its more upright branches and uniformly brown sheath teeth. Both *E. arvense* and *E. pratense* can be distinguished from *E. palustre* by having the first internodes of their branches equal to or longer than the stem sheaths.

Equisetum pratense (meadow horsetail) is a common species with a circumpolar distribution that typically grows in moist woodlands and meadows. It can be easily confused with *E. arvense* and *E. palustre* (see paragraph above).

There are two other species of *Equisetum* subg. *Equisetum* in Ontario. *Equisetum fluviatile* (water horsetail) has rarely branching stems and typically grows in standing water. *Equisetum palustre* (marsh horsetail; mentioned above) has often sparsely branched stems and grows in marshes and swamps.

E. pratense

CRYPTIC BOTANY

**subgenus *Hippochaete* (Scouring Rushes)
Species Descriptions**

Equisetum scirpoides (dwarf scouring rush) is a common species with a more northerly circumpolar distribution. It can be found growing in forests, peat bogs, and tundra. It can be easily identified by its small wiry stems with a solid centre and sheaths with 3 teeth.

Equisetum variegatum ssp. *variegatum* (variegated scouring rush) is also a common species with a more northerly distribution. It can be found growing in riverbanks, forests, and tundra. It can be identified as having 3–12 white-margined sheath teeth and a similar number of stem ridges.

Equisetum hyemale ssp. *affine* (common scouring rush) is a common and sometimes weedy species with a circumpolar distribution. It can be found growing on roadsides, riverbanks, and forest slopes, often in sandy soils. It can be easily identified by its stem sheaths that are constricted at the base and have distinctive black bands above and/or below, and by its numerous deciduous sheath teeth.

There is one other species of *Equisetum* subg. *Hippochaete* in Ontario. *Equisetum laevigatum* (smooth scouring rush) has unbranched annual stems and typically grows in open areas including riverbanks, roadsides, and prairies. It is the only *Equisetum* species that is endemic to North America.

E. scirpoides

E. variegatum

E. hyemale

CRYPTIC BOTANY

Ferns (Pteridophyta)

Most species of ferns have finely divided compound leaves (fronds), although a few species have simple leaves. Fern fronds are typically found growing in either circular tufts or scattered along an undergound stem. All fern species produce spores inside sporangia. The sporangia are sometimes produced on seperate fertile fronds that are distinctly different from the regular fronds are typically brown at maturity, but in most species the sporangia are produced in clusters (called sori) on the underside of regular fronds. The sori may be exposed, protected by a translucent piece of tissue called an indusium, or protected by the inrolled margins of the frond (termed 'a false indusium').

There are 60 species and 22 hybrids of true ferns that can be found in Ontario plus an additional 18 species of Ophioglossoid ferns. The Ophioglossoid ferns (e.g. *Botrychium*, pg. 77) are often grouped with the true ferns but the evolutionary relationships between the Ophioglossoid ferns, horsetails, and true ferns are poorly understood. They are included here for simplicity. On the following pages 32 of the most common woodlot fern species in southern and central Ontario are profiled, including representatives from most of the genera you are likely to find in an Ontario woodlot.

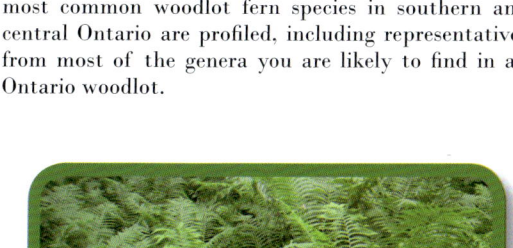

CRYPTIC BOTANY

Important Fern Characters

Fern biology has its own unique set of terms to describe important characteristics of fern leaves. The entire leaf is called a frond. The expanded part of the leaf is called the blade. The leaf stalk is either called the stipe or the rachis, depending on if it is the section below or above the start of the leaf blade, respectively. The first division of the frond is called a pinna and the second division is called a pinnule. Some ferns have a small piece of tissue called an indusium that covers each sorus and the shape and size of the indusium are often important for identification (a hand lens will be helpful here!).

Figure 14. Parts of a fern frond

Figure 15. Magnified fern sorus

Figure 15. Underside of a fern frond

Figure 16. Sorus without indusium

Figure 17. Sorus with false indusium formed by the leaf margin

Figure 18. Sorus with true indusium

CRYPTIC BOTANY

Another important character for fern identification is the degree to which the fronds are subdivided into leaflets. A frond that has leaflets arranged along the primary stalk is once-divided or pinnate. A frond that has secondary stalks arranged along the primary stalk with the leaflets placed on the secondary stalk is two times divided or bipinnate. The leaflets may also be lobed; for example, a bipinnate frond with deeply lobed leaflets is described as 2.5 times divided (see third image below). Another useful character for fern identification is whether the ferns grow in circular tufts or grow scattered along a creeping stem or rhizome.

Figure 19. Frond once-divided (pinnate)

Figure 20. Frond two times divided (bipinnate)

Figure 21. Frond 2.5 times divided

Figure 20. Fronds growing in a circular tuft

Figure 20. Fronds scattered along a creeping rhizome

CRYPTIC BOTANY

Genus Key to the Common Woodlot Ferns of Southern Ontario

1(a) Spores produced on separate stalks or leaf segments that are typically brown at maturity and distinctly different than the regular leafy green tissue 2

2(a) Sporangia not covered by stiff brown tissue; fertile stalk deciduous; leaf stalk with either numerous vascular bundles or a single U-shaped vascular bundle 3

3(a) Fronds occurring singly; leaf blade 3 times divided and distinctly triangular; leaf stalk with numerous vascular bundles ....*Botrychium virginianum*, pg. 77

3(b) Fronds growing in circular tufts; leaf blade 1.5 to 2 times divided, not triangular; leaf stalk with a U-shaped vascular bundle 4

4(a) Sporangia borne on a separate stalk with no green leafy tissue; leaf blade with a tuft of woolly hairs at the base of each segment ....*Osmundastrum cinnamomeum*, pg. 92

4(b) Sporangia borne on the same stalk as regular green leafy tissue; leaf blade without a tuft of woolly hairs at the base of each segment ....*Osmunda*, pg. 91

2(b) Sporangia covered by stiff brown tissue; fertile stalk often persisting through winter; leaf stalk with two vascular bundles at the base, sometimes uniting into a single U-shaped bundle farther up 5

5(a) Leaf blade once-divided, triangular; fronds scattered along a creeping rhizome ....*Onoclea sensibilis*, pg. 90

CRYPTIC BOTANY

5(b) Leaf blade 1.5 times divided, broadest at the middle and tapering to the base and apex; fronds growing in circular tufts ....*Matteuccia struthiopteris*, pg. 89

1(b) Spores produced in clusters on the underside of regular green leafy tissue 6

6(a) Sori occuring along margins of leaf segments, partly covered by the inrolled leaf margins; fronds always scattered along an underground stem, often in dense colonies but never in distinctive circular tufts 7

7(a) Plants small with two distinct types of fronds; fertile fronds longer than sterile, noticeably different in shape; plants typically growing on calcareous rock ....
..................................*Cryptogramma stelleri*, pg. 78

7(b) Plants large with only one type of frond; plants typically growing in soil 8

8(a) Leaf blade triangular, leathery, with oblong segments; stalk with numerous vascular bundles; sori forming a continuous band ....*Pteridium aquilinum*, pg. 96

8(b) Leaf blade semi-circular, membranous, with fan-shaped lobed segments; stalk with a single U-shaped vascular bundle; sori close together but separated on distinct lobes ....*Adiantum pedatum*, pg. 73

6(b) Sori not ocurring along margins of leaf segments, or if sori along margins then not partly covered by inrolled leaf margins; frond arrangement various 9

CRYPTIC BOTANY

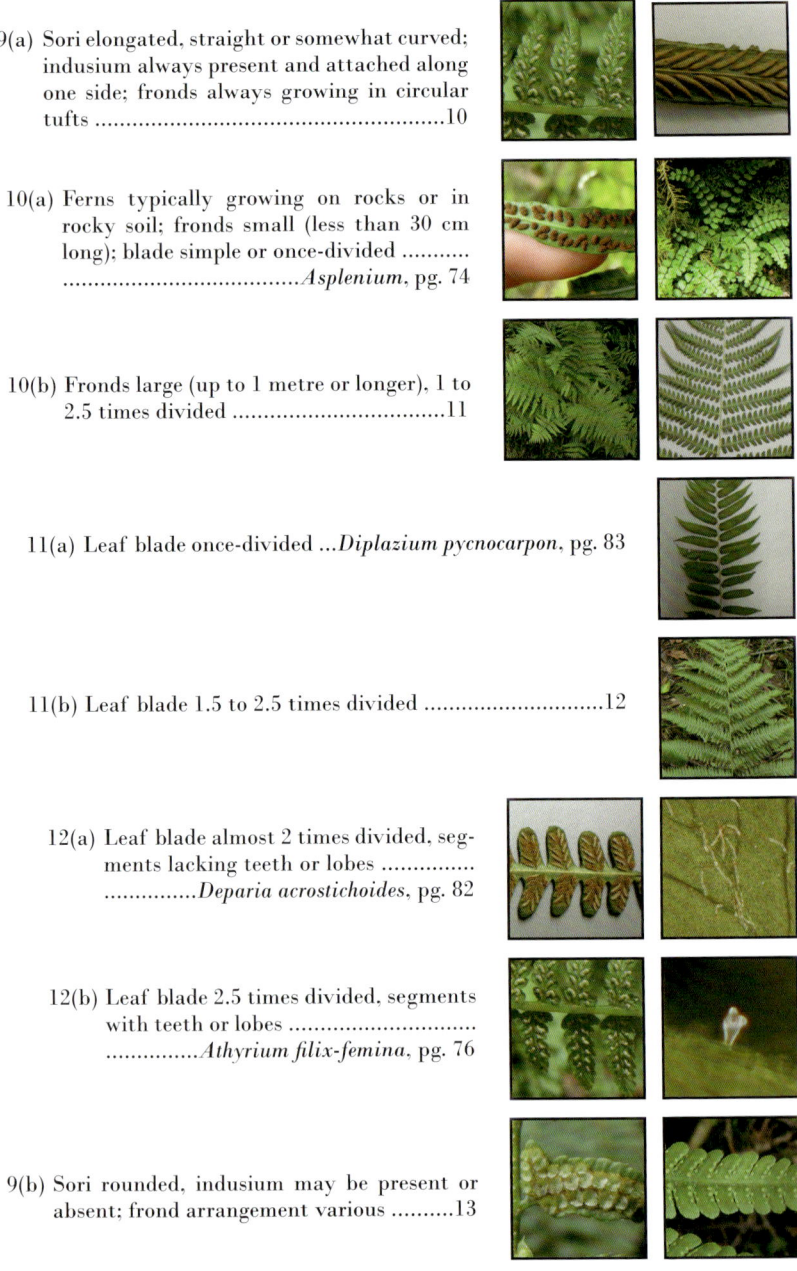

9(a) Sori elongated, straight or somewhat curved; indusium always present and attached along one side; fronds always growing in circular tufts 10

10(a) Ferns typically growing on rocks or in rocky soil; fronds small (less than 30 cm long); blade simple or once-divided .... *Asplenium*, pg. 74

10(b) Fronds large (up to 1 metre or longer), 1 to 2.5 times divided 11

11(a) Leaf blade once-divided ...*Diplazium pycnocarpon*, pg. 83

11(b) Leaf blade 1.5 to 2.5 times divided 12

12(a) Leaf blade almost 2 times divided, segments lacking teeth or lobes ....*Deparia acrostichoides*, pg. 82

12(b) Leaf blade 2.5 times divided, segments with teeth or lobes ....*Athyrium filix-femina*, pg. 76

9(b) Sori rounded, indusium may be present or absent; frond arrangement various 13

CRYPTIC BOTANY

13(a) Sori lacking an indusium; fronds always scattered along a creeping underground stem, never in circular tufts 14

14(a) Leaf blade deeply lobed and appearing nearly once-divided; sori with glandular paraphyses scattered amongst the sporangia ...*Polypodium virginianum*, pg. 94

14(b) Leaf blade more than once-divided; sori without glandular paraphyses 15

15(a) Leaf blade 1.5 times divided, moderately hairy; leaf stalk with wings connecting the blade segments .... *Phegopteris connectilis*, pg. 93

15(b) Leaf blade strongly triangular, 2.5 times divided, lacking hairs; leaf stalk without wings .... *Gymnocarpium dryopteris*, pg. 88

13(b) Sori with an indusium (sometimes withering and falling off at maturity); frond arrangement various 16

16(a) Fronds scattered along a creeping rhizome; leaf stalk, veins, and leaf margins with easily visible non-glandular hairs ... . *Thelypteris*, pg. 97

16(b) Fronds growing in circular tufts; leaf stalk, veins, and blade glabrous or with tiny glandular hairs 17

71

CRYPTIC BOTANY

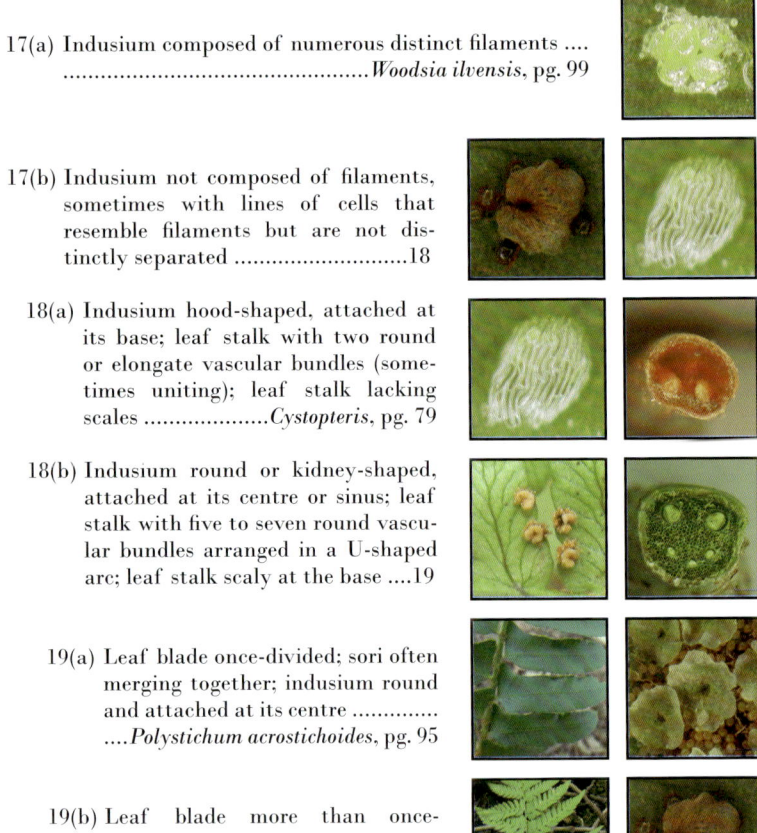

17(a) Indusium composed of numerous distinct filaments ....
....*Woodsia ilvensis*, pg. 99

17(b) Indusium not composed of filaments, sometimes with lines of cells that resemble filaments but are not distinctly separated 18

18(a) Indusium hood-shaped, attached at its base; leaf stalk with two round or elongate vascular bundles (sometimes uniting); leaf stalk lacking scales ....*Cystopteris*, pg. 79

18(b) Indusium round or kidney-shaped, attached at its centre or sinus; leaf stalk with five to seven round vascular bundles arranged in a U-shaped arc; leaf stalk scaly at the base 19

19(a) Leaf blade once-divided; sori often merging together; indusium round and attached at its centre ....
....*Polystichum acrostichoides*, pg. 95

19(b) Leaf blade more than once-divided; sori distinct; indusium kidney-shaped, attached at it sinus ....*Dryopteris*, pg. 84

CRYPTIC BOTANY

Northern maidenhair fern
Adiantum pedatum

Family: *Pteridaceae*

Vegetative Characteristics: Fronds many times divided, with a distinctive semicircular blade; pinnules thin and membranous, fan-shaped, with several blunt lobes; stipe stiff, shiny, and purplish, with a U-shaped vascular bundle; fronds scattered along a creeping rhizome.

Reproductive Characteristics: Sori elongate, found along the outer margin of each leaf segment, covered by a translucent false indusium formed by the inrolled margins of the leaf segment.

Notes: *Adiantum pedatum* (northern maidenhair fern) is a less common diploid species that is typically found growing in humus-rich soils in moist woodlands, often on slopes. This is an attractive fern with a unique appearance and is one of the most distinctive ferns in eastern Canada. Its membranous semi-circular blades make it easily recognizable even from a distance. The genus name *Adiantum* is from the Greek *adiantos*, meaning 'unwetted', a reference to the way the fan-shaped leaves shed raindrops.

CRYPTIC BOTANY

A. trichomanes ssp. *quadrivalens*

Spleenworts
Asplenium spp.

Family: *Aspleniaceae*

Vegetative Characteristics: In our flora, all species have distinctive fronds that are simple or once divided (2 times divided in *Asplenium ruta-muraria*); stipes usually with two small round vascular bundles, sometimes appearing as a single bundle; fronds growing in circular tufts.

Reproductive Characteristics: Sori linear, found on the underside of fronds; indusium opening along one side, attached along its outer edge.

Notes: Our *Asplenium* species are habitat-specific and locally abundant species that are frequently found growing on rocks. The genus name *Asplenium* is from Greek *splen*, meaning 'spleen', as these species were thought to be useful for treating spleen diseases.

Key to Some Common *Asplenium* Species

1(a) Leaf blade simple, long-tapering, rooting at the tip ....*Asplenium rhizophyllum*

1(b) Leaf blade once divided, not rooting at the tip 2

2(a) Leaf stalk strongly reddish-brown ....*Asplenium trichomanes*

2(b) Leaf stalk green....................*Asplenium trichomanes-ramosum*

CRYPTIC BOTANY

Asplenium Species Descriptions

Asplenium trichomanes (maidenhair spleenwort; also pictured on previous page) is a locally common species that grows in sheltered rock crevices and on mossy boulders in forests. In Ontario it has been traditionally divided into two subspecies based on cytotype: ssp. *trichomanes* is a diploid that grows exclusively on acidic rocks, whereas ssp. *quadrivalens* is an autotetraploid that grows exlusively on calcareous rocks.

Asplenium trichomanes-ramosum (green spleenwort) is a less common diploid species that grows on calcareous rocks in shady locations. *Asplenium trichomanes-ramosum* is similar to *Asplenium trichomanes* but can be distinguised by its distinctive green stipe.

Asplenium rhizophyllum (walking fern) is a less common diploid species that is frequently found growing on sheltered mossy rocks in moist forests. It can be easily distinguished by its simple fronds that elongate and root at the tip to form new plantlets (hence the name 'walking fern').

There are three other uncommon and rare species in Ontario: *A. platyneuron* (ebony spleenwort), which is considered to be rare in Ontario; *A. ruta-muraria* (wall-rue), a rare species that is typically only found growing as isolated individuals on calcareous rocks along the Bruce Peninsula and on Manitoulin Island; and *A. scolopendrium* (hart's-tongue fern), a species of special concern in Ontario that is only found growing in calcareous soils along the Niagara Escarpment.

A. trichomanes ssp. *quadrivalens*

A. trichomanes-ramosum

A. rhizophyllum

A. rhizophyllum sori

CRYPTIC BOTANY

Lady fern
Athyrium filix-femina

Family: *Athyriaceae*

Vegetative Characteristics: Fronds 2.5 times divided; blade lance-shaped, broadest near the middle and somewhat to moderately narrowed at base; pinnae lanceolate, tapering to a slender point; pinnules toothed and often lobed halfway to midrib; stipe with two strap-like vascular bundles near the base; veins often with distinctive bulbous-headed glands; fronds up to 1 m tall with blades 10–35 cm wide, growing in circular tufts rather than in loose patches.

Reproductive Characteristics: Sori elongated, produced on underside of regular leaves; indusium curved to half-moon shaped, attached on one side.

Notes: *Athyrium filix-femina* var. *angustum* (lady fern) is a very common diploid species typically found growing in moist woods, thickets, and occasionally in swamps. The genus name *Athyrium* means 'doorless', indicating the way in which the sporangia push back the outer edge of the indusium and the species name *filix-femina* is Latin for 'female fern'.

CRYPTIC BOTANY

Rattlesnake fern
Botrychium virginianum

Family: *Ophioglossaceae*

Vegetative Characteristics: Fronds triangular, 3 times divided; blade membranous; stipe fleshy; pinnule segments lanceolate and either toothed or lobed; stipe with numerous round vascular bundles; fronds occurring singly.

Reproductive Characteristics: Sporangia are borne on a long stalk that extends from the first division of the leafy tissue; sporangia exposed, containing thousands of spores.

Notes: *Botrychium virginianum* (rattlesnake fern) is a very common tetraploid species that is found in rich shaded forests. *Botrychium virginianum* is the most common species of *Botrychium* in Ontario. There are 16 other species of *Botrychium* in Ontario, almost all of which are uncommon or rare. The common name 'rattlesnake fern' refers to the resemblance between its sporangial clusters and the tail of a rattlesnake. While often grouped with the true ferns, the evolutionary relationships between the Ophioglossoid ferns, horsetails, and true ferns are poorly understood. The genus name *Botrychium* is from Latin *botry*, 'bunch of grapes'; the common name for most species in the genus is grape fern, which refers to how the sporangial clusters resemble a bunch of grapes.

CRYPTIC BOTANY

Slender cliff-brake
Cryptogramma stelleri

Family: *Pteridaceae*

Vegetative Characteristics: Fronds dimporphic, fertile fronds 2 times divided, sterile fronds 1.5 times divided; fertile fronds longer than sterile; blades thin and fragile, turning brown later in the season; pinnules on fertile fronds lanceolate with inrolled margins; pinnules on sterile fronds oblong, ovate, or obovate, minutely lobed; stipe with two round vascular bundles at the base; fronds scattered along a creeping rhizome.

Reproductive Characteristics: Sori elongated, found along the margins of fertile fronds, covered by a false indusium formed by the inrolled frond margins.

Notes: *Cryptogramma stelleri* (slender cliff-brake) is an uncommon diploid species that can be found growing in shaded crevices on calcareous rocks, typically in coniferous forests. The genus name *Cryptogramma* is from Greek *cryptos*, 'hidden', and *gramme*, 'line', referring to the linear sori covered by the inrolled fron margins. The species was named after George Wilhelm Steller, an 18th century naturalist.

CRYPTIC BOTANY

Bladder ferns
Cystopteris spp.

C. bulbifera

Family: *Cystopteridaceae*

Vegetative Characteristics: Fronds 1.5 to 2.5 times divided (3.5 in *C. montana*); stipe with two round or elongate vascular bundles at the base, often uniting into a single U-shaped vascular bundle farther up; fronds typically growing in tufts (scattered along a creeping rhizome in *C. protrusa*).

Reproductive Characteristics: Sori round, found on the underside of regular fronds; indusium hood-shaped and attached on one side, typically withering and falling off at maturity.

Notes: Hybridization and polyploidy are common in the genus and the species are linked in a complex reticulate network. *Cystopteris laurentiana*, *C. protrusa*, and *C. tenuis* were formerly considered to be varieties of *C. fragilis* but genetic and genomic analyses have revealed that these are four distinct species. The genus name *Cystopteris* is derived from Greek *kystos*, 'bladder', and *pteris*, 'fern', referring to the indusium which is inflated and bladder-like.

Key to Some Common *Cystopteris* Species

1(a) Leaf blades tapering to a long-attenuate tip; blades and indusia densely covered in glandular hairs; blades often bearing distinctive bulblets beneath ....*Cystopteris bulbifera*, pg. 80

1(a) Blades and indusia glabrous or with scattered glandular hairs; blades lacking bulblets 2

2(a) Blades and indusia with scattered glandular hairs; blade usually broadest near the base; veins ending in both points and notches ....
........................*Cystopteris laurentiana*, pg. 81

2(b) Blades and indusia glabrous; blade usually broadest near the middle; venation various 3

79

CRYPTIC BOTANY

3(a) Pinnae perpendicular to rachis; veins ending only in points, never in notches ....*Cystopteris fragilis*, pg. 80

3(b) Pinnae often at an acute angle to rachis and curving upwards toward blade tip; veins ending in both points and notches ... .*Cystopteris tenuis*, pg. 81

Cystopteris Species Descriptions

C. bulbifera

Cystopteris bulbifera (bulblet bladder fern; also pictured on previous page) is a very common diploid species that can be found growing in moist woods, in shaded ravines, and on calcareous rocks. It is a distinctive fern that can be easily identified by its long, tapering fronds with a dense covering of glandular hairs and by the presence of bulblets. Immature specimens can be difficult to identify.

Cystopteris fragilis (fragile fern) is a circumpolar tetraploid species found throughout Ontario that grows primarily in rock crevices but also grows in thin soil over rock. *Cystopteris fragilis* can be easily confused with *C. tenuis* and *C. laurentiana* where their ranges overlap, but *C. fragilis* can be distinguished by the absence of glandular hairs in combination with pinnae that are perpendicular to the rachis and veins ending only in points.

C. fragilis

Cystopteris fragilis veins

CRYPTIC BOTANY

Cystopteris tenuis (Mackay's fragile fern) is a common tetraploid species that grows in rock crevices, on rocky slopes, and on boulders, but also grows occasionally in soil. It is much more restricted in range than *C. fragilis* and can be found in the southern and eastern parts of Ontario. *Cystopteris tenuis* can be distinguished by the absence of glandular hairs in combination with pinnae at an acute angle to the rachis and veins ending either in points or notches.

Cystopteris laurentiana (Laurentian fragile fern) is an uncommon hexaploid species that can be found in abundance on Manitoulin Island and also occurs along the north shore of Lake Superior. It is mainly found on calcareous rocks and rocky slopes. *Cystopteris laurentiana* can be distinguished from *C. fragilis* and *C. tenuis* by the presence of sparse glandular hairs, particularly on fresh indusia. On mature specimens, the indusia are frequently withered and/or missing, but *C. laurentiana* can be distinguished by having pinnae perpendicular to the rachis (*C. tenuis* has pinnae at an angle) and veins ending in both notches and points (*C. fragilis* has veins ending only in points). *Cystopteris laurentiana* is also typically broader toward the base and has a longer tip than *C. fragilis*.

There are two other species of *Cystopteris* in Ontario. *Cystopteris montana* (mountain bladder fern) is rare and found only along the north shore of Lake Superior. *Cystopteris protrusa* (southern bladder fern) is a southern species, known from only a few locations in the southernmost part of Ontario.

C. tenuis

C. tenuis veins

C. laurentiana

C. laurentiana veins

CRYPTIĆ BOTANY

Silvery glade fern
Deparia acrostichoides

Family: *Athyriaceae*

Vegetative Characteristics: Fronds elongate, 1.5 times divided; blades broadest at the middle, gradually narrowing to the base and apex; pinnae lobes blunted and finely toothed; stipe with two elongate vascular bundles near the base, uniting farther up the stem; veins with multicellular hairs; fronds growing in circular tufts.

Reproductive Characteristics: Sori linear, elongated; indusia becoming silvery at maturity, opening along one side.

Notes: *Deparia acrostichoides* (silvery glade fern) is a less common diploid species that is typically found in rich woodlands. The name 'silvery glade fern' refers to the sori which turn silver at maturity. The genus name *Deparia* is from the Greek *depas*, 'saucer', a reference to the saucer-shaped indusium of the type species *D. prolifera*.

CRYPTIC BOTANY

Narrow-leaved glade fern
Diplazium pycnocarpon

Family: *Athyriaceae*

Vegetative Characteristics: Fronds elongate, once-divided; pinnae membranous, tapering to a pointed tip with entire or sometimes crenulate margins; veins lacking hairs or glands; stipe with two elongate vascular bundles near the base, uniting farther up the stem; fronds growing in tufts.

Reproductive Characteristics: Sori linear, elongated; opening along one side.

Notes: *Diplazium pycnocarpon* (narrow-leaved glade fern) is a less common diploid species that grows in rich moist forests. It is an attractive species and at first glance could be mistaken for *Polystichum acrostichoides* (pg. 95) but it can be easily distinguished by its membranous non-evergreen fronds which are much thinner than those of *P. acrostichoides*. The genus name *Diplazium* is from the Greek *diplazein*, 'double', in reference to a double sorus.

CRYPTIC BOTANY

D. carthusiana

Wood ferns
Dryopteris spp.

Family: *Dryopteridaceae*

Vegetative Characteristics: Fronds 1.5 to 3 times divided; stipe often scaly, especially at the base; stipe with 5 to 7 round vascular bundles arranged in a U-shaped arc; fronds usually large, growing in circular tufts.

Reproductive Characteristics: Sori rounded, found on the underside of fronds; indusium kidney-shaped, attached at its sinus.

Notes: This is a difficult genus. Hybridization and polyploidy are common. Recent phylogenetic studies have revealed that the species are related in a complicated reticulate network. In Ontario alone, 15 of 36 possible hybrid combinations have been reported. The genus name *Dryopteris* is from Greek *drys*, a 'tree', and *pteris*, a 'fern'.

Key to *Dryopteris* Species

1(a) Blades 1.5 to 2 times divided 2

2(a) Sori located near margins of pinnules; blade leathery, paler beneath; tips of pinnules with distinctly rounded lobes or teeth ....
..........................*Dryopteris marginalis*, pg. 86

84

CRYPTIC BOTANY

2(b) Sori located between midrib and margins of pinnules; pinnules distinctly sharp-toothed 3

3(a) Fertile fronds noticeably different than sterile; fertile fronds taller and more erect than sterile fronds; pinnae of fertile fronds nearly horizontal; ferns growing in wetlands ....*Dryopteris cristata*, pg. 86

3(b) Fertile fronds not noticeably different than sterile; pinnae of fertile fronds nearly vertical; ferns growing in wetlands or not 4

4(a) Blades gradually reduced to tip; pinnae broadest at base; ferns growing in wetlands .....*Dryopteris clintoniana*, pg. 86

4(b) Blades abruptly reduced to tip; pinnae broadest near middle; ferns growing in rich forest soils ....*Dryopteris goldiana*, pg. 86

1(b) Blades 2 to 3 times divided 5

5(a) Basal pinnules on basal pinnae distinctly shorter than adjacent pinnules; blades and indusia glandular ....*Dryopteris intermedia*, pg. 87

5(b) Basal pinnules on basal pinnae distinctly longer than adjacent pinnules; blades and indusia glabrous ....*Dryopteris carthusiana*, pg. 87

85

CRYPTIC BOTANY

Dryopteris Species Descriptions

Dryopteris marginalis (marginal wood fern) is a common diploid species that is frequently found growing in rocky soil, on rocky forested slopes, or in sheltered rock crevices. It can be easily distinguished by its leathery fronds and distinctive sori positioned near the margins of pinnules.

Dryopteris goldiana (Goldie's wood fern) is a less common diploid species that grows primarily in rich, moist forests. It has distinctive, large, attractive fronds that abruptly taper to a pointed tip.

Dryopteris cristata (crested wood fern) is a common tetraploid species that grows primarily in swampy woods and wetlands. It can be identified by its distinctive fertile fronds that have pinnae rotated horizontally from the blade.

Dryopteris clintoniana (Clinton's wood fern) is a common hexaploid species that grows in woody swamps. *Dryopteris clintoniana* can be identified by its large, long-tapering fronds with pinnae not rotated horizontally. Poorly developed specimens can be difficult to identify. *Dryopteris clintoniana* is believed to have arisen from a fertile cross between *D. goldiana* and *D. cristata*.

CRYPTIC BOTANY

Dryopteris carthusiana (spinulose wood fern; also pictured on pg. 84) is a very common tetraploid species that grows in moist to wet or swampy woods but also occasionally on drier sites. *Dryopteris carthusiana* is frequently confused with *D. intermedia*. *Dryopteris carthusiana* can be distinguished from *D. intermedia* by having less lacy yellow-green fronds, glabrous blades and indusia, and inner lower pinnules that are disctinctly longer than the adjacent pinnules.

Dryopteris intermedia (evergreen wood fern) is a very common diploid species that grows in moist woods and in rocky forest soils. *Dryopteris intermedia* can be distinguished from *D. carthusiana* by having lacy blue-green fronds, densely glandular blades and indusia, and inner lower pinnules that are distinctly shorter than the adjacent pinnules. *Dryopteris intermedia* and *D. carthusiana* readily hybridize to form a triploid hybrid (*Dryopteris* × *triploidea*), which can be identified by the presence of aborted and misshapen spores and characteristics intermediate between its parent species.

There are three other species of *Dryopteris* in Ontario. *Dryopteris expansa* (northern wood fern) is a northern and western species, *D. filix-mas* (male fern) is an uncommon species found along the Niagara Escarpment, and *D. fragrans* (fragrant wood fern) is a circumpolar northern species.

D. carthusiana

D. intermedia

D. cristata (fertile frond)

D. intermedia (basal)

CRYPTIC BOTANY

Oak fern
Gymnocarpium dryopteris

Family: *Cystopteridaceae*

Vegetative Characteristics: Blade triangular, 2.5 times divided; uppermost pinnules bluntly lobed, basal pinnules more deeply cut; stipe with two round vascular bundles near the base; blade yellow-green, triangular, horizontal to ground; fronds up to 30 cm tall, scattered along a rhizome rather than in distinct clusters.

Reproductive Characteristics: Sori small and rounded, found on the underside of pinnules; indusium not present.

Notes: *Gymnocarpium dryopteris* (oak fern) is a common diploid species that occurs in patches in moist to wet mixed forests, wet conifer forests, and also in rocky woods. It is an attractive fern, easily recognized by its small triangular blades positioned horizontal to the ground. There are two other species of *Gymnocarpium* in Ontario that are much less common: *G. robertianum* (limestone oak fern) and *G. jessoense* ssp. *parvalum* (Nahanni oak fern). The genus name *Gymnocarpium* means 'naked fruit', alluding to the lack of an indusium on the sori, and the species name *dryopteris* is from Latin *dryo*, 'oak', and *pteris*, 'fern'.

CRYPTIC BOTANY

Ostrich fern
Matteuccia struthiopteris

Family: *Onocleaceae*

Vegetative Characteristics: Fronds 1.5 times divided, strongly tapered to the base and abruptly tapered at tip, broadest above the middle; pinnae deeply cut, blades appearing nearly bipinnate; pinnules entire, oblong, blunted; stipe with two elongated vascular bundles near the base; fronds very large, up to 2 m long, growing in dense circular tufts.

Reproductive Characteristics: Sori borne on separate and distinct fertile fronds; rolled-up margins of pinnae enclose sori and form pod-like structures, turning brown at maturity.

Notes: *Matteuccia struthiopteris* var. *pensylvanica* (ostrich fern) is a common and widespread diploid species that often forms dense stands in shaded hardwood forests, moist hardwood swamps, and in ditches and along streambanks. One of the largest ferns in our area, its distinctive crown resembles a large headdress. The genus is named for Carlo Matteucci, a 19th century Italian physicist.

CRYPTIC BOTANY

Sensitive fern
Onoclea sensibilis

Family: *Onocleaceae*

Vegetative Characteristics: Fronds once-divided; blade broadly triangular, widest at base; rachis with 'wings' that are narrow at the base of the blade and broaden towards the apex; pinnae with wavy margins, most pronounced on basal pinnae; stipe and veins with multicellular hairs; stipe with two elongated vascular bundles near the base, sometimes uniting; fronds 50–80 cm long, loosely clustered along a creeping rhizome.

Reproductive Characteristics: Fertile fronds shorter than sterile, blackened at maturity, persisting through the winter; sporangia are contained in berry-like clusters (modified pinnules) that split open at maturity to release spores.

Notes: *Onoclea sensibilis* (sensitive fern) is a very common diploid species that is found in moist to wet conifer and hardwood swamps, low-lying woodland areas, wet meadows, and along the edges of ponds. The genus name *Onoclea* means 'closed cup' and refers to the way sori are enclosed by the leaflets of fertile fronds. The species name *sensibilis* refers to the sensitivity of the fronds, which will blacken following the first frost.

CRYPTIC BOTANY

Osmunda

O. claytoniana

Family: *Osmundaceae*

Vegetative Characteristics: Fronds large, 1.5 or 2 times divided; pinnae lacking tufts of hairs at the base; stipe winged at the base; stipe with a distinctive U-shaped vascular bundle curled at both ends; fronds large, growing in large circular tufts.

Reproductive Characteristics: Spores produced on a fertile frond that also contains green leafy tissue above and/or below the fertile pinnae; fertile pinnae lacking leafy tissue and consisting of masses of exposed sporangia.

Notes: The genus name *Osmunda* is thought to be derived from 'Osmunder', the Saxon god of war.

Key to *Osmunda* Species

1(a) Fronds 1.5 times divided; fertile tissue interrupts the frond with green leafty tissue above and below .... *Osmunda claytoniana*

1(b) Fronds 2 times divided; fertile tissue occurs at the frond apex with green leafy tissue below .... *Osmunda regalis*

O. regalis

Osmunda Species Descriptions

Osmunda claytoniana (interrupted fern) is a common diploid species that typically grows in swamps and marshes. It can be identified by having fronds that are 1.5 times divided, pinnae that lack a tuft of woolly hairs at the base, and fertile pinnae that 'interrupt' the fertile frond.

Osmunda regalis (royal fern), like *O. claytoniana*, is a common diploid species that typically grows in swamps and marshes. It can be identified by having fronds that are 2 times divided and fertile pinnae at the apex of the fertile frond.

CRYPTIC BOTANY

Cinnamon fern
Osmundastrum cinnamomeum

Family: *Osmundaceae*

Vegetative Characteristics: Fronds 1.5 times divided; pinnae each with a tuft of hairs at the base; pinnae segments short, oblong, blunt-tipped, with entire margins; stipe not winged at the base, with a single U-shaped vascular bundle curled at the ends; fronds large, growing in circular tufts.

Reproductive Characteristics: Spores produced on a seperate fertile frond with no green tissue; fertile pinnae lacking leafy tissue and consisting of masses of naked sporangia.

Notes: *Osmundastrum cinnamomeum* (cinnamon fern; syn. *Osmunda cinnamomea*) is a common diploid species that typically grows in swamps and marshes. It was formerly included in the genus *Osmunda* (previous page) but was segregated out when phylogenetic studies revealed that the rest of the genus *Osmunda* was more closely related to *Todea* then to the section containing *Osmundastrum cinnamomeum*. The sterile fronds of *Osmunastrum cinnamomeum* can be easily confused with sterile fronds of *Osmunda claytoniana* (previous page) that lack fertile pinnae but *Osmundastrum cinnamomeum* can be distinguished by the tuft of hairs found at the base of each pinna. The common name 'cinnamon fern' refers to the rusty colour of the fertile pinnae.

CRYPTIC BOTANY

Northern beech fern
Phegopteris connectilis

Family: *Thelypteridaceae*

Vegetative Characteristics: Blades bipinnatifid, triangular, broadest at base; 'wings' present on rachis and connect all but the lowest pair of pinnae; lowermost pinnae bent downwards and outwards; pinnules oblong, blunted, moderately covered in hairs; stipe long, somewhat scaly, with two strap-like vascular bundles; fronds up to 40 cm tall, occurring singly along a creeping rhizome.

Reproductive Characteristics: Sori small and round, found near the margins on the underside of pinnules; indusium not present.

Notes: *Phegopteris connectilis* (northern beech fern) is a common triploid species that occurs in moist conifer and hardwood swamps, near the edge of ponds, and in moist rocky crevices. This is a distinctive fern, easily recognized by the winged rachis and downward-pointing basal pinnae. *Phegopteris connectilis* is an apogamous species, generating unreduced spores, and a sporophyte is produced directly from the gametophyte without fertilization. The genus name *Phegopteris* is derived from Greek *phegos*, 'beech', and *pteris*, 'fern'.

CRYPTIC BOTANY

Rock polypody
Polypodium virginianum

Family: *Polypodiaceae*

Vegetative Characteristics: Fronds deeply lobed and appearing nearly once-divided; blade oblong, somewhat leathery; pinnae oblong with entire or crenulate margins; stipe with three round vascular bundle near the base, often appearing as a single bundle; fronds scattered along a creeping rhizome.

Reproductive Characteristics: Sori bulbous, round, located near the apex on the underside of fronds; indusium not present but sori contain numerous glandular paraphyses (modified sporangia) scattered amongst the sporangia.

Notes: *Polypodium virginianum* (rock polypody) is a very common tetraploid species that is typically found growing in moss or shallow layers of organic matter on rocks in wooded areas. This is a distinctive fern with small evergreen fronds that can be found in abundance on rocks in mixed forests. In older texts you will find refererence to a diploid cytotype that has a more eastern distribution but this taxon is now recognized as a separate species, *P. appalachianum*. Their sterile triploid hybrid is also common. The genus name *Polypodium* is from Greek *poly*, 'many' and *podion*, 'little foot'.

CRYPTIC BOTANY

Christmas fern
Polystichum acrostichoides

Family: *Dryopteridaceae*

Vegetative Characteristics: Blade narrow, once-divided, evergreen; pinnae oblong, toothed, each with a prominent auricle ('ear') on the upper side; stipe short, densely scaly, containing 5 vascular bundles arranged in a U-shaped arc; fronds 35–65 cm long, growing in a circular tuft.

Reproductive Characteristics: Sori borne on reduced upper pinnae, usually merging and completely covering the lower surfaces of pinnae; indusium round, attached at its centre.

Notes: *Polystichum acrostichoides* (Christmas fern) is a very common and distinctive diploid species found growing in rich woods and on humus-rich rocky slopes. The name Christmas fern refers to the evergreen fronds, which can be found underneath the snow at Christmas time. There are two other *Polystichum* species in Ontario: *P. braunii* (Braun's holly fern) and *P. lonchitis* (holly fern). The genus name *Polystichum* is from Greek *poly*, 'many', and *stichos*, 'row', and refers to the dense rows of sori on the backs of the fertile pinnae.

95

CRYPTIC BOTANY

Bracken fern
Pteridium aquilinum

Family: *Dennstaedtiaceae*

Vegetative Characteristics: Fronds very large, triangular, 3 to 3.5 times divided; blade leathery; pinnule segments oblong, entire, often with revolute margins; stipe with numerous scattered vascular bundles; fronds scattered along an underground stem.

Reproductive Characteristics: Sori continuous along the margins of leaf segments with a false indusium formed by the revolute margins of the leaf segments.

Notes: *Pteridium aquilinum* var. *latiusculum* (bracken fern) is a very common diploid species that is found worldwide in dry forests, open clearings, and ditches. This is a highly variable and often weedy species that typically forms large colonies. Its fiddleheads are often consumed as a green vegatable but recent studies have shown that bracken contains a number of mutagenic and carcinogenic compounds. The genus name is from the Greek *pteridion*, meaning a 'small fern'.

CRYPTIC BOTANY

Thelypteris

Family: *Thelypteridaceae*

Vegetative Characteristics: Fronds nearly 2 times divided; pinnules oblong, blunted, with entire or weakly crenulate margins; stipe with two vascular bundles near the base uniting into a single U-shaped bundle farther up; stipe and veins with long hairs visible without magnification; fronds scattered along a creeping rhizome, not growing in circular tufts.

Reproductive Characteristics: Sori round, located on the underside of pinne segments; indusium present, rounded.

Notes: This is a very large genus, with over 800 species worldwide, although most are tropical and only a few occur in our area. The genus name *Thelypteris* is from Greek *thelys*, 'female', and *pteris*, 'fern'.

Thelypteris palustris

Key to *Thelypteris* Species

1(a) Leaf blade tapering toward the apex and base; ferns typically growing in moist forests and along roadsides ....
....*Thelypteris noveboracensis*

1(b) Leaf blade scarcely tapering toward the base; ferns typically growing in marshes and swamps ....*Thelypteris palustris*

97

CRYPTIC BOTANY

Thelypteris palustris sori

Thelypteris Species Descriptions

Notes: *Thelypteris noveboracensis* (New York fern) is a common diploid species typically found growing in large colonies in moist woods and along roadsides. *Thelypteris noveboracensis* can be identified by having fronds that are tapered toward the apex and base of the blade.

Notes: *Thelypteris palustris* var. *pubescens* (marsh fern; also pictured on previous page) is a common diploid species that can be found in ditches, marshes, swamps, wet forests, and along riverbanks, lakeshores and streambanks. This is a very soft, thin, and fragile fern and often has a mildly distorted appearance. *Thelypteris palustris* var. *pubescens* can be identified by having fronds that are only weakly tapered at the base.

Thelypteris noveboracensis sori

Thelypteris noveboracensis

Thelypteris noveboracensis sorus

CRYPTIC BOTANY

Rusty woodsia
Woodsia ilvensis

Family: *Woodsiaceae*

Vegetative Characteristics: Fronds nearly 2 times divided; blade narrow and elongated; pinnae oblong and somewhat elongated, with a blunt rounded tip; stipe and blade usually covered with abundant scales and hairs; fronds growing in tufted clumps; old frond bases persistent and all of equal length.

Reproductive Characteristics: Sori round and closely spaced; indusium composed of numerous narrow filaments, attached below and extending over the sporangia.

Notes: *Woodsia ilvensis* (rusty woodsia) is a common diploid species that is typically found growing on exposed acidic rocks but can be sometimes found in forested areas. This is a distinctive species that can be confused with *Dryopteris fragrans* where their ranges overlap, but *W. ilvensis* can be indentified by its filamentous indusia and the presence of stubble from old frond bases (*D. fragrans* can be easily identified by the curled-up dead fronds at the base of the plant). There are six species of *Woodsia* in Ontario but *W. ilvensis* is the most common. The genus *Woodsia* was named after the English botanist Joseph Woods.

The Key to Floral Diversity

Steven Newmaster & Carole Ann Lacroix

Introduction
The key to plant identification is attained through an understanding of family concepts. Any plant can be quickly placed within a family using a handful of characters such as leaf arrangement or the number of petals. Descriptions of vegetative characters (e.g. leaf shape, leaf arrangement, etc.) are provided in this guide within the plant descriptions and botanical glossary. Although vegetative characters are useful for plant identification, floral characters provide the basis for classification. This is because floral structures are generally less variable and more consistent. A good understanding of floral morphology is critical for understanding the family concepts that follow.

The Basic Parts of a Flower
Variation in the basic parts of a flower forms the basis for classification. At the base of the flower is a receptacle where the stalk (peduncle/pedicel) is attached. This is the point of insertion of four alternating series of parts (Fig. 1): the sepals (calyx), the petals (corolla), the stamens (androecium – male parts), and the carpels (gynoecium – female parts). The stamens or male parts (androecium) are each made up of a stalk, called a filament, and a sac-like structure that contains the pollen, called an anther. The carpels or female parts (gynoecium) are each made up of a basal ovary located in the centre of the flower, a stalk called a style, and an upper portion receptive to pollen called the stigma. The gynoecium forms the fruit (pericarp) and protects the ovules that give rise to seeds. The gynoecium may have nectaries that function to attract pollinators.

The stalk of a single flower (e.g. a tulip) is called a peduncle. If the stalk branches to more than one flower, then the primary stalk is called a peduncle and each of the secondary stalks is a called a pedicel (Fig. 2). Shown on the next page are some common arrangements of flowers (inflorescences) (Figs. 3–8).

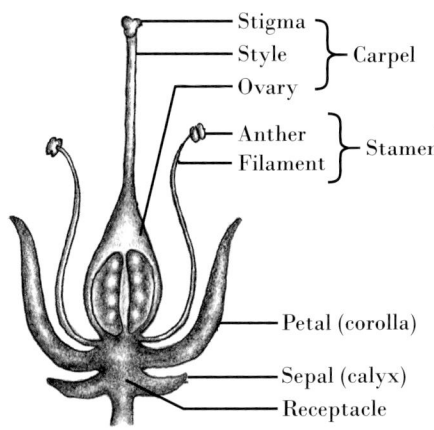

Figure 1. *Parts of a flower*

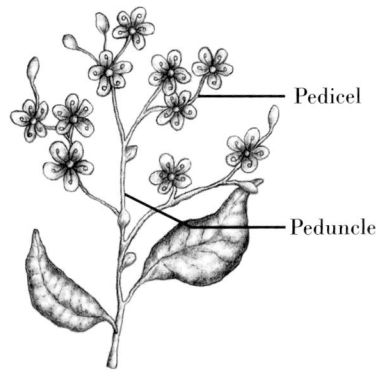

Figure 2. *Parts of an inflorescence*

Figure 3. *Spike*
Flowers sessile on the stem

Figure 4. *Raceme*
Flowers with pedicels in an unbranced inflorescence

Figure 5. *Panicle*
Flowers in a multi-branched inflorescence

Figure 6. *Umbel*
Pedicels arising from one point on the stem

Figure 7. *Corymb*
Pedicels of different lengths resulting in a flat-topped inflorescence

Figure 8. *Cyme*
Terminal or center flower the oldest

Regular/Irregular Flowers

Together, the sepals and petals make up the perianth. In a regular perianth (actinomorphic) the petals are similar to one another, as are the sepals, all of which radiate from the center of the flower so that it is radially symmetric and could be sectioned/divided many ways to give equal parts (Fig. 9). An irregular perianth (zygomorphic) is bilaterally symmetric and cannot be divided into more than two equal parts (Fig. 10).

Figure 9. Regular flower   *Figure 10. Irregular flower*

Counting Floral Parts

Interpreting the variability in the number of floral parts is a key character in determining families. Carefully count the number of sepals, petals, stamens, and carpels. A complete flower has all four series; an incomplete flower is missing one or more series (e.g. an apetalous flower is one without petals). The number of parts are referred to using the term 'merous', e.g. a flower with three sepals is 3-merous. Most monocots (e.g., grasses, rushes, irises, orchids, palms, etc.) are 3-merous with floral parts in series of three or multiples thereof (3 petals and 6 stamens), while most dicots have floral parts in multiples of four or five and are 4- or 5-merous. The series number is rather uniform within a flower so a flower with four petals will likely have four stamens/carpels or multiples thereof (e.g. 8 stamens, 12 carpels). On a piece of fruit there may be clues to the series number. For example, a blueberry has five residual sepals (5-merous).

Sexual Orientation

Determining sexual orientation is difficult but critical in developing a family concept. A flower can be perfect (bisexual) or imperfect (unisexual). Unisexual flowers with only male parts are called staminate and those with only female parts are called pistillate. A species with unisexual flowers that has both staminate and pistillate flowers on the same individual plant is termed monoecious (Greek – 'in one house'). A species with unisexual flowers that has staminate and pistillate flowers on separate individual plants is termed dioecious (Greek – 'in two houses').

Typically the stamens are inserted above and alternate with the petals. However, variability in the androecium among plant families provides many good characters, such as: stamens united to petals; filaments fused into one or two groups; two long and two short stamens (mint family – Lamiaceae); or four long and two short stamens (mustard family – Brassicaceae).

Variation in the carpels among plant families also provides many good characters for classification. Although determining carpel number is difficult, three broad categories of gynoecia can be recognized based on the degree of fusion among the carpels, namely:

Figure 11. Superior ovary

1) monocarpous/unicarpellate – gynoecium has a single carpel (e.g. cannabis, most legumes)
2) apocarpous – gynoecium has multiple distinct (free, unfused) carpels (e.g. buttercup, magnolia)
3) syncarpous – gynoecium has multiple carpels fused into a single structure (e.g. maples, most flowers)

The location of the ovary in relation to the other floral parts also provides a good character. If the insertion of the perianth and stamens is <u>below</u> the ovary, then the ovary is said to be superior (and the flower is called hypogynous) (Fig. 11). In a perigynous flower the receptacle has elongated, elevating the calyx, corolla, and androecium so that they are attached onto an open cup-like structure called a hypanthium (e.g. roses and cherries – Rosaceae) (Fig. 12). If the insertion of the perianth and stamens is <u>above</u> the ovary, then the ovary is said to be inferior (and the flower is epigynous) (Fig. 13). Some epigynous flowers have a hypanthium that is completely fused to the ovary (e.g. apple – Rosaceae).

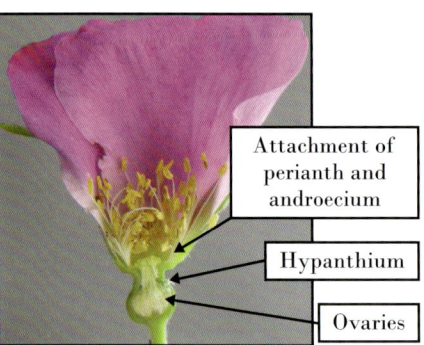

Figure 12. Perigynous flower

Figure 13. Inferior ovary

Key to Flowering Plant Families

Flower Aster-like with ray and disc flowers in a head				Asteraceae
Monocots:	Perianth a spathe and spadix (zygomorphic - irregular)			Araceae
flower parts in multiples of three; major leaf veins parallel	Perianth of 3 sepals and 3 petals	Zygomorphic (irregular)		Orchidaceae
		Actinomorphic (regular)		Melanthiaceae
	Perianth of 6 similar tepals	Flowers bisexual	Flowers single, yellow	Liliaceae
			Flowers multiple, greenish/white/blue	Asparagaceae
			Flowers multiple, yellow	Colchicaceae
		Flowers unisexual		Smilacaceae
Dicots: flower parts in multiples of four or five (sometimes three); major leaf veins reticulated				
Epigynous (inferior ovaries) or perigynous (hypanthium present)				
Imperfect (unisexual)	Male flowers in catkins	Female flower without perianth		Betulaceae
		Female flower with 4 sepals	Fruit with an upper cupule	Fagaceae
			Fruit without cupule	Juglandaceae
	Flowers not in catkins			Nyssaceae
Perfect (bisexual)	Sepals fused at least basally	3 perianth parts visible (no petals)		Aristolochiaceae
		4–5 petals	Stamens 4, petals 4	Hamamelidaceae
			Stamens 5, petals 5	Grossulariaceae
			Stamens 8–10	Ericaceae
			Stamens 20 or more	Hydrangeaceae
	Sepals separate	Petals fused	Stipules present	Rubiaceae
			Stipules absent, style short	Adoxaceae
			Stipules absent, style long	Caprifoliaceae
		Petals separate	Stamens many	Rosaceae
			Stamens 4–5	Cornaceae
Hypogynous (superior ovaries)				
Zygomorphic (irregular)	Imperfect (unisexual)		Corolla absent	Ulmaceae
			Corolla with 4–5 petals	Rutaceae
	Perfect (bisexual)	Corolla tube-like with 5 lobes		Bignoniaceae
		Corolla with 3 petals		Polygalaceae
		Corolla with 4–5 petals	Stamens 5	Violaceae
			Stamens 6–8	Sapindaceae
			Stamens 10	Fabaceae

Actinomorphic (regular)					
Flowers in dense spherical heads					Platanaceae
Stamens many (>15)	Sepals separate	Apocarpous; trees	Flowers greenish or white		Magnoliaceae
			Flowers red		Annonaceae
		Apocarpous; herbs or woody vines			Ranunculaceae
		Syncarpous			Papaveraceae
	Sepals fused	Ovary 4-lobed			Boraginaceae
		Ovary unlobed			Malvaceae
Stamens few (<12)	Petals absent	Sepals absent; flowers in catkins	Stamens 2		Salicaceae
			Stamens 3+ (rarely 2)		Myricaceae
		Stamens 4 or 8; sepals fused with 4 lobes	Leaves covered in scales		Elaeagnaceae
			Leaves without scales		Thymelaeaceae
			Milky latex		Moraceae
		Stamens usually 8; fruit a winged key			Sapindaceae
		Stamens 5; sepals 5			Cannabaceae
	Petals only fused at tip				Vitaceae
	Petals fused	Stamens 2			Oleaceae
		Stamens 5, united around style			Solanaceae
		Stamens 5, epipetalous			Polemoniaceae
		Stamens 5			Primulaceae
		Stamens 8–10			Ericaceae
		Stamens 10, on a disc			Anacardiaceae
	Petals 3	Stamens 9			Lauraceae
	Petals 4–5	Stamens 4–5, on a disc			Celastraceae
		Stamens usually 8, on a disc; fruit a winged key			Sapindaceae
		Stamens 4			Aquifoliaceae
		Stamens 4 long, 2 short			Brassicaceae
		Stamens 5	Stamens opposite petals		Rhamnaceae
			Stamens alternate petals, anthers pink		Montiaceae
			Stamens alternate petals, anthers yellow		Staphyleaceae
			Stamens alternate petals with double anthers		Geraniaceae
		Stamens 5–7	Leaves woolly below		Ericaceae
		Stamens 8–10	Flowers greenish		Fabaceae
			Flowers white		Saxifragaceae
	Petals 6	Stamens 6–12			Berberidaceae

Index to Angiosperm Families

Adoxaceae 324–325, 356–366
Anacardiaceae 293–297, 343–346
Annonaceae 159
Aquifoliaceae 218–219
Araceae 379
Aristolochiaceae 380
Asparagaceae 395–397, 401–402, 407
Asteraceae 408, 414
Berberidaceae 160–161, 384, 399
Betulaceae 154–155, 162–166, 168, 189–190, 249
Bignoniaceae 167, 175
Boraginaceae 394
Brassicaceae 377, 382–383
Cannabaceae 177
Caprifoliaceae 193, 233–241
Celastraceae 176, 200–202
Colchicaceae 415–416
Cornaceae 183–188, 386
Elaeagnaceae 195–198, 327
Ericaceae 157, 180, 199, 212–214, 225–226, 229, 354–355
Fabaceae 179, 215–216, 304
Fagaceae 174, 203, 275–288
Geraniaceae 390
Grossulariaceae 298–303
Hamamelidaceae 217
Hydrangeaceae 252
Juglandaceae 169–173, 220–221
Lauraceae 231, 326
Liliaceae 388
Magnoliaceae 232, 242
Malvaceae 342
Melanthiaceae 412–413
Montiaceae 385
Moraceae 244–245
Myricaceae 182, 246–247
Nyssaceae 248
Oleaceae 204–211, 230, 339
Orchidaceae 387
Papaveraceae 406
Platanaceae 261
Polemoniaceae 398
Polygalaceae 400
Primulaceae 411
Ranunculaceae 181, 375–376, 378, 381, 393, 403–405, 409
Rhamnaceae 289–292
Rosaceae 156, 158, 191, 243, 253, 267–273, 305–322, 331–337, 389, 391–392, 425
Rubiaceae 178
Rutaceae 274, 368
Salicaceae 262–266, 323
Sapindaceae 141–153
Saxifragaceae 410
Smilacaceae 328–329
Solanaceae 330
Staphyleaceae 338
Thymelaeaceae 192, 194
Ulmaceae 348–353
Violaceae 417–424
Vitaceae 250–251, 367

107

WOODY PLANTS

Woody Plants

Aron Fazekas, Brian Lacey, Thomas Henry & Steven Newmaster

Introduction

There is considerable diversity of woody plants in Ontario woodlots, with many species of trees and shrubs as well as several species of woody vines that can be commonly found. Although there are a great variety of forms, woody plants all possess tough fibres, i.e. rigid composites of cellulose and lignin, that allow individual plants to persist for many years and hold themselves upright under the pressure of their own weight.

The distinction between a tree and a shrub is difficult to define. Many species have forms that bridge these two categories, and examples can found that contradict most definitions. Nevertheless, trees are most often defined as woody plants that have a main trunk with secondary branches that form a crown. Mature trees are typically taller (>10 m) than mature shrubs (<10 m). Some biologists will use a minimum trunk diameter of 10 cm (30 cm girth) as the distinction between trees and shrubs, but this is a only general rule. Some trees, such as the cedars of the Bruce Peninsula, have stunted growth, and some shrubs may occasionally exceed this size. Shrubs are usually defined as woody plants that have multiple stems and no central trunk. Some shrubs are tall (1–10 m; e.g. dogwood, pg. 183) and others are short (<1 m; e.g. blueberry, pg. 354). Woody vines (technically called lianas) such as grapevines are not stiff enough to hold their long stems upright and often climb on other structures, such as trees, poles, or fences, to achieve vertical growth.

Compared with most other plants, trees are long-lived. Some reach over a thousand years old such as the dwarf eastern white cedars on the Niagara escarpment (1300+ years old). It is estimated that there may be as many as 100,000 tree species across the globe, representing approximately 25% of all living plants. In Ontario there are over 500 species of woody plants, of which 20% are rare or uncommon. This includes approximately 170 species of trees and 340 species of shrubs. Many of these woody plants are found in woodlots, but we do not have quantitative surveys with exact numbers of species and their distribution at a regional scale. Students and woodlot owners can play an important role is exploring and documenting the woody plant diversity in Ontario woodlots.

Woody plants, particularly trees, play important ecological roles in woodlots. They are the major plant life form in woodlots, contributing large amounts of biomass. All plants play a critical role in producing oxygen, but woody plants have the additional function of sequestering carbon in the form of woody biomass for long periods of time. The decay of this woody biomass is a critical part of nutrient cycling in soil. Trees and shrubs serve as a food source for many wildlife species and create structural habitat for many other plant species (e.g. epiphytic cryptogams: lichens and mosses). They are important regulators of hydrological processes, especially those that involve groundwater hydrology. A mature forest with a closed canopy can affect local patterns of precipitation and evaporation by intercepting rainfall and moderating ground temperatures. These factors aid in the prevention of erosion and provide a weather-sheltered ecosystem.

WOODY PLANTS

The woody plants in our region can be divided into two natural groups: the conifers (belonging to the gymnosperms) and the flowering plants (the angiosperms). The main characteristic separating these groups is that the ovules and seeds of gymnosperms are borne on a scale rather than being enclosed by a carpel as they are in angiosperms. The conifers, therefore, do not have flowers, but analogous structures called strobili, which are often cone-shaped and commonly referred to as 'cones'. The pollen-producing stobili often disintegrate after pollination, whereas the ovule-bearing strobili develop into persistent, hard, scaly cones. All the conifers are wind pollinated whereas many angiosperms have evolved various adaptations to promote animal-mediated pollination. Reproduction is much slower in the conifers: fertilization can take up to a year, and seeds may not fully develop for up to 3 years. This contrasts with angiosperms, some of which can complete their life cycle in a few weeks. Almost all the conifers have evergreen leaves (often modified into needles) that remain on the tree throughout the year, creating the familiar evergreen appearance. The needles are replaced on an ongoing basis, with the average needle lasting one to several years depending on the species. Most woody angiosperms (especially in temperate areas) are deciduous, losing their leaves every year in the autumn.

Winter Key to the Woody Plants of Southern and Central Ontario

The following pictures illustrate a number of specific characters used by this key that may be unfamiliar to many readers.

Thorns vs. Spines vs. Prickles – While many people use these terms interchangeably, in botany they have distinct meanings and are not interchangeable. Thorns develop from modified branches and are typically found at the end of twigs (e.g. buckthorn, pg. 289) or where a twig would be (e.g. hawthorns, pg. 191). Spines are formed from modified leaves or stipules and are found only at nodes, either subtending a leaf or replacing it entirely. Prickles are outgrowths from the bark or epidermis of a plant and often densely cover the stem (e.g. raspberries and roses have prickles, not thorns).

110

WOODY PLANTS

False Terminal Buds – Most woody plants have terminal buds that arrest shoot growth each year and are distinctly different than the lateral buds (e.g. in size and shape). However, some species have false terminal buds: the 'end bud' is actually a lateral bud and the end of the twig bears the shrivelled remnant of a twig or a 'twig scar' opposite the leaf scar where the twig fell off.

buds subopposite

terminal bud
lateral bud
buds opposite

buds alternate

Twig scar
False terminal bud

bud scale
leaf scar
bundle scar

naked bud (no scales)

Leaf scar
Twig scar
False terminal bud

persistent leaf base

lenticels

111

WOODY PLANTS

Winter Key

1. Plants evergreen, with green leaves or needles persisting throughout the winter 2
2. Conifers; leaves needle-like or scale-like; seeds produced in woody cones (most species), berry-like blue cones (*Juniperus*), or fleshy red arils (*Taxus*) ....**go to summer key lead 2**, pg. 126
 2. Ericaceous shrubs; leaves usually broad (*Kalmia* has narrow linear leaves); seeds produced in capsules or berry-like fruits 3
  3. Leaves opposite or whorled ....***Kalmia***, pg. 225
  3. Leaves alternate 2
   4. Upright shrub up to 1 m; leaves usually 3–4 times longer than wide; leaf undersides covered with dense woolly hairs or small white or rusty scales 3
    5. Leaf undersides covered with small white or rusty scales; leaves becoming progressively smaller towards the branch tips ....***Chamaedaphne calyculata***, pg. 180
    5. Leaf undersides covered with white or rusty woolly hairs; leaves all of similar size ....***Ledum groenlandicum***, pg. 229
   4. Low or trailing shrub, under 30 cm; leaves usually less than 3 times longer than wide; leaf undersides hairless or with scattered hairs 4
    6. Leaf margins with brown hairs; petiole half the length of the leaf blade ... .***Epigaea repens***, pg. 199
    6. Leaf margins not hairy; petiole short or absent 5
     7. Leaves often widest above the middle; leaf tips rounded or notched ....***Arctostaphylos uva-ursi***, pg. 157
     7. Leaves widest at or below the middle; leaf tips pointed or with a small spine ....***Gaultheria***, pg. 212
1. Plants deciduous, leaves not persisting through the winter (but sometimes dead yellow leaves can remain on the plant) 8
 8. Buds opposite, subopposite or whorled ....**Key B**, pg. 114
 8. Buds alternate 9
  9. Twigs and buds densely covered with silver and/or brown scales, or silver stellate hairs ....***Elaeagnus***, pg. 195
  9. Twigs and buds not as above 10
   10. Climbing vine 11
    11. Climbing by adhesive aerial roots (not strictly originating from nodes) (DANGER: poison ivy!) ....***Toxicodendron radicans*** ssp. ***radicans***, pg. 346
    11. Climbing by tendrils (originating at nodes) or twining around a support 12
     12. Tendrils absent, climbing by twining 13
      13. Stems round; bundle scar U-shaped; bud scales hairless ....***Celastrus scandens***, pg. 176
      13. Stems ridged; bundle scar circular; bud scales with short hairs ....***Solanum dulcamara***, pg. 330
     12. Tendrils present 14
      14. Tendrils borne in pairs on persistent leaf bases; stems green, with prickles ....***Smilax***, pg. 328
      14. Tendrils borne singly (may be branched) on twigs; stems grey or brownish, unarmed 15
       15. Stems reddish-brown, minutely ridged; leaf and bundle scars indistinct; buds fully evident ....***Vitis***, pg. 367

WOODY PLANTS

     15. Stems light grey, smooth; leaf scars elliptical; bundle scars many, arranged in a closed ring; buds somewhat sunken in twig .... ***Parthenocissus***, pg. 250
10. Shrub or tree 16
   16. Twigs armed with spines, thorns, or prickles 17
     17. Leaf scars indistinct on torn and shrivelled persistent petiole bases ... . ***Rubus***, pg. 313
     17. Leaf scars distinct or absent (buds subtended instead by spines) 18
       18. Spines paired, occurring on either side of the leaf scar 19
         19. Buds inconspicuous, covered by leaf scar which often develops 3 irregular cracks; large tree .. ***Robinia pseudoacacia***, pg. 304
         19. Buds clearly visible; shrub or small tree 20
           20. Buds orange-red with a woolly appearance; leaf scars broad, triangular; young twigs brown or grey .... ***Zanthoxylum americanum***, pg. 368
           20. Buds red or green, smooth; leaf scars narrow, linear; young twigs green or red .... ***Rosa***, pg. 305
       18. Spines solitary or scattered along the twig 21
         21. Some leaf scars lacking, spines immediately subtending buds 22
           22. Bud scales thin, papery; prickles often scattered along stem between nodes; bark often shredding; inner bark not bright yellow .... ***Ribes***, pg. 298
           22. Bud scales thick, pointed; prickles never occurring between nodes; bark not shredding; inner bark bright yellow .... ***Berberis***, pg. 160
         21. Leaf scars always present 23
           23. Thorns terminal, occurring only at the ends of branches including short dwarf shoots; prickles absent; medium-sized trees 24
             24. Buds bright red, somewhat hairy; twigs often hairy towards the tips; terminal bud present .... ***Malus***, pg. 243
             24. Buds grey or reddish-brown, hairless; twigs hairless; terminal bud absent or false .... ***Prunus***, pg. 268
           23. Spines and/or prickles growing laterally from twig; growth forms various 25
             25. Spines located directly above leaf scars, often branched; prickles absent; buds inconspicuous, sunken in twig; twig strongly zig-zagging; medium to large tree .... ***Gleditsia triacanthos***, pg. 215
             25. Spines and/or prickles various, but not only directly above leaf scars; buds clearly visible; twigs straight or slightly zig-zagging; shrub or small tree 26
              26. Buds buff, brown, or dull orange ..... ***Ribes***, pg. 298
              26. Buds bright red or green 27
                27. Spines red, sometimes branched, longer than 2.5 cm .... ***Crataegus***, pg. 191
                27. Spines buff, brown, or grey, not branched, shorter than 2.5 cm .... ***Rosa***, pg. 305
   16. Twigs unarmed .... **Key C**, pg. 116

WOODY PLANTS

Key B – Buds opposite, subopposite, or whorled
1. Buds whorled..2
   2. Leaf scars triangular; buds conspicuous, oblong; twigs with 2 or 4 longitudinal lines below the nodes ....***Diervilla lonicera***, pg. 193
   2. Leaf scars round or elliptical; buds inconspicuous, often somewhat sunken in twig; twigs without longitudinal lines 3
     3. Shrub, up to 3 m high; twig slender; bundle scar one, U-shaped ....
....***Cephalanthus occidentalis***, pg. 178
     3. Tree, up to 30 m high; twig stout; bundle scars many, arranged in a ring ....
....***Catalpa speciosa***, pg. 175
1. Buds opposite or subopposite 4
   4. Twigs ending in sharp thorns ....***Rhamnus cathartica***, pg. 291
   4. Twigs not ending in thorns 5
     5. Climbing vine 6
       6. Climbing by adhesive aerial roots ....***Campsis radicans***, pg. 167
       6. Climbing by coiling leaf petioles ....***Clematis***, pg. 181
     5. Shrub or tree 7
       7. Shrub 8
         8. Stems soft, twigs easily compressed between fingertips ....
....***Sambucus***, pg. 324
         8. Stems firm, twigs not easily compressed 9
           9. Bundle scars 3 10
             10. Buds covered by a single scale (cap-like), often flattening toward the tip to resemble a duck's bill ....***Salix (purpurea)***, pg. 323
             10. Buds naked or covered by more than one scale; not resembling a duck's bill 11
               11. Terminal bud absent 12
                 12. Buds inconspicuous, covered by leaf scar or breaking through it; opposing leaf scars connected by a line; stipule scars absent ....***Philadelphus***, pg. 252
                 12. Buds clearly visible above leaf scar; opposing leaf scars not connected by a transverse line; stipule scars conspicuous ....
....***Staphylea trifolia***, pg. 338
               11. Terminal bud present 13
                 13. Buds naked, or with more than 2 scales 14
                   14. Buds naked and covered with stellate hairs, or buds with 4–6 scales (the lower 2 may be quite small) ....
....***Viburnum***, pg. 356
                   14. Buds with more than 6 scales 15
                     15. Twigs with 2 or 4 longitudinal lines below the nodes; fruit a brown, vase-shaped capsule with a 5-pointed tip ....***Diervilla lonicera***, pg. 193
                     15. Twigs round or slightly angled, but lacking longitudinal lines; fruit a round berry ....***Lonicera***, pg. 233
                 13. Buds with 2 scales 16
                   16. Young twigs red, green, purple, or pink 17
                     17. Buds completely hairless and trunk solid grey or brown; twigs 6 sided; fruit bright red berry-like drupes, in large hanging clusters, often persistent through winter ....***Viburnum opulus***, pg. 362

WOODY PLANTS

17. Buds hairy at least at the tips, or if hairless then trunk white-striped; twigs round; fruits various, but if red berry-like drupes, not persisting through winter in abundance **18**
    18. Leaf scars U-shaped, raised on persistent leaf bases which are connected by a transverse line; fruit a berry-like drupe ....***Cornus***, pg. 183
    18. Leaf scars V-shaped, nearly meeting around the twig, not significantly raised; fruit a papery-winged seed ....***Acer***, pg. 141
  16. Young twigs buff, brown, or grey **19**
    19. Buds red and hairy; leaf scars on raised, red petiole bases; fruit white ....***Cornus racemosa***, pg. 187
    19. Buds not red, or if red then hairless; leaf scars scarcely or not raised; fruit red or blue-black ....
....***Viburnum***, pg. 356
9. Bundle scars 1, more than 3, or indistinguishable **20**
  20. Bundle scars 1 ....**21**
    21. Twigs 4-sided, bright green or red; fruit a capsule bearing red or orange arils ....***Euonymus***, pg. 200
    21. Twigs round or angled (but not 4-sided), brown or grey; fruits various **22**
      22. Buds and twigs covered in brown and silver scales ....
....***Shepherdia canadensis***, pg. 327
      22. Buds and twigs lacking brown and silver scales **23**
        23. Leaf scars round, connected by transverse lines; buds inconspicuous, often sunken into twig ....
................................***Cephalanthus occidentalis***, pg. 178
        23. Leaf scars semi-circular, not connected; buds clearly visible, ovoid **24**
          24. Buds and twigs with short hairs; terminal bud present; fruit a black berry-like drupe ....
................................***Ligustrum vulgare***, pg. 230
          24. Buds and twigs hairless; terminal bud often absent; fruit a dry, crescent-shaped capsule ....
................................***Syringa***, pg. 339
  20. Bundle scars numerous or indistinguishable **25**
    25. Terminal bud absent; stipule scars present ....
................................***Staphylea trifolia***, pg. 338
    25. Terminal bud present; stipule scars absent **26**
      26. Twigs bright green or red; fruit a papery-winged seed ....
................................***Acer***, pg. 141
      26. Twigs buff, brown, or grey; fruit a berry ....***Lonicera***, pg. 233
7. Tree **27**
  27. Bundle scars 1; fruit a dry, crescent-shaped capsule ......***Syringa***, pg. 339
  27. Bundle scars 3 or more; fruits various **28**
    28. Bundle scars 3 ....**29**
      29. Young axillary buds often covered by persistent leaf bases; some twigs terminating in a long-stalked, bulbous flower bud; fruit a red berry-like drupe ....***Cornus florida***, pg. 186

WOODY PLANTS

    29. Young axillary buds visible; all twigs terminating in normal buds; fruit a papery-winged seed or large, globose, nut-like capsule ...**30**
        30. Terminal bud 1.5 cm long or longer; fruit a large, globose, nut-like capsule ....*Aesculus*, pg. 152
        30. Terminal bud less than 1.5 cm long; fruit a papery-winged seed ....*Acer*, pg. 141
  28. Bundle scars more than 3 ....**31**
    31. Bundle scars arranged in a closed ring; terminal bud absent; buds inconspicuous, often somewhat sunken in twig ....
....*Catalpa speciosa*, pg. 175
    31. Bundle scars arranged in a U-shape; terminal bud present; buds clearly visible **32**
        32. Terminal buds pyramidal, with 6 or fewer scales; bundle scars more than 9; fruit a winged seed ....*Fraxinus*, pg. 204
        32. Terminal buds ovoid, with more than 6 scales; bundle scars 3–9, sometimes arranged in 3 groups; fruit a large, globose, nut-like capsule ....*Aesculus*, pg. 152

Key C – Buds alternate; plant a shrub or tree; twigs unarmed
1. Coniferous tree, cones often persistent; twigs with longitudinal furrows; leaf scars extremely small, of two types: clustered in rings on bud-like warty spur-shoots, and singly on main stem; bundle scars 1 ....*Larix*, pg. 227
1. Not coniferous (*Alnus* bears cone-like fruits); twigs lacking furrows; leaf scars various; bundle scars various **2**
  2. Leaf scars indistinct on a torn persistent petiole base; bark shedding................**3**
    3. Petiole base with 2 erect, linear stipules; stem straight; bark shedding in thin longitudinal shreds ....*Potentilla fruticosa*, pg. 267
    3. Petiole base lacking stipules; stem somewhat zig-zagging; bark splitting longitudinally and shedding in larger sheets ....*Rubus*, pg. 313
  2. Leaf scars distinct; bark various **4**
    4. Bundle scars more than 3 (including numerous bundle scars collected in 3 groups) ....*Key D*, pg. 117
    4. Bundle scars 1 or 3 ....**5**
      5. Bundle scars 1 ....**6**
        6. Tree **7**
          7. Terminal bud lacking (false); all buds of similar size; buds hairy, deltoid; stipule scars present ....*Celtis occidentalis*, pg. 177
          7. Terminal bud many times larger than axillary buds; buds hairless, ovoid; stipule scars absent ....*Sassafras albidum*, pg. 326
        6. Shrub **8**
          8. Buds hairy, of two types: buds near twig tips large with more than 20 scales, and buds lower on twig smaller with 4 or fewer scales; twigs hairy, dotted with small yellow resin glands ...*Comptonia peregrina*, pg. 182
          8. Buds hairy or not, all of one type; twigs hairy or not, lacking yellow resin glands **9**
            9. Many stems ending in persistent dry flowers or flower-like fruits; buds hairy; leaf scars on much raised petiole bases; bark shedding **10**
              10. Leaf scar bearing a pair of erect, linear stipules; stems ending in a fewer than 5 dry flowers ....*Potentilla fruticosa*, pg. 267

WOODY PLANTS

    10. Leaf scar without stipules; stems ending in more than 5 dry flower-like fruits ....*Spiraea*, pg. 334
  9. Stems ending in normal buds; buds hairless; leaf scars not raised or slightly raised; bark shredding or smooth **11**
    11. Erect shrub up to 4 m tall; terminal bud present ......*Ilex*, pg. 218
    11. Erect or sprawling shrub up to 1 m; terminal bud absent or false **12**
        12. Buds bright purple, almost perpendicular to the twig; leaf scars almost perfectly flush with surface of twig; twig straight ....*Daphne mezereum*, pg. 192
        12. Buds green or red, more closely appressed to twig; leaf scars somewhat raised; twig zig-zagging or curving **13**
            13. Twigs green or red, warty; young twigs not resinous ....*Vaccinium*, pg. 354
            13. Twigs brown, not warty; young twigs densely resinous ....*Gaylussacia baccata*, pg. 214
  5. Bundle scars 3 ....**14**
    14. Tree ....**Key E**, pg. 119
    14. Shrub ....**Key F**, pg. 121

Key D – From Key C; bundle scars more than 3
1. Bundle scars not arranged in a crescent-shaped line **2**
  2. Multiple buds clustered at the tips of the twigs ....*Quercus*, pg. 275
  2. Buds not clustered at the tips of twigs **3**
    3. Twig completely encircled by a line emanating from the leaf scar **4**
      4. Buds scales 10–20 ....*Fagus grandifolia*, pg. 203
      4. Bud scales 1 or 2 ....**5**
        5. Buds dark red or purple, with white speckles; leaf scar beneath bud ....*Liriodendron tulipifera*, pg. 232
        5. Buds covered in long white hairs; leaf scar half encircling bud ....*Magnolia acuminata*, pg. 242
    3. Twig not encircled by a line emanating from the leaf scar **6**
      6. Stipule scars present **7**
        7. Bud scales 4 or more **8**
          8. Catkins often present; bundle scars not projecting; shrub ....*Corylus*, pg. 189
          8. Catkins absent; bundle scars often projecting; tree ....*Morus*, pg. 244
        7. Bud scales 2 or 3 ....**9**
          9. Buds ovoid; twig round ....*Tilia americana*, pg. 342
          9. Buds deltoid; twig angled ....*Castanea dentata*, pg. 174
      6. Stipule scars absent **10**
        10. Leaf scars raised, nearly circular; twigs covered with short hairs; buds towards tip of twig oblong, with more than 10 hair-fringed scales; medium-sized shrub ....*Rhus aromatica*, pg. 293
        10. Leaf not significantly raised, somewhat triangular or 3-lobed; twigs variously hairy; buds with fewer than 10 scales; shrubs or trees **11**
          11. Axillary buds solitary; twigs mottled; fruit a greenish-white berry-like drupe; shrubs or small trees; growing in wetlands (DANGER: poison sumac!) ....*Toxicodendron vernix*, pg. 345

WOODY PLANTS

      11. Axillary buds often superposed; twigs of uniform colour; fruit a nut; large trees; growing in various sites **12**
          12. Buds scales uniformly woolly ....***Juglans***, pg. 220
          12. Buds scales either hairless, with outer scales noticeably less hairy than inner scales, or with glandular yellow dots ....***Carya***, pg. 169
1. Bundle scars arranged in a crescent-shaped line **13**
   13. Leaf scar on dark red petiole base, which contrasts sharply with grey or brown twig; buds conical, dark red, often gummy or woolly; bark of main trunk dark grey with conspicuous light-coloured horizontal lenticels; small or medium-sized trees ....***Sorbus***, pg. 331
   13. Leaf scars not on a sharply contrasting red petiole base; buds various; bark not as above; shrubs or trees **14**
      14. Leaf scar encircling at least half of bud **15**
          15. Twig completely encircled by a line emanating from the leaf scar; buds with a single sac-like scale **16**
              16. Buds hairless ....***Platanus occidentalis***, pg. 261
              16. Buds covered in white hairs ....***Magnolia acuminata***, pg. 242
          15. Twig not encircled by a line emanating from the leaf scar; buds naked or with multiple scales **17**
              17. Buds inconspicuous, somewhat sunken in twig, often superposed; leaf scars roughly heart-shaped; twigs stout; medium to large tree ....***Gymnocladus dioicus***, pg. 216
              17. Buds clearly visible, superposed or solitary; leaf scars triangular, U-shaped, or crescent-shaped; twigs various; shrub or small tree **17**
                18. Bud hairs dark coloured **18**
                    19. Leaf scar encircling ¾ or less of the bud; buds taller than wide; terminal bud present; young twigs reddish-brown ....***Asimina triloba***, pg. 159
                    19. Leaf scar almost completely encircling the bud; buds wider than tall; terminal bud absent or false; young twigs olive ....***Dirca palustris***, pg. 194
                18. Bud hairs light coloured **19**
                    20. Buds stalked; twigs slender; fruit a greenish-white berry-like drupe (DANGER: poison ivy!) ....***Toxicodendron radicans*** ssp. ***rydbergii***, pg. 345
                    20. Buds sessile; twigs stout; fruit a red berry-like drupe ....***Rhus***, pg. 293
          15. Leaf scar encircling less than half the bud or positioned entirely below it ..**20**
            21. Shrub **21**
              22. Buds bright red; woody cone-like catkins persisting through winter .. ..***Alnus***, pg. 154
              22. Buds brown or buff; cone-like catkins absent **22**
                23. Leaf scars distinctly raised with 3 descending lines; bark shredding in papery strips; fruit a pendulous cluster of 2–5-parted dry follicles ....***Physocarpus opulifolius***, pg. 253
                23. Leaf scars not significantly raised; bark not shredding; fruit a greenish-white berry-like drupe (DANGER: poison ivy or poison sumac!) ....***Toxicodendron***, pg. 343
            21. Tree **23**

WOODY PLANTS

24. Lowest bud scale of axillary buds (i.e. outermost scale on the bottom of the bud) positioned directly above leaf scar; bark smooth and whitish or greenish when young ....*Populus*, pg. 262
24. Lowest bud scale of axillary buds to one side of the leaf scar; bark various 24
   25. Buds covered in white or buff silky hairs (DANGER: poison sumac!) ....*Toxicodendron vernix*, pg. 345
   25. Buds hairless, or with dark or rust-coloured hairs 25
      26. Bud with 2 or 3 scales, bright green or red ....*Tilia americana*, pg. 342
      26. Bud with more than 3 scales, rust-coloured or dark brown ....*Ulmus*, pg. 348

Key E – From Key C; bundle scars 3; trees
1. Lowest bud scale of axillary buds (i.e. outermost scale on the bottom of the bud) positioned directly above leaf scar; bark smooth and whitish or greenish when young ....*Populus*, pg. 262
1. Lowest bud scale of axillary buds to one side of the leaf scar; bark various 2
   2. Buds covered by a single scale (cap-like), often flattening toward the tip to resemble a duck's bill ....*Salix*, pg. 323
   2. Buds naked or with more than one scale 3
      3. Buds inconspicuous, somewhat sunken in twig, often superposed 4
      4. Buds covered in silky light-coloured hairs; shrub or small tree ....*Ptelea trifoliata*, pg. 274
      4. Buds hairless or with dark-coloured hairs; medium to large tree 5
         5. Buds in hairy pits; leaf scars large; twigs fairly straight, stout ....*Gymnocladus dioicus*, pg. 216
         5. Buds hairless; leaf scars moderate; twigs strongly zig-zagging, slender with swollen nodes ....*Gleditsia triacanthos*, pg. 215
      3. Buds clearly visible, superposed or not 6
         6. Buds scales 4 or fewer 7
         7. Terminal bud absent or false 8
            8. Buds often multiple above each leaf scar; leaf scars with an even fringe of hairs on upper margin; twigs with three ridges descending from leaf scars; stipule scars absent ....*Cercis canadensis*, pg. 179
            8. Buds solitary; leaf scars hairless or with a few scattered hairs on upper margin; twigs with or without three ridges descending from leaf scars; stipule scars present 9
               9. Bark with conspicuous horizontal lenticels or lines, in some species peeling in papery strips; pendulous catkins often present through winter ....*Betula*, pg. 162
               9. Bark lacking horizontal lenticels or lines; catkins absent 10
                  10. Buds hairy; twigs dark brown or grey, often mottled; stipule scars equal ....*Celtis occidentalis*, pg. 177
                  10. Buds hairless; twigs various colours, not mottled; stipule scars unequal 11
                     11. Buds bright red or green; twigs round, zig zagging ....*Tilia americana*, pg. 342

119

11. Buds yellow-orange to orange-brown; twigs often with ridges descending from leaf scars, not strongly zig-zagging ....
....*Castanea dentata*, pg. 174
7. Terminal bud present 12
    12. Buds densely woolly ....*Juglans*, pg. 220
    12. Buds hairless or with a few scattered hairs 13
        13. Twigs green or purple, main twig often exceeded in length by laterally branching twigs; large shrub or small tree; often with a few dead, bright yellow branches ...*Cornus alternifolia*, pg. 186
        13. Twigs yellowish-brown, reddish-brown, or grey; main twig extending beyond laterally branching twigs; medium to large tree; dead branches, if present, not distinctly coloured 14
            14. Bundle scars of the same colour or darker than leaf scars; bark of trunk smooth or peeling horizontally in papery strips, with conspicuous horizontal lenticels; pendulous catkins often present through winter ....*Betula*, pg. 162
            14. Bundle scars noticeably lighter than leaf scar; bark of trunk corky, forming irregular vertical ridges or blocky plates, lacking conspicuous lenticels; catkins absent ....
            ....*Nyssa sylvatica*, pg. 248
6. Buds scales 5 or more 15
    15. Buds with 10 or more scales; bark grey, smooth 16
        16. Buds red, hairy, ovoid; terminal bud less than 1 cm long; trunk deeply rippled, appearing muscle-like ....*Carpinus caroliniana*, pg. 168
        16. Buds orange-brown, hairless, spine-like; terminal bud almost 2 cm long; trunk more or less round ....*Fagus grandifolia*, pg. 203
    15. Buds with less than 10 scales; bark various 17
        17. Leaf scars in more than two rows along the length of the twig 18
            18. Buds usually with more than 7 scales; stipule scars present ....
            ....*Prunus*, pg. 268
            18. Buds with 7 or fewer scales; stipule scars absent 19
                19. Buds less than 3 times longer than wide; bark scaly or shedding ....*Malus*, pg. 243
                19. Buds 3 or more times longer than wide; bark smooth, often with vertical lines ....*Amelanchier*, pg. 156
        17. Leaf scars 2-ranked (i.e. in two rows along the twig, the third scar directly below the first) 20
            20. Buds often multiple above each leaf scar; leaf scars with a even fringe of hairs on upper margin; twig with three ridges descending from leaf scars ....*Cercis canadensis*, pg. 179
            20. Buds solitary or rarely superposed; leaf scars hairless or with a few scattered hairs on upper margin; twig without three ridges descending from leaf scars 21
                21. Buds pink or red; bud scales elongated, sometimes twisted; stipule scars absent ....*Amelanchier*, pg. 156
                21. Buds various colours but not pink or red; bud scales not twisted; stipule scars present (may be difficult to see in *Betula*) 22
                    22. Bud scales 2-ranked; catkins absent; bark splitting into vertical ridges, but not peeling and not papery ....
                    ....*Ulmus*, pg. 348

WOODY PLANTS

22. Bud scales not 2-ranked (i.e. not in two rows); pendulous catkins in groups of 2 or 3 present in winter; bark smooth or peeling in papery strips 23
23. Buds scales with thin longitudinal lines; bark peeling vertically, without conspicuous lenticels; spur shoots not in abundance ....*Ostrya*, pg. 249
23. Bud scales not lined; bark smooth or peeling horizontally, with conspicuous horizontal lenticels; spur shoots often in abundance ....*Betula*, pg. 162

Key F – From Key C; bundle scars 3; shrubs
1. Buds naked, or bud scales entirely covered in dense woolly hairs 2
   2. Buds wider than long; leaf scars U-shaped, partly encircling bud; terminal bud absent or false ....*Ptelea trifoliata*, pg. 274
   2. Buds longer than wide; leaf scars crescent shaped or semi-circular, entirely below bud; terminal bud present 3
      3. Terminal bud stalked; buds buff or greyish .....*Hamamelis virginiana*, pg. 217
      3. Terminal bud not stalked; buds rusty orange ....*Rhamnus frangula*, pg. 292
1. Bud scales hairless or with some hair but surface of scale still visible 4
   4. Buds covered by a single scale (cap-like), often flattening toward the tip to resemble a duck's bill ....*Salix*, pg. 323
   4. Buds with more than 1 scale 5
      5. Terminal bud absent or false 6
         6. Buds often multiple above each leaf scar, reddish- or yellowish-green; twigs green or olive, with white lenticels ....*Lindera benzoin*, pg. 231
         6. Buds single above each leaf scar, colours various; twigs red, purple, grey or brown with or without lenticels 7
            7. Buds of two types: buds near twig tips large with more than 20 scales, and buds lower on twig smaller with 4 or fewer scales; twigs dotted with small yellow resin glands 8
               8. Buds and twigs hairy ....*Comptonia peregrina*, pg. 182
               8. Buds and twigs hairless ....*Myrica gale*, pg. 247
            7. Buds of one type, with 4–6 scales; twigs lacking yellow resin glands 9
               9. Stipule scars conspicuous; pendulous catkins present; bud scars 2 ranked; twigs zig-zagging ....*Corylus*, pg. 189
               9. Stipule scars inconspicuous; catkins absent; buds scars in more than 2 rows on the twig; twigs not zig-zagging ......*Rhamnus alnifolia*, pg. 291
      5. Terminal bud present 10
         10. Twigs with 2 or 3 distinct lines descending from leaf scars; bud scales 5 or 6; bark papery, shredding 11
            11. Leaf scar much raised on persistent petiole base; buds hairy; stipule scars present, but inconspicuous; fruit a pendulous cluster of 2–5-parted dry follicles ....*Physocarpus opulifolius*, pg. 253
            11. Leaf scar only slightly raised; stipule scars absent; buds hairy or hairless; fruit a berry ....*Ribes*, pg. 298
         10. Twigs without distinct lines descending from leaf scars; bud scales various; bark not papery or shredding 12
            12. Stipule scars present 13
               13. Buds with 4 or fewer scales; twigs roughly 3-sided; pendulous catkins present ....*Alnus*, pg. 154

WOODY PLANTS

      13. Buds with more than 4 scales; twigs round; catkins absent ....*Prunus*, pg. 272
   12. Stipule scars absent **14**
      14. Buds purple or greenish; lateral bud scales 2; plants often with a few dead, bright yellow branches ....*Cornus alternifolia*, pg. 186
      14. Buds pink or red; bud scales 3 or more; dead branches, if present, not distinctly coloured **15**
         15. Buds elongated, more than 2 times longer than wide **16**
            16. Bud scales somewhat spreading, not twisted, often notched at tip; second bud scale half the length of the bud or more ....*Aronia melanocarpa*, pg. 158
            16. Bud scales closely appressed, often twisted, not notched at tip; second bud scale usually less than half the length of the bud ....*Amelanchier*, pg. 156
         15. Buds globose or ovoid, less than 2 times longer than wide **17**
            17. Buds pinkish; twigs brown, hairy, often dotted with yellow resin glands; fruits white, warty berry-like ....*Myrica pensylvanica*, pg. 247
            17. Buds bright red; twigs red, hairless, resin glands absent; fruits orange or red hips ....*Rosa blanda*, pg. 312

WOODY PLANTS

Summer Key to the Woody Plants of Southern and Central Ontario

The following diagrams illustrate a number of specific characters used by this key that may be unfamiliar to many readers.

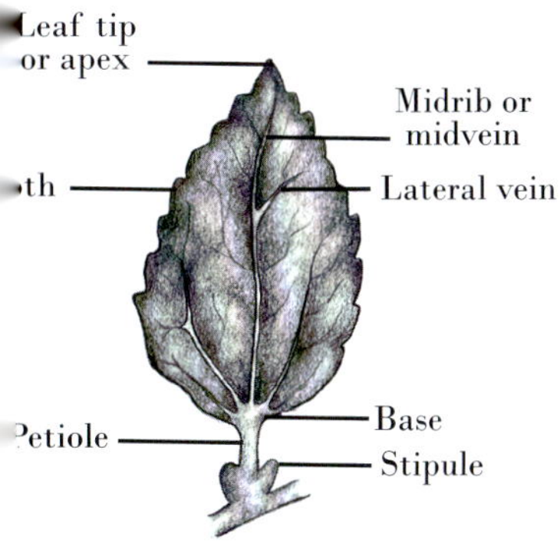

Parts of a leaf	Linear	Oblong
Oval	Elliptic	Lanceolate
Oblanceolate	Ovate	Obovate

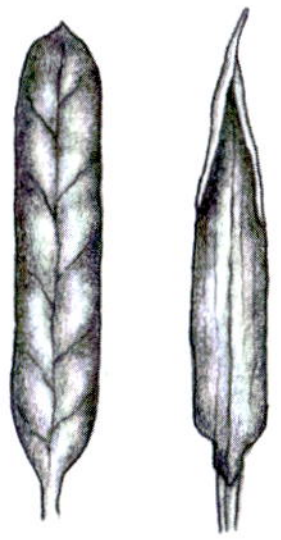

Labels on "Parts of a leaf": Leaf tip or apex, Midrib or midvein, Lateral vein, Tooth, Petiole, Base, Stipule

123

WOODY PLANTS

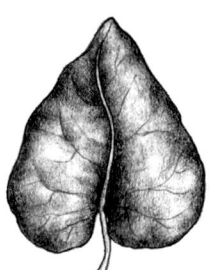
Cordate

Obcordate

Simple

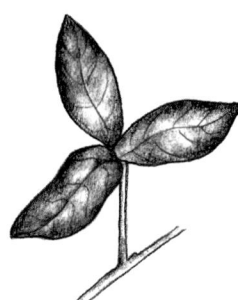
Trifoliate (ternate)

Trifoliate (terminal leaflet stalked)

Biternate (two times ternate)

Palmately compound

Pinnately compound

WOODY PLANTS

Pinnate/once-divided

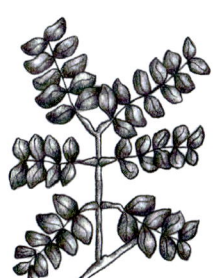
Bipinnate/two times divided

Tripinnate/three times divided

Alternate

Opposite

Whorled

Solitary

Axillary

Terminal

Axillary and terminal

WOODY PLANTS

Summer Key

1. Conifers; leaves evergreen (except *Larix*) and needle-like or scale-like; seeds produced in woody cones (most species), berry-like blue cones (*Juniperus*), or fleshy red arils (*Taxus*) 2
   2. At least some leaves scale-like; leaves opposite, short, and concealing the twigs ... 3
      3. Leaves all scale-like; leaves strongly flattened on the twig; cones yellowish or brown, woody .... ***Thuja occidentalis***, pg. 341
      3. Leaves both needle-like and scale-like; leaves appressed to twig but not strongly flattened; cones fleshy, dark blue at maturity .... ***Juniperus***, pg. 222
   2. Leaves all needle-like; leaves alternate, whorled, or in clusters, but not concealing the twigs 4
      4. Plant a low shrub; seeds in berry-like blue cones or fleshy red arils 5
         5. Leaves in whorls of 3; seeds in berry-like blue cones .... ***Juniperus communis***, pg. 223
         5. Leaves alternate; seeds in fleshy red arils .... ***Taxus canadensis***, pg. 340
      4. Plant a small to large tree; seeds in woody cones 6
         6. Needles in clusters 7
            7. Needles evergreen, in bundles of 2, 3, or 5 .... ***Pinus***, pg. 258
            7. Needles deciduous, densely clustered on dwarf shoots .... ***Larix***, pg. 227
         6. Needles occurring singly on the twigs 8
            8. Twigs not grooved; needles with rounded bases ... ***Abies balsamea***, pg. 140
            8. Twigs grooved; needles with peg-like bases that persist after the needles fall 9
               9. Needles 4-sided (roll between fingers), with pointed tips ... ***Picea***, pg. 254
               9. Needles flat, with blunt tips .... ***Tsuga canadensis***, pg. 347
1. Angiosperms; leaves usually broad, deciduous (evergreen in some ericaceous shrubs and some introduced species of *Ilex* and *Ligustrum*); seeds not produced in cones (*Alnus* has cone-like catkins) 10
   10. Leaves subopposite, opposite, or whorled .... **KEY B**, pg. 127
   10. Leaves alternate (in some species, leaves may be densely packed on dwarf shoots and appear whorled) 11
      11. Leaves compound .... **KEY C**, pg. 129
      11. Leaves simple 12
         12. Plant a climbing vine or twining shrub 13
            13. Stems with prickles; veins distinctive, arising from a single point at the base of the leaf, curving outwards, and uniting again at the tip ... . ***Smilax***, pg. 328
            13. Stems without prickles; veins ending near margins, not uniting at the tip 14
               14. Leaves unlobed; fruit an orange capsule .... ***Celastrus scandens***, pg. 176
               14. At least some leaves lobed; fruit a berry 15
                  15. Plants with tentrils; all leaves lobed; leaves toothed; fruit a blue or purple berry .... ***Vitis***, pg. 367
                  15. Plants lacking tentrils; lower leaves unlobed, upper leaves lobed at the base; leaves entire; fruit a red berry .... ***Solanum dulcamara***, pg. 330
         12. Plant a creeping shrub, upright shrub, or tree 16

WOODY PLANTS

 16. Leaves lobed (in *Sassafras* and *Morus*, both lobed and unlobed leaves can occur on the same plant) ....**Key D**, pg. 131
 16. Leaves not lobed **17**
  17. Plants armed with spines or thorns **18**
   18. Spines often 3-pronged, found at the nodes; flowers yellow or orange, 6-parted; fruit an oblong red berry ...***Berberis***, pg. 160
   18. Thorns single, found on the stem but not at the nodes; flowers white, 5-parted; fruit resembling an apple or a plum **19**
    19. Leaf tip long, slender, and pointed; fruit a plum, with a single stone ....***Prunus***, pg. 268
    19. Leaf tip usually pointed but short, not long and slender; fruit resembling an apple (but may be quite small) **20**
     20. Thorns smooth; mature bark shredding ....
....***Crataegus***, pg. 191
     20. Thorns rough, often leafy, with leaf scars or buds; mature bark scaly ....***Malus***, pg. 243
  17. Plants unarmed **21**
   21. Leaves with distinct teeth ....**Key E**, pg. 132
   21. Leaf margins entire or obscurely toothed ....**Key F**, pg. 134

Key B – Angiosperms; leaves opposite, subopposite, or whorled
1. Leaves compound **2**
 2. Plant a trailing or climbing woody vine **3**
  3. Leaves with 3 leaflets; flowers white, blue, or purple; fruit a distinctive cluster of small seeds each with a long feathery plume ....***Clematis***, pg. 181
  3. Leaves with 7–13 pinnately compound leaflets; flowers orange to red, trumpet-shaped, showy; fruit a long pod ....***Campsis radicans***, pg. 167
 2. Plant a shrub or tree **4**
  4. Leaves with 3–7 leaflets, at least some leaves usually with 3 leaflets **5**
   5. Leaves always with 3 leaflets; leaflets finely toothed (5 to many teeth per cm); fruit a large inflated capsule with 3 lobes at the tip ....
....***Staphylea trifolia***, pg. 338
   5. Leaves with 3–7 leaflets; leaflets irregularly toothed (1 tooth per cm or less) or sometimes lobed; fruit a dry winged 'maple key', occurring in pairs ....
....***Acer negundo***, pg. 147
  4. Leaves with 5–15 leaflets, never with 3 leaflets **6**
   6. Leaves palmately compound with 5–7 leaflets ....***Aesculus***, pg. 152
   6. Leaves pinnately compound with 5–15 leaflets **7**
    7. Erect shrubs (up to 4 m tall); twigs with warty spots; pith large and spongy; fruit red to black, berry-like ....***Sambucus***, pg. 324
    7. Small to large trees; twigs lacking warty spots; fruit a winged seedcase ..
....***Fraxinus***, pg. 204
1. Leaves simple **8**
 8. Plant a vine; flowers in terminal clusters above a distinctive leafy disk; uppermost stem leaves also united to form distinctive disks ....***Lonicera***, pg. 233
 8. Plant not a vine; flowers and leaves various but not as above **9**
  9. Lateral leaf veins curving toward leaf tip **10**

WOODY PLANTS

10. Leaves always opposite; leaf margins entire; when the leaves are gently broken, thin strands remain to connect the veins; twigs lacking thorns at the tips ....*Cornus*, pg. 183
10. Leaves opposite or subopposite; leaf margins finely toothed; when leaves are broken, no strands remain to connect the veins; branches and twigs with thorns at the tips ....*Rhamnus cathartica*, pg. 291
9. Lateral leaf veins not distinctly curved toward leaf tip, directed more or less toward leaf margin **11**
  11. Leaves palmately lobed with 3–7 lobes **12**
    12. Leaves short-stalked; leaves always with 3 lobes; flowers white, in clusters broader than long; fruit berry-like ....*Viburnum*, pg. 356
    12. Leaves long-stalked; leaves with 3–7 lobes (depending on species); flowers greenish, yellowish, or reddish, in clusters longer than broad; fruit a dry winged 'maple key', occurring in pairs ....*Acer*, pg. 141
  11. Leaves not lobed **13**
    13. Some leaves subopposite; leaves long and narrow; flowers in catkins ....*Salix (purpurea)*, pg. 323
    13. Leaves all opposite or whorled; leaf shape various; flowers not in catkins **14**
      14. Leaf margins distinctly toothed **15**
        15. Leaves coarsely toothed (less than 25 teeth per side) **16**
          16. Flowers citrus-scented, with 4 petals; fruit a small capsule; uncommon escapees from cultivation ...*Philadelphus*, pg. 252
          16. Flowers with 5 petals; fruit berry-like; native shrubs ....
          ....*Viburnum*, pg. 356
        15. Leaves finely toothed (more than 25 teeth per side) **17**
          17. Twigs somewhat 4-sided; flowers greenish-white, greenish-purple or purple; fruit a lobed pink capsule, splitting open at maturity to reveal orange or red arils ......*Euonymus*, pg. 200
          17. Twigs rounded; flowers white or yellow; fruit a brown capsule or a berry-like drupe **18**
            18. Low shrub, usually less than 1 m tall; flowers yellow; fruit a brown capsule, often persisting through winter; fruit clusters both terminal and axillary ....
            ....*Diervilla lonicera*, pg. 193
            18. Larger shrub, up to 6 m tall; flowers white; fruit berry-like; fruit clusters only terminal ....*Viburnum*, pg. 356
      14. Leaf margins entire, sometimes wavy but not distinctly toothed ...**19**
        19. Leaves leathery, evergreen; plants typically growing in swamps, bogs, or wet woods; flowers pink, with fused petals, hanging in clusters; fruit a capsule with a persistent style ....*Kalmia*, pg. 225
        19. Leaves deciduous (some introduced *Ilex* and *Ligustrum* species have evergreen or semi-evergreen leaves); flowers and fruit not as above **20**
          20. Leaves and twigs covered with rusty brown scales; fruit red and berry-like ....*Shepherdia canadensis*, pg. 327
          20. Leaves and twigs not covered with rusty brown scales **21**
            21. Leaves heart-shaped; fruit a capsule **22**

WOODY PLANTS

22. Large trees; leaves opposite or whorled; flowers with 5 fused petals; petals white and streaked with yellow and purple; fruit a long tube-like capsule, large and conspicuous .... *Catalpa speciosa*, pg. 175
22. Shrubby trees; leaves always opposite; flowers fragrant, with 4 fused petals; petals solid white, pink, or purple; fruit a small flattened capsule ....
.... *Syringa*, pg. 339
21. Leaves various, but not heart-shaped; fruit a berry, a berry-like drupe, or a spherical head of nutlets **23**
23. Leaves opposite or whorled, bright and glossy; flowers and fruit in distinctive spherical heads on long stalks .... *Cephalanthus occidentalis*, pg. 178
23. Leaves always opposite; flowers and fruit not as above **24**
24. Mature bark shredding; flowers and fruits in distinctive pairs from the leaf axils; flowers white, pink, yellow, or orange-red; fruit a berry ....
.... *Lonicera*, pg. 233
24. Mature bark not shredding; flowers and fruit growing in clusters at the end of stems and branches; flowers always white; fruit a berry-like drupe (i.e. with a pit) **25**
25. Leaf margins always entire; leaves smaller, usually less than 4 cm long and 1.5 cm wide; inflorescence usually longer than wide; introduced shrubs, commonly escaped from cultivation .... *Ligustrum vulgare*, pg. 230
25. Leaf margins entire or with irregular wavy/rounded teeth; leaves larger, up to 9 cm long and 5 cm wide; inflorescence usually wider than long; native shrubs ....
..*Viburnum nudum* var. *cassinoides*, pg. 366

Key C – Angiosperms; leaves alternate, compound
1. Medium to large trees; fruit a large pod (legume) or a large nut enclosed in a husk (walnuts and hickories) **2**
2. Leaflets entire; fruit a large pod (legume) **3**
   3. Leaves always singly compound; terminal leaflet present; two spines present at the base of each leaf .... *Robinia pseudoacacia*, pg. 304
   3. At least some leaves doubly compound; terminal leaflet absent; leaves without a pair of spines at the base (but twigs and bark may have large branched thorns) **4**
      4. Leaves always doubly compound; leaflets ovate with a pointed tip, up to 8 cm long; thorns never present; twigs distinctive, very thick; fruit a thick, leathery pod .... *Gymnocladus dioicus*, pg. 216
      4. Leaves both singly and doubly compound; leaflets oblong with a rounded tip, less than 4 cm long; twigs, branches, and trunk sometimes with prominent branched thorns; fruit a flat, twisted pod ..... *Gleditsia triacanthos*, pg. 215

WOODY PLANTS

    2. Leaflets toothed; fruit a large nut enclosed in a husk (walnuts and hickories) **5**
        5. Leaflets 5–11; husk splitting into 4 sections (hickories) ....***Carya***, pg. 169
        5. Leaflets 11–23; husk not splitting into sections (walnuts) ....***Juglans***, pg. 220
1. Vines, shrubs, or small trees; fruit various but neither a large pod or a large nut enclosed in a husk **6**
    6. Leaflets with distinct teeth (or rarely lobes)...................**7**
        7. Plant a climbing vine...................**8**
            8. Leaves with 3 leaflets; plant climbing with aerial roots; fruit white, berry-like (DANGER: poison ivy!) ....***Toxicodendron radicans* ssp. *radicans***, pg. 346
            8. Leaves with 5 leaflets; plant climbing with tendrils; fruit a blue or purple berry ....***Parthenocissus***, pg. 250
        7. Plant a shrub or small tree, not a climbing vine (some *Rubus* species have trailing vine-like stems but never climb) **9**
            9. Leaves with 3–9 leaflets **10**
                10. Leaves lacking stipules; stems lacking bristles and prickles; flowers yellow or greenish-yellow; fruit white and smooth or red and hairy, berry-like **11**
                    11. Terminal leaflet distinctly stalked; fruit smooth and white (DANGER: poison ivy!) ....***Toxicodendron radicans* ssp. *rydbergii***, pg. 345
                    11. Terminal leaflet narrowed at the base but not distinctly stalked; fruit red and hairy ....***Rhus aromatica***, pg. 296
                10. Leaves usually with stipules (sometimes falling off as the leaves mature); most species with bristles/prickles on the stem; flowers white, pink, or purple; fruit a rose hip or resembling a raspberry **12**
                    12. Stipules persistent, fused to the leafstalk; flowers usually pink or rose-coloured, sometimes white; fruit a rose hip ....***Rosa***, pg. 305
                    12. Stipules deciduous, not fused to the leafstalk; flowers usually white, sometimes pink to purple; fruit resembling a raspberry ....***Rubus***, pg. 313
            9. Leaves with 11–31 leaflets (rarely 9) **13**
                13. Leafstalks exuding a milky latex when removed; flowers small and greenish-yellow; fruit hairy, in clusters longer than wide ....***Rhus***, pg. 293
                13. Leafstalks without a milky latex; flowers white and showy; fruit hairless, in clusters wider than long ....***Sorbus***, pg. 331
    6. Leaflet margins entire, wavy, or obscurely toothed...................**14**
        14. Leaves with stipules; stem and leaves often covered in silky hairs; flowers large, bright yellow, and showy ....***Potentilla fruticosa***, pg. 267
        14. Leaves lacking stipules; stems and leaves without silky hairs; flowers small and greenish or greenish-yellow **15**
            15. Leaves with 3 leaflets **16**
                16. Plant a vine or low shrub; terminal leaflet with a distinct stalk; fruit a smooth white drupe (DANGER: poison ivy!) ....***Toxicodendron radicans***, pg. 343
                16. Plant a tall shrub or very small tree; terminal leaflet narrowed at the base but without a distinct stalk; fruit a flat, winged seedcase ....***Ptelea trifoliata***, pg. 274
            15. Leaves with 5–13 leaflets **17**
                17. Leaves with prickles on the leafstalk; stem with paired spines at each node; fruit a small capsule ....***Zanthoxylum americanum***, pg. 368

WOODY PLANTS

17. Leafstalk and stem lacking prickles and spines; fruit white or red, berry-like 18
18. Leafstalk with wings between leaflets; fruit red and hairy ....*Rhus copallina*, pg. 296
18. Leafstalk without wings; fruit white and smooth (DANGER: poison sumac!) ....*Toxicodendron vernix*, pg. 345

Key D – Angiosperms; leaves alternate, simple, lobed
1. Plants with thorns, spines, or bristles 2
   2. Some branches modified into large, thick thorns; fruit resembling an apple (but may be quite small) 3
      3. Thorns smooth; mature bark shredding ....*Crataegus*, pg. 191
      3. Thorns rough, often leafy, with leaf scars or buds; mature bark scaly ....*Malus*, pg. 243
   2. Plants with smaller, thinner spines at the nodes and/or bristles on the stem; fruit a red or black berry, covered with glands or prickles in some species ...*Ribes*, pg. 298
1. Plants unarmed 4
   4. Plants with unlobed, 2-lobed, and 3-lobed leaves on the same individual 5
      5. Leaves with entire margins; twigs and leaves fragrant when bruised; fruit solitary, dark-blue, on a long red stalk ....*Sassafras albidum*, pg. 326
      5. Leaves with toothed margins; twigs exude a milky sap when cut; fruit in a red or white berry-like cluster ....*Morus*, pg. 244
   4. All leaves similarly lobed 6
      6. Small to large trees 7
         7. Leaves with 4 or 6 lobes; leaf tip notched and flattened, very distinctive; flowers large, yellowish-green and showy; fruit an erect cone-like structure ... .*Liriodendron tulipifera*, pg. 232
         7. Leaves with 3, 5, or more lobes; leaf tip pointed; flowers small; fruit an acorn or a ball-like cluster of seeds 8
            8. Bark very distinctive, smooth and flaking off in large irregular pieces; leaves with 3 (sometimes 5) palmate lobes; fruit a dense round cluster of seeds ....*Platanus occidentalis*, pg. 261
            8. Bark furrowed, not flaking; leaves with 5 or more pinnate lobes; fruit an acorn, i.e. a nut enclosed in a tough, rounded shell with a prominent cap ....*Quercus*, pg. 275
      6. Small to medium-sized shrubs, less than 3 m tall 9
         9. Leaves with numerous pinnate lobes; lobes lacking teeth; leaves with a pleasant aromatic fragrance when crushed ...*Comptonia peregrina*, pg. 182
         9. Leaves with 3–5 palmate lobes; lobes usually toothed; leaves unscented or with an unpleasant odour when crushed 10
            10. Leaves large (10–20 cm long); flowers large (3–5 cm in diameter); fruit a raspberry ....*Rubus*, pg. 313
            10. Leaves smaller (3–10 cm long); flowers smaller (less than 1 cm in diameter); fruit a simple berry or small pod 11
               11. Leaves without stipules at the base; fruit a red or black berry ....*Ribes*, pg. 298
               11. Leaves with a pair of small stipules at the base; fruit a reddish-brown pod, in densely packed clusters, often persistent ....*Physocarpus opulifolius*, pg. 253

WOODY PLANTS

Key E – Angiosperms; leaves alternate, simple, toothed
1. Lateral leaf veins curving toward leaf tip 2
  2. Plant a tall shrub or small tree; veins usually indistinct at the tips; fruit greenish, yellowish, or reddish ....***Malus***, pg. 243
  2. Plant a low shrub (usually less than 1 m tall), mainly found at wet sites; veins always distinct at the tips; fruit black ....***Rhamnus alnifolia***, pg. 291
1. Lateral leaf veins not distinctly curved toward leaf tip, directed more or less toward leaf margin 3
  3. Leaf base asymmetrical, i.e. leaf tissue at the base of the leaf extends farther on one side than the other 4
    4. Leaf margin wavy or with coarse rounded teeth ....
    ....***Hamamelis virginiana***, pg. 217
    4. Leaf margin with numerous sharp, fine teeth 5
      5. Leaf margin double-toothed (teeth have teeth); fruit a flat, dry, winged seedcase ....***Ulmus***, pg. 348
      5. Leaf margin single-toothed; fruit a round nut-like capsule or a berry-like drupe 6
        6. Leaves tapering to a long slender point; fruit red to dark purple, berry-like, solitary ....***Celtis occidentalis***, pg. 177
        6. Leaves heart-shaped, with a short point; fruit a small brown spherical capsule, in clusters, with green leafy bracts ....***Tilia americana***, pg. 342
  3. Leaf base symmetrical, i.e. leaf tissue at the base of the leaf similar on both sides 7
    7. Leaves distinctly heart-shaped; fruit fruit a small brown spherical capsule, in clusters, with green leafy bracts ....***Tilia americana***, pg. 342
    7. Leaves not heart-shaped; fruits various 8
      8. Teeth only at the end of lateral veins 9
        9. Leaves with stellate hairs below; fruit fruit an acorn, i.e. a nut enclosed in a tough, rounded shell with a prominent cap ....***Quercus***, pg. 275
        9. Leaves usually glabrous below, occasionally with non-stellate hairs along the veins; fruit a nut, with several nuts enclosed in a spiny or bristly husk 10
          10. Veins extending beyond teeth to form short bristles; bark rough and brown, ridged when mature ....***Castanea dentata***, pg. 174
          10. Teeth usually lacking bristles; bark smooth and grey, very distinctive ....***Fagus grandifolia***, pg. 203
      8. At least some teeth not at the end of veins 11
        11. Leaves usually double-toothed (but can appear single-toothed in some species); lateral veins always reaching the margin and ending in teeth; intervening teeth often (but not always) smaller than those at the end of veins; male flowers always in catkins, often present throughout the winter 12
          12. Fruit a woody cone-like catkin, often remaining on the plant all year ....***Alnus***, pg. 154
          12. Fruit not a woody cone-like catkin 13
            13. Bark with distinctive elongated horizontal lenticels or lines, in some species mature bark distinctively curled and peeling or shredded; fruit in densely packed clusters, usually more than 20 fruits per cluster ....***Betula***, pg. 162

WOODY PLANTS

13. Bark without elongated horizontal lenticels or lines; fruit in loosely arranged clusters, less than 20 fruits per cluster **14**
14. Plant a low or medium shrub, <4 m tall; leaves usually around 1.5 times as long as wide with a short point at the tip; leaf margins sometimes wavy or irregularly shaped; fruit clusters not drooping; bracts around nut forming a distinctive fringe or beak ....***Corylus***, pg. 189
14. Plant a small tree; leaves usually 2 times as long as wide (or longer) with a long point at the tip; leaf margins regularly shaped; fruit in drooping clusters; nut in an inflated sac or in the axil of a 3-lobed leafy bract................................**15**
15. Bark smooth, light grey, often with muscle-like ridges; lateral veins rarely forking; nut in the axil of a 3-lobed leafy bract ....***Carpinus caroliniana***, pg. 168
15. Bark rough, brown, splitting into vertical strips that peel at both ends; some lateral veins forking near the tip; nut enclosed in an inflated sac ....
....***Ostrya virginiana***, pg. 249
11. Leaves always single-toothed; teeth usually all the same size; veins usually not reaching the margin or ending in teeth (except in some *Amelanchier* species, but *Amelanchier* never has flowers in catkins); flowers in catkins or not **16**
16. Buds with a single cap-like scale; flowers in catkins ...***Salix***, pg. 323
16. Buds with more than one scale; flowers in catkins or not **17**
17. Leaves with a pleasant aromatic fragrance when crushed; leaves toothed at the tip; fruit a tiny nutlet, in dense sessile clusters ....***Myrica***, pg. 246
17. Leaves not fragrant when crushed; leaves toothed along all or most of their length; fruits various **18**
18. Small glands present on leaves, either on the leafstalk near the base of the blade or on the midrib and leaf margins ...
....**19**
19. Small glands on the leafstalk near the base of the blade; shrubs or small trees up to 12 m tall ....
....***Prunus***, pg. 268
19. Small glands on the midrib and leaf margins; shrubs up to 2.5 m tall ....***Aronia melanocarpa***, pg. 158
18. Small glands not present on leaves **20**
20. Leaves sessile or nearly so, leafstalk indistinct or very short; plant a low erect shrub up to 1.5 m tall **21**
21. Leaves with minute teeth; flowers in small clusters (sometimes solitary); petals fused; fruit a blueberry ....***Vaccinium angustifolium***, pg. 355
21. Leaves with fine to coarse teeth; flowers densely packed in erect terminal clusters; petals not fused; fruit a persistent capsule ....***Spiraea***, pg. 334
20. Leaves distinctly stalked, leafstalk usually at least 1 cm long; plant a shrub or tree **22**

133

WOODY PLANTS

22. Leaf distinctly tapered at the base; fruit red, berry-like, often remaining on the shrub into the winter; plant a medium-sized shrub up to 4 m tall ....***Ilex verticillata***, pg. 219
22. Leaf usually rounded or flat at the base (sometimes somewhat tapered in *Amelanchier*); fruit various; plant a shrub or tree **23**
23. Bark smooth and whitish, greyish, or greenish, older bark sometimes ridged; leaves usually triangular or egg-shaped; teeth always rounded; flowers and fruit in catkins ...***Populus***, pg. 262
23. Bark scaly (*Malus*) or smooth with distinctive vertical lines (*Amelanchier*); leaves rounded; teeth usually sharp (sometimes crenate in *Malus*); flowers showy and white, fruit an apple-like or berry-like pome **24**
24. Bark with distinctive vertical lines; lateral veins straight ....***Amelanchier***, pg. 156
24. Bark scaly; lateral veins curved ....***Malus***, pg. 243

Key F – Angiosperms; leaves alternate, simple, entire

1. Lateral leaf veins curving toward leaf tip; when the leaves are gently broken, thin strands remain to connect the veins ....***Cornus alternifolia***, pg. 186
1. Lateral veins not curving toward leaf tip (*Rhamnus frangula* sometimes has veins weakly curved towards leaf tip but never has thin strands connecting the veins) **2**
2. Leaves leathery, evergreen; plant a creeping or upright shrub usually less than 1 m tall **3**
3. Plant a prostrate or trailing shrub less than 20 cm tall **4**
4. Stems all trailing **5**
5. Leaves large (up to 7 cm long) and rounded or heart-shaped at the base; leafstalk up to 1/2 the length of the blade; leaf margins and stems densely covered in bristly brown hairs ....***Epigaea repens***, pg. 199
5. Leaves small (less than 1 cm long) and somewhat tapered at the base; leafstalk short and indistinct; leaf margins and stems with scattered appressed hairs ....***Gaultheria hispidula***, pg. 213
4. Stems both trailing and upright; upright stems up to 20 cm tall **6**
6. Leaves with a wintergreen scent when crushed; leaf margins often obscurely toothed, the teeth tipped with bristle-like hairs; leaves few and clustered at the tips of upright stems ....***Gaultheria procumbens***, pg. 213
6. Leaves not scented when crushed; leaf margins entire; leaves more numerous and usually spread along upright stems ....***Arctostaphylos uva-ursi***, pg. 157
3. Plant a low erect shrub up to 1 m tall **7**
7. Leaves getting progressively smaller toward the tip of the stem; lower surface of the leaf densely covered in white to brownish scales; leaf margins not inrolled ....***Chamaedaphne calyculata***, pg. 180
7. Leaves not progressively smaller toward the tip of the stem; lower surface of leaf densely covered in white to orange woolly hairs; leaf margins strongly inrolled ....***Ledum groenlandicum***, pg. 229

WOODY PLANTS

2. Leaves deciduous; plant a shrub or tree 8
   8. Leaves and twigs covered in silvery or brownish scales .... *Elaeagnus*, pg. 195
   8. Leaves and twigs without scales 9
      9. Leaves heart-shaped to somewhat rounded; main veins of leaf radiating from the base of the leaf; fruit a pod (legume) .... *Cercis canadensis*, pg. 179
      9. Leaves not heart-shaped; main veins of leaf pinnately arranged; fruit various 10
         10. Buds with a single cap-like scale; flowers in catkins .... *Salix*, pg. 323
         10. Buds without scales or with more than one scale; flowers not in catkins 11
            11. Leaves with an aromatic fragrance when crushed 12
               12. Larger leaves often toothed at the apex; flowers in catkins; fruit in dense sessile clusters, covered in a bluish-white wax .... *Myrica pensylvanica*, pg. 247
               12. Leaves always entire; flowers not in catkins; fruit bright red and berry-like .... *Lindera benzoin*, pg. 231
            11. Leaves without an aromatic fragrance 13
               13. Leaves large (usually more than 15 cm long) 14
                  14. Leaves widest above the middle; veins distinctively looped together near the margins of the leaf; fruit banana-like, turning yellowish-green when mature .... *Asimina triloba*, pg. 159
                  14. Leaves widest around the middle; veins not looped together; fruit cone-like, turning dark pink when mature and opening to expose orange seeds .... *Magnolia acuminata*, pg. 242
               13. Leaves smaller (usually less than 15 cm long) 15
                  15. Stems flexible, bark tough and pliable; nodes conspicuously swollen; end of the petiole covers the following season's buds .... *Dirca palustris*, pg. 194
               15. Stems and bark rigid; nodes not swollen 16
                  16. Leafstalks and also usually the young twigs purple; fruit purple to red and berry-like .... *Ilex mucronata*, pg. 219
                  16. Leafstalks and young twigs not purple (*Nyssa* has a reddish leafstalk); fruit various 17
                     17. Plant a low shrub usually less than 1 m tall 18
                        18. Flowers tubular, with 4 large lobes; fruit bright red .... *Daphne mezereum*, pg. 192
                        18. Flowers urn-shaped, with 5 small lobes; fruit reddish-purple to blue 19
                           19. Leaves densely velvety-hairy; leaves lacking resin dots .... *Vaccinium myrtilloides*, pg. 355
                           19. Leaves glabrous or minutely hairy; leaves with yellow-orange resin dots .... *Gaylussacia baccata*, pg. 214
                     17. Plant a medium to large shrub or small tree 20
                        20. Leaves often crowded near the tips of shoots; buds with 5 scales; fruit blue to black, plum-like, 1 to 3 cm long .... *Nyssa sylvatica*, pg. 248
                        20. Leaves spread out along shoots; buds lacking scales; fruit purplish-black, berrylike, less than 1 cm long .... *Rhamnus frangula*, pg. 292

WOODY PLANTS

Silhouettes of Some Common and Unique Woody Plants

Eastern white pine
Pinus strobus

Red pine
Pinus resinosa

Scots pine
Pinus sylvestris

Jack pine
Pinus banksiana

Eastern hemlock
Tsuga canadensis

Black spruce
Picea mariana

Sugar maple
Acer saccharum

Silver maple
Acer saccharinum

American beech
Fagus grandifolia

WOODY PLANTS

American basswood
Tilia americana

White ash
Fraxinus americana

Red oak
Quercus rubra

White elm
Ulmus americana

Slippery elm
Ulmus rubra

Rock elm
Ulmus thomasii

Butternut
Juglans cinerea

Shagbark hickory
Carya ovata

WOODY PLANTS

Horsechesnut
Aesculus hippocastanum

Black locust
Robinia pseudoacacia

Honey locust
Gleditsia triacanthos

Willow (tree form)
Salix sp.

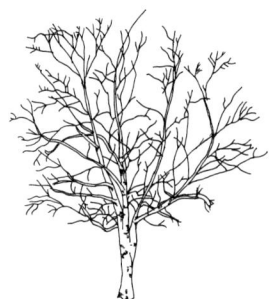

White birch
Betula papyrifera

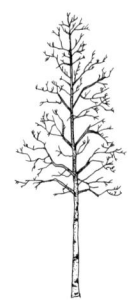

Trembling aspen
Populus tremuloides

Ironwood
Ostrya virginiana

Pin cherry
Prunus pensylvanica

American chesnut
Castanea dentata

WOODY PLANTS

Alternate-leaved dogwood
Cornus alternifolia

Hawthorn
Crataegus sp.

Redbud
Cercis canadensis

Mountain maple
Acer spicatum

Burning bush
Euonymus atropurpureus

Poison sumac
Toxicodendron vernix

Willow (shrub form)
Salix sp.

Honeysuckle
Lonicera sp.

WOODY PLANTS

Abies Fir
1 species in Ontario

Family: *Pinaceae*

Abies balsamea (balsam fir) is the only native fir found in southern Ontario, although it has a wide distribution in the boreal forest from the Rocky Mountains to the Atlantic and extending south into the hardwood forests of the Great-Lakes and Acadian regions. Being a shade-tolerant species it is often present in the understory with spruce and hemlock, but it is sometimes found in pure stands. It is a medium-sized tree with a tall and narrow symmetrical crown. The needles are 1–2.5 cm long, flat, and in two ranks along the twig, sometimes curving towards the light. Like *Tsuga* (hemlock), the needles have lines of stomata beneath, appearing as white stripes on either side of the midrib. The seed-producing cones are 5–10 cm long and in erect clusters towards the ends of the branches. The cone scales fall off the central axis in autumn, leaving the bare upright axis remaining on the tree. It can be readily distinguished from *Tsuga canadensis* (eastern hemlock; pg. 347) by its symmetrical form with regular branches and stiff twigs.

Abies balsamea — Balsam fir — Medium-sized tree

WOODY PLANTS

Acer Maple
Family: *Sapindaceae*   13 species in Ontario

Acer is perhaps one of the best recognized trees in Canada. Its iconic leaf shape is featured on the national flag and its fruit is the distinctive 'maple key'. The various species of *Acer* vary in size from shrubs to tall trees. Maples have deciduous leaves in opposite pairs on stalks about the same length as the leaf. The leaves are simple and palmately lobed (with the exception of *Acer negundo* [Manitoba maple], which has compound leaves). The terminal buds have up to eight pairs of bud scales. Clusters of small flowers with five sepals (some species also have five petals) usually appear before the leaves emerge. The fruit is composed of a seed case with a long wing (2–5 cm) and are joined in pairs on a single stalk.

In southern Ontario there are six native species: *Acer negundo* (Manitoba maple), *Acer saccharum* (sugar maple), *Acer nigrum* (black maple), *Acer rubrum* (red maple), *Acer saccharinum* (silver maple), and *Acer spicatum* (mountain maple). *Acer pensylvanicum* (striped maple) and *Acer spicatum* are more prominent to the north of our area. *Acer saccharum* and *Acer nigrum* can intergrade, and *Acer nigrum* has been considered as a variety of *Acer saccharum*. *Acer* × *freemanii* is the hybrid of *Acer rubrum* and *Acer saccharinum*; it can occur naturally where the two species occur together and has characteristics intermediate between the two species. Two European species, *Acer platanoides* (Norway maple) and *Acer pseudoplatanus* (sycamore maple), are widely planted in urban areas and can escape to woodlots. Maples often represent a large proportion of many woodlots and natural forests in Ontario (as well as Quebec, the Maritimes, and the northeastern United States) and are thus important both ecologically and commercially.

Winter Key to *Acer*

1(a) Buds with only 2 scales 2

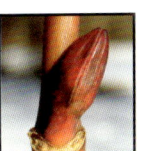 

2(a) Buds and twigs with short white hairs; terminal buds ~5 mm in length ....

Acer spicatum   Mountain maple   pg. 148

WOODY PLANTS

2(b) Buds and twigs hairless; terminal buds 8–10 mm in length..............................

Acer pensylvanicum   Striped maple   pg. 148

  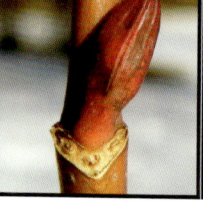 

1(b) Buds with more than 2 scales 3

3(a) Buds brown; scales 8 or more 4

4(a) Twigs glossy, reddish-brown, hairless ....

Acer saccharum   Sugar maple   pg. 150

4(b) Twigs dull, buff or tan, slightly hairy ....

Acer nigrum   Black maple   pg. 150

WOODY PLANTS

3(b) Buds red or green; scales 8 or fewer 5

5(a) Opposite leaf scars touching 6

6(a) Buds hairless or with a minute fringe of hairs on the scale margins; twigs reddish-brown or greenish-brown ....

Acer platanoides Norway maple   pg. 149

6(b) Buds covered in dense white hairs; twigs purple or green ....

Acer negundo Manitoba maple   pg. 147

5(b) Opposite leaf scars not touching, connected by a transverse line 7

 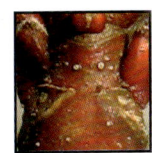

143

WOODY PLANTS

7(a) Buds green, 5mm or more; buds solitary throughout winter ....

Acer pseudoplatanus   Sycamore maple   pg. 149

7(b) Buds red or orange, 3–4 mm; twig tips developing clusters of spherical flower buds towards spring 8

8(a) Twigs with an unpleasant odour when bruised; mature bark splitting into wide strips that peel from the ends ....

Acer saccharinum   Silver maple   pg. 151

8(b) Twigs lacking an unpleasant odour; mature bark splitting into wide strips that do not peel from the ends ....

Acer rubrum   Red maple   pg. 151

WOODY PLANTS

Summer Key to Acer

1(a) Leaves pinnately compound with 3–9 leaflets ....*Acer negundo*, pg. 147

1(b) Leaves simple 2

2(a) Leaf central lobe tapering toward tip; shrub or small tree 3

3(a) Leaves coarsely single-toothed, 2–3 teeth per cm; leaves and twigs hairy; inflorescence erect; bark solid brown ....*Acer spicatum*, pg. 148

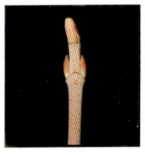 

3(b) Leaves finely double-toothed, 7–12 teeth per cm; leaves and twigs hairless; inflorescence pendant; bark brown or green, with vertical white stripes ....*Acer pensylvanicum*, pg. 148

2(b) Leaf central lobe widening toward tip or with parallel sides; medium to large tree 4

4(a) Leaves coarsely single-toothed.....................5

5(a) Sap from broken twigs and leaves milky; fruit with wings almost 180° apart............. ....*Acer platanoides*, pg. 149

145

WOODY PLANTS

5(b) Sap from broken twigs and leaves watery; fruit with wings roughly parallel 6

6(a) Leaves wrinkled; leaf sinuses V-shaped; central lobe with more than 10 teeth; bark scaly, shedding ....
................*Acer pseudoplatanus*, pg. 149

6(b) Leaves smooth; leaf sinuses U-shaped; central lobe with 3–7 teeth; bark ridged, not shedding 7

7(a) Leaves and petioles densely hairy; leaves usually 3-lobed, drooping at the margins; stipules often present ....
...........................*Acer nigrum*, pg. 150

7(b) Leaves and petioles hairless or with few scattered hairs; leaves usually 5-lobed, margins not drooping; stipules absent ..... *Acer saccharum*, pg. 150

4(b) Leaves double-toothed 8

8(a) Leaf lobes separated by an angle of approximately 90°, central lobe with roughly parallel sides ....
..............................*Acer rubrum*, pg. 151

8(b) Leaf lobes separated by an angle of less than 45°, central lobe with sides diverging towards the tip 9

WOODY PLANTS

9(a) Leaves smooth; lobes separated by U-shaped notches; leaf undersides white or silvery ....*Acer saccharinum*, pg. 151

9(b) Leaves rough; lobes separated by V-shaped notches; leaf undersides light green ....*Acer pseudoplatanus*, pg. 149

Acer negundo   Manitoba maple   Medium-sized tree

147

WOODY PLANTS

Acer spicatum — Mountain maple — Large shrub

Acer pensylvanicum — Striped maple — Large shrub

WOODY PLANTS

Acer platanoides — Norway maple — Large tree

Acer pseudoplatanus — Sycamore maple — Large tree

WOODY PLANTS

Acer nigrum   Black maple   Large tree

Acer saccharum   Sugar maple   Large tree

WOODY PLANTS

Acer rubrum   Red maple   Large tree

Acer saccharinum   Silver maple   Large tree

WOODY PLANTS

Aesculus Horsechestnut, Buckeye   2 species in Ontario
Family: *Sapindaceae*

Aesculus is represented in southern Ontario by only one native species, *Aesculus glabra* (Ohio buckeye). *Aesculus glabra* is restricted to a single natural site in Ontario, on Walpole Island in Lake St. Clair, although it is widespread in the United States. Much more common in southern Ontario is *Aesculus hippocastanum* (horsechestnut), a European native that is widely planted and has naturalized in some areas. Both species are medium-sized trees when mature, with deciduous leaves in opposite pairs on long stalks. The leaves are easily identifiable by their palmately compound leaflets. The large flowers appear after the leaves emerge and are bell- or cone-shaped in erect clusters. The terminal buds are quite large, with many pairs of overlapping scales. The fruit is a large globose or ovoid capsule, 2.5–5 cm in length.

Winter Key to *Aesculus*
1(a) Buds dull, light brown, 1.5–2 cm long, not gummy ....

Aesculus glabra   Ohio buckeye

1(b) Buds shiny, dark brown or black, 2–4 cm long, gummy ....

Aesculus hippocastanum   Horsechestnut

Summary Key to *Aesculus*
1(a) Leaflets usually 5, widest in the middle, 8–12 cm long; flowers yellowish green; buds matte, not sticky ....*Aesculus glabra*

1(b) Leaflets usually 7, widest above the middle, 10–25 cm long; flowers white; buds shiny, sticky ....*Aesculus hippocastanum*

WOODY PLANTS

Aesculus glabra   Ohio buckeye   Small tree

Aesculus hippocastanum   Horsechesnut   Medium-sized tree

WOODY PLANTS

Alnus Alder
Family: *Betulaceae*

3 species in Ontario

Alnus is composed of shrubs or trees that often form dense thickets and occur predominantly in temperate and boreal regions of the Northern hemisphere. Alders possess the unusual, and greatly beneficial, ability to form symbiotic relationships with nitrogen-fixing soil bacteria. As a result they are able to colonize newly exposed or nutrient-poor soils and improve soil fertility for later successional species. The leaves are simple, alternate, deciduous, and ovate to broadly oval in outline. The leaf margins are sharp-toothed (often double-toothed) and the venation is prominent. Pollen and seed flowers occur in separate catkins on the same plant. Male catkins are narrow, elongated and pendulous while female catkins are more egg-shaped. When mature, the female catkins form hard woody cone-like structures that often persist through the winter and provide a highly distinctive character with which to identify the alders. Ontario has two native species, *Alnus viridis* (green alder) and *Alnus incana* spp. *rugosa* (speckled alder), and a naturalized escape of British origin, *Alnus glutinosa* (black alder).

Winter Key to Alnus
1(a) Buds stalked, with 2–3 scales; woody catkins longer than their stalks ....

Alnus incana spp. *rugosa*  Speckled alder

1(b) Buds sessile, with 3–6 scales; woody catkins roughly equal to, or shorter than, their stalks ....*Alnus viridis* ssp. *crispa*

Summer Key to Alnus
1(a) Leaves not sticky; leaf margins coarsely double-toothed; woody catkins longer than their stalks; shrub up to 9 m high ...*Alnus incana* spp. *rugosa*

1(b) Leaves sticky when young; leaf margins finely toothed; woody catkins roughly as long as, or shorter than, their stalks; shrubs up to 3 or 4 m high ....*Alnus viridis* ssp. *crispa*

WOODY PLANTS

Alnus incana ssp. *rugosa* Speckled alder — Large shrub

Alnus viridis ssp. *crispa* Green alder — Large shrub

155

WOODY PLANTS

Amelanchier   Shadbush, Serviceberry   9 species in Ontario
Family: *Rosaceae*

Amelanchier is comprised of shrubs and small trees, with about 5 species in southern Ontario. Like many other genera of the rose family, species delimitation is difficult within *Amelanchier* due to the large overlapping variation in traits and hybridization. The leaves are alternate, deciduous, oval-shaped, and simple, with toothed margins. The flowers appear before the leaves are mature and are arranged on stalks from a central stem. The fruit is a small red or purple pome.

Amelanchier spp.   Serviceberry

Amelanchier spp.   Serviceberry   Shrubs or small trees

WOODY PLANTS

Arctostaphylos Bearberry
3 species in Ontario

Family: *Ericaceae*

Arctostaphylos is comprised of low, spreading shrubs associated with nutrient-poor acidic soils. The flowers are white or pale pink, urn-shaped, and hang in clusters from the tips of the branches. The fruit is a red or purplish-black drupe roughly 6–10 mm in diameter. Two of Ontario's three species are restricted to the tundra in the far north of the province. In southern Ontario, *Arctostaphylos uva-ursi* (common bearberry) can be found growing in open areas on sandy or rocky soil, especially along shorelines. Its leaves are alternate, evergreen, leathery, and simple, with entire margins.

Arctostaphylos uva-ursi   Common bearberry   Spreading shrub

WOODY PLANTS

Aronia Chokeberry

1 species in Ontario

Family: *Rosaceae*

A common wetland shrub of up to 2.5 m, *Aronia melanocarpa* (black chokeberry) can be found in bogs, swamps, and wet woods throughout southern Ontario. The finely toothed leaves are alternate, simple, and oval to obovate. A series of small black glands, barely visible to the naked eye, can be found along the midrib on the upper surface of the leaf and on the tips of the teeth. The flowers are white, 5-parted, and form flat-topped or rounded terminal clusters of 5–15 flowers. The berry-like fruits are purplish-black and, though bitter, are edible and were consumed by First Nations people. *Aronia melanocarpa* may soon be more extensively cultivated, as recent research has found extremely high levels of cancer-fighting antioxidants in the fruits.

Aronia melanocarpa   Black chokeberry

Aronia melanocarpa   Black chokeberry   Medium-sized shrub

WOODY PLANTS

Asimina   Pawpaw, Indian-banana   1 species in Ontario
Family: *Annonaceae*

Asimina triloba is a small tree (<10 m tall) found in Carolinian forests in southern Ontario along the shores of Lake Erie. The solitary leaves are distinctively large and egg-shaped with the end tapered toward the base. The characteristic veins are looped at the ends, before the leaf margins. The terminal bud on winter twigs is elongated and flattened without any scales, and the twigs have solid whitish pith. The flower buds are large and globular with reddish hairs. The flowers are 4 cm across, drooping, and reddish-purple with three parts (petals/sepals). The fleshy fruits are reminiscent of stubby bananas that are green and turn yellow when ripe.

Asimina triloba   Pawpaw

Asimina triloba   Pawpaw   Small tree

WOODY PLANTS

Berberis Barberry   2 species in Ontario
Family: *Berberidaceae*

Berberis is comprised of spiny shrubs or sub-shrubs with yellow inner bark and wood. The stems are of two types: long primary stems on which leaves have been modified into single or 3-pronged spines, and short axillary shoots 1–2 mm long bearing several normal leaves. The leaves are simple, entire or toothed, and narrowly elliptic, oblanceolate, or obovate. Though the branching is alternate, the dense growth of leaves on the dwarf shoots may appear opposite or whorled. The flowers are yellow or orange, 6-parted, and produced singly or in racemes of up to 20. The fruits are edible, may be spherical or ellipsoidal, are red or black, and have 1–10 seeds.

Winter Key to Berberis
1(a) Low spreading shrub 0.5–1.5 m high; twigs reddish-brown; spines single or with 2 smaller lateral branches ....

Berberis thunbergii   Japanese barberry

1(b) An erect or arching shrub 2–3 m high; twigs grey or buff; spines single or with 2 lateral branches equal in length to the central branch ....

Berberis vulgaris   European barberry

Summer Key to Berberis
1(a) Leaves entire; twigs reddish-brown; spines single or with 2 much smaller lateral branches; low shrub 0.5–1.5 m high; flowers and fruit occur in clusters of 1–4 ....*Berberis thunbergii*

1(b) Leaves toothed; twigs grey or buff; spines single or with 2 lateral branches equal in size to the central branch; medium shrub 2–3 m high; flowers in drooping racemes of 10–20 ....*Berberis vulgaris*

WOODY PLANTS

Berberis thunbergii — Japanese barberry — Low shrub

Berberis vulgaris — European barberry — Small shrub

WOODY PLANTS

Betula Birch
12 species in Ontario

Family: *Betulaceae*

The paper-like white bark of *Betula papyrifera* (white birch) is a common sight in many southern Ontario woodlots. The birches range in size from shrubs to tall trees. Birches are shade-intolerant, which results in a open crown and twigs of two distinct types. Twigs located toward the exterior of the tree receive the most light and develop normally, whereas twigs that are towards the interior of the crown receive less light and do not elongate. The leaves are deciduous and alternate, but on the dwarf interior twigs they can be so closely spaced as to appear opposite. The flowers develop in catkins before the leaves appear. The fruits, borne in ovoid or cylindrical catkins, are small (3–4 mm) flat nuts with two thin wings.

Winter Key to *Betula*

1(a) Twigs smelling of wintergreen; bark grey-black or yellow-bronze 2

2(a) Older bark bronze coloured, peeling in thin strips; twigs and buds often hairy; twigs greenish-yellow ....

Betula alleghaniensis   Yellow birch   pg. 165

2(b) Older bark grey-black with conspicuous horizontal lenticels, not peeling; twigs and buds hairless; twigs brown ....

Betula lenta   Cherry birch   pg. 165

WOODY PLANTS

1(b) Twigs not smelling of wintergreen; bark chalky
     white 3

3(a) Bark peeling in large horizontal strips, separating into papery layers; twigs with
     sparse hairs ....

Betula papyrifera   White birch   pg. 166

3(b) Bark not peeling or separating into papery layers; twigs hairless ....

Betula populifolia   Grey birch   pg. 166

WOODY PLANTS

Summer Key to Betula

1(a) Bark bronze or charcoal grey to black and tinged with red; twigs smelling of wintergreen; leaves with 9–12 lateral veins per side 2

2(a) Bark bronze, peeling in thin horizontal curls; twigs hairy ....*Betula alleghaniensis*, pg. 165

2(b) Bark grey to black and tinged with red, smooth with prominent horizontal lenticels; twigs hairless or nearly so ....*Betula lenta*, pg. 165

1(b) Bark white; twigs not smelling of wintergreen; leaves with 6–9 lateral veins per side 3

3(a) Bark peeling in horizontal strips to reveal orange-pink inner bark; leaves oval to ovate with a short, pointed tip; leaves with hair in the vein axils beneath ....*Betula papyrifera*, pg. 166

3(b) Bark not conspicuously peeling; leaves triangular to ovate with an elongated tip; leaves hairless beneath ....*Betula populifolia*, pg. 166

WOODY PLANTS

Betula alleghaniensis Yellow birch Medium-sized tree

Betula lenta Cherry birch Medium-sized tree

WOODY PLANTS

Betula papyrifera — White birch — Medium-sized tree

Betula populifolia — Grey birch — Small tree

WOODY PLANTS

Campsis   Trumpet creeper   1 species in Ontario
Family: *Bignoniaceae*

Campsis radicans is a perennial vine that can be found in southwestern Ontario in old fields, fencerows, and open forests/woodlots. It is pollinated by a variety of birds (hummingbirds, orioles) and moths (sphinx moth). As a weedy species it grows very aggressively, outcompeting many other native species in full sun or partially shaded habitats. The shiny, dark green, compound leaves have 7-11 leaflets with large teeth on the margins. The twigs are light brown with solid pith and small buds. The bark of older vines becomes light brown and scaly. The orange to red tubular flowers are in clusters and very showy; these are often planted in gardens. The fruit is a dry pod 8-15 cm long, containing large numbers of winged seeds.

Campsis radicans   Trumpet creeper

Campsis radicans   Trumpet creeper   Woody vine

167

WOODY PLANTS

Carpinus Hornbeam
Family: *Betulaceae*

1 species in Ontario

Only one species, *Carpinus caroliniana* (American hornbeam, blue beech), occurs naturally in Ontario. Although the leaves are similar in shape to those of *Ostrya virginiana* (ironwood; pg. 249) and the birches (pg. 162), *C. caroliniana* is readily distinguished by its growth form. *Carpinus caroliniana* is usually a small tree with multiple low branches, thin bark, and distinctive thick ridges along the trunk that give it a muscle-like appearance. The leaves are deciduous and alternate in two rows along the twig. The teeth along the leaf margins are of two sizes: a larger tooth at the end of each vein with one or two smaller teeth in between. The flowers develop before the leaves appear, forming clusters of tight stalkless unisexual flowers called catkins. The fruit is a small (< 1 cm) nut that lies at the base of a leafy bract.

Carpinus caroliniana Blue beech

Carpinus caroliniana Blue beech Small tree

WOODY PLANTS

Carya Hickory
4 species in Ontario
Family: *Juglandaceae*

Carya consists of medium-sized trees commercially valued for their nuts (which include the pecan) and their strong, shock-resistant wood for use in axe handles, baseball bats, and a variety of other tools. Hickories are native to eastern North America and eastern Asia. The leaves are deciduous, alternate, and pinnately compound, with 5–11 finely toothed leaflets. The leaflets vary in size with the terminal leaflet being the largest and successive pairs of leaflets getting smaller towards the leaf axil. The flowers are small and yellow-green, appearing along with the leaves in spring. The pollen flowers grow in groups of three catkins while the seed flowers grow in small bunches from the branch tips. The globose or egg-shaped nuts are enclosed within a four-parted husk that splits open as the fruit matures.

Winter Key to Carya
1(a) Buds covered with yellow glandular dots; terminal bud scales elongated, wavy .....
    Carya cordiformis   Bitternut hickory   pg. 173

1(b) Buds without bright yellow dots; terminal buds ovate 2

2(a) Terminal buds 10 mm long or less; mature bark not peeling from trunk ....
    Carya glabra   Red hickory   pg. 173

WOODY PLANTS

2(b) Terminal buds over 10 mm long; mature bark extremely shaggy, peeling in long vertical strips 3

    3(a) Twigs orange or buff, dull; terminal bud 20–25 mm long; leaf petioles commonly persisting on twig ....

Carya laciniosa   Shellbark hickory   pg. 172

    3(b) Twig reddish-brown or grey, shiny; terminal bud 12–18 mm long; leaf petioles rarely persisting on twig ....

Carya ovata   Shagbark hickory   pg. 172

WOODY PLANTS

Summer Key to Carya

1(a) Bark peeling in wide vertical strips giving a shaggy appearance; fruit husk splitting at maturity along its full length; twigs stout 2

2(a) Leaflets 7 or 9 (rarely 5); leaflets with hairs arising singly (not in tufts) on the teeth; twigs yellow-brown to orange; fruit mostly 4–6 cm long ....*Carya laciniosa*, pg. 172

2(b) Leaflets 5 (rarely 7); leaflets with hairs tufted near the apex of the teeth; twigs reddish-brown; fruit mostly 2–4 cm long ....*Carya ovata*, pg. 172

1(b) Bark smooth or ridged, but not peeling in wide strips; fruit husk splitting at maturity along half its length; twigs slender 3

3(a) Leaflets 7–9; buds bright yellow; fruit more or less globose, with raised ridges along sutures ....*Carya cordiformis*, pg. 173

3(b) Leaflets 5 (rarely 7); buds pale reddish-yellow to brown; fruit pear-shaped, lacking raised ridges ....*Carya glabra*, pg. 173

WOODY PLANTS

Carya laciniosa   Shellbark hickory   Medium-sized tree

Carya ovata   Shagbark hickory   Medium-sized tree

WOODY PLANTS

Carya cordiformis Bitternut hickory Medium-sized tree

Carya glabra Red hickory Medium-sized tree

WOODY PLANTS

Castanea Chestnut   1 species in Ontario
Family: *Fagaceae*

Now extremely rare, *Castanea dentata* (American chestnut) once constituted up to 1/4 of the hardwood trees in eastern North America. Giants of the forest, mature chestnut trees could grow up to 35 m in height and 1 m in diameter. Their abundant nut crops were enjoyed by people and wildlife alike, and their straight-grained, durable lumber was used to build fences, barns, and homes. In the early 20th century a fungus known as the chestnut blight was accidentally introduced from Asia, killing roughly 4 billion trees in a few decades and dramatically changing the composition of eastern forests. Today, the presence of *Castanea dentata* in the province is limited to a few specimens growing in the Carolinian zone at the southernmost tip of Ontario. The leaves are simple, alternate, deciduous and elliptic. The nuts, contained in a spiny husk, hang in clusters of 1–5 from the leaf axils.

Castanea dentata   American chestnut

Castanea dentata   American chesnut   Usually small trees

WOODY PLANTS

Catalpa Catalpa   3 species in Ontario
Family: *Bignoniaceae*

Though not native to Ontario, *Catalpa* is often planted as an ornamental tree outside of its natural range due to its showy flowers and fruit. *Catalpa speciosa* (northern catalpa) can occasionally be found as an escape in disturbed areas such as roadsides and forest edges in the Carolinian zone, the southernmost part of the province. The leaves are large, heart-shaped, and opposite or whorled. The fruits are conspicuous elongated capsules that resemble bean pods and can be almost two feet long. The pods often persist on the tree through the winter, turning from green to brown as the season progresses.

Catalpa speciosa   Northern catalpa

Catalpa speciosa   Northern catalpa   Medium-sized tree

175

WOODY PLANTS

Celastrus Staff-tree, Bittersweet   2 species in Ontario
Family: *Celastraceae*

The genus *Celastrus* is composed of about 30 species of woody shrubs and vines, of which only one, *Celastrus scandens*, is native to Ontario. This species is a vine or twining shrub with deciduous, alternate, simple leaves. The small white or yellow-green flowers are borne in terminal clusters 3–8 cm long. The fruits are orange 3-valved capsules about 1 cm across. The valves split open when mature, each releasing 1–2 seeds enclosed in a red aril.

Celastrus scandens   Bittersweet

Celastrus scandens   Bittersweet   Woody vine

WOODY PLANTS

Celtis Hackberry
2 species in Ontario

Family: *Cannabaceae*

Sparsely distributed throughout the Carolinian region of southern Ontario, *Celtis occidentalis* (common hackberry) is a medium-sized tree growing primarily in rich, moist woods, in ravines, or along streams. Its most unique feature is its bark, which is composed of irregular stratified ridges with a corky texture. Hackberry is also known for its tendency to develop 'witch's broom', a dense, tangled mass of twigs growing from a single point. The leaves are simple, alternate, evenly toothed, and ovate to ovate-lanceolate, with an elongated, often curved, tip and an asymmetrical cordate base. The size and hairiness of the leaves are known to be highly variable. The flowers are small, green, and inconspicuous. Male and female reproductive parts can occur in separate flowers on the same tree or together in the same flower. The fruit is a sweet-tasting, cranberry-sized, red or dark-purple drupe with a large pit.

Celtis occidentalis   Hackberry

Celtis occidentalis   Hackberry   Medium-sized tree

177

WOODY PLANTS

Cephalanthus Buttonbush   1 species in Ontario
Family: *Rubiaceae*

The name *Cephalanthus* stems from the Greek *kephalos*, meaning 'head', and *anthos*, meaning 'flower', a reference to this shrub's eye-catching spherical inflorescence. The inflorescence is composed of 100–200 small 4-parted white flowers emanating from the same point to create a dense globe with a diameter of 2–4 cm. The fruit is a spherical collection of tiny cone-shaped brown nutlets borne on a long stalk. The leaves are simple, entire, deciduous, and may be opposite or whorled. Buttonbush favours wet environments including marshes, ditches, and the edges of streams or ponds. In spring it can be found with its roots in standing water.

Cephalanthus occidentalis   Buttonbush

Cephalanthus occidentalis   Buttonbush   Spreading shrub

WOODY PLANTS

Cercis Redbud

1 species in Ontario

Family: *Fabaceae*

Ontario's only representative of this genus is *Cercis canadensis* (eastern redbud), a small tree with a short, twisted trunk, found in rich soils in forest understories and along riverbanks. Its natural range includes only the southernmost tip of Ontario, but it is often planted farther north as an ornamental. Early in the spring, while the tree is still leafless, the bright red buds and pink flowers cover the branches and twigs in abundance, producing a beautiful display. Like other members of the bean family (Fabaceae), redbud's flowers are zygomorphic with five petals forming a 'banner', 'wings', and a 'keel'. Its fruit is a flat reddish-brown pod, 5–10 cm long, that regularly persists on the tree into winter.

Cercis canadensis Redbud

Cercis canadensis Redbud Small tree

WOODY PLANTS

Chamaedaphne Leatherleaf — 1 species in Ontario
Family: *Ericaceae*

A low, spreading, evergreen shrub, *Chamaedaphne calyculata* (leatherleaf) can be found in almost any bog in Ontario. In the wet, acidic conditions found in sphagnum bogs, leatherleaf reproduces vigorously by vegetative shoots to produce dominant clonal colonies. It can also be found growing along the banks of streams, swamps, and lakes. The leaves are thick and leathery (hence the common name), alternate, entire to irregularly toothed, oval to elliptic, and get progressively smaller towards the tips of the branches. Whitish to brownish scales cover the twigs, flower buds, and leaf blades (especially the undersides). The flowers are 5-parted, white, and urn-shaped with a slightly projecting style.

Chamaedaphne calyculata   Leatherleaf

Chamaedaphne calyculata   Leatherleaf   Low shrub

WOODY PLANTS

Clematis Clematis   10 species in Ontario
Family: *Ranunculaceae*

With only two native species of *Clematis* in Ontario – *Clematis occidentalis* (purple clematis) and *Clematis virginiana* (virgin's-bower) – most of the diversity in the province is due to escapes of European or Asian origin that were initially planted as ornamentals for their attractive flowers. The fragrant flowers attract bees, butterflies, and other pollinators and reward them with an abundance of pollen and nectar. Our native species are woody vines that climb trees or fences to a height of 7 m using elongated petioles as tendrils to wrap around supports. The leaves are opposite, coarsely toothed to entire, and ternately compound (rarely with 5 leaflets). The fruits are distinctive clusters of small seeds each with a long, feathery, white plume.

Clematis virginiana   Virgin's-bower

Clematis virginiana   Virgin's-bower   Woody vine

WOODY PLANTS

Comptonia Sweet-fern 1 species in Ontario
Family: *Myricaceae*

Easily identified by its combination of woody stems and fern-like leaves, *Comptonia peregrina* (sweet-fern) is a low shrub bearing no relation to the true ferns. It grows in sandy to rocky soil in open conifer forests and is commonly associated with jack pine. Like other nitrogen fixing plants, sweet-fern does well in disturbed or nutrient-poor soils, which, in combination with its intolerance for shade, makes it well suited to being an early colonizer. The long, narrow leaves are alternate and simple with rounded, fern-like lobes. The leaves and shoots are dotted with glands that give off a pleasant fragrance when bruised. Male flowers grow in catkins at the tips of shoots, while the female flowers grow in bright red, bristly, bur-like clusters lower down on the shoot.

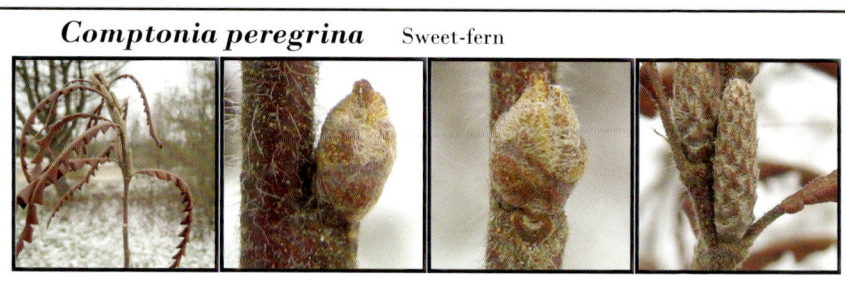

Comptonia peregrina Sweet-fern

Comptonia peregrina Sweet-fern   Low shrub

WOODY PLANTS

Cornus Dogwood   9 species in Ontario
Family: *Cornaceae*

Most members of *Cornus* are shrubs or small trees with deciduous, opposite, simple leaves with entire margins. One species is herbaceous (*Cornus canadensis*, pg. 386), and one has an alternate arrangement (*Cornus alternifolia*, pg. 186). The leaf venation of most *Cornus* species is distinctive: leaf veins begin parallel at an angle from the midrib but soon curve towards the leaf tip. The small flowers are 4-parted, often clustered in heads. The fruit is a red, white, blue, or black drupe.

Winter Key to Cornus
1(a) Branching alternate; often with dead bright yellow twigs ....

Cornus alternifolia   Alternate-leaved dogwood   pg. 186

1(b) Branching opposite; without dead bright yellow twigs 2

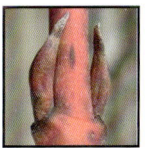 

2(a) Axillary buds sometimes concealed by vestigial petiole bases; some branches tipped with large, round flower buds ....

Cornus florida   Eastern flowering dogwood   pg. 186

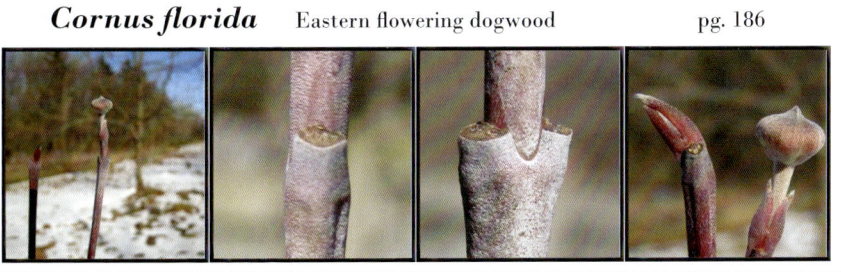

2(b) Axillary buds clearly visible; all terminal buds similar 3

 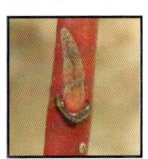

WOODY PLANTS

3(a) Twigs green with purple streaks or pink with dark spots ....

Cornus rugosa   Round-leaved dogwood   pg. 187

3(b) Twigs red or brown 4

4(a) Older twigs bright red; vestigial inflorescence flat-topped ....

Cornus sericea   Red-osier dogwood   pg. 188

4(b) Older twigs light grey or brown; vestigial inflorescence vertically elongated ..

Cornus racemosa   Grey dogwood   pg. 187

WOODY PLANTS

Summer Key to Cornus

1(a) Branching alternate ....*Cornus alternifolia*, pg. 186

1(b) Branching opposite 2

2(a) Leaves often clustered at the ends of the branches; flowers lacking stalks, surrounded by four white petal-like bracts 2–5 cm long; fruit red, ellipsoid; mature bark with small, 'alligator-skin' scales ....*Cornus florida*, pg. 186

2(b) Leaves more or less evenly spaced along new growth; flowers stalked, lacking bracts; fruit, white or blue, globose; mature bark smooth, warty, or vertically splitting 3

3(a) Leaves elliptic to lanceolate, 2–3 times as long as wide; twigs with pale brown pith; older twigs light brown ....
...........................*Cornus racemosa*, pg. 187

3(b) Leaves elliptic to oval, usually 2 times as long as wide or less; twigs with white pith; older twigs red, yellow, or green 4

4(a) Twigs pink or green, with purple streaks; leaves with 6–8 pairs of lateral veins; leaves usually 1.5 times as long as wide or less; fruit pale blue ....
.............................*Cornus rugosa*, pg. 187

4(b) Twigs red, without purple streaks; leaves with 4–5 (rarely 6) pairs of lateral veins; leaves usually 1.5 times as long as wide or more; fruit white ....*Cornus sericea*, pg. 188

WOODY PLANTS

Cornus alternifolia — Alternate-leaved dogwood — Large shrub

Cornus florida — Eastern flowering dogwood — Large shrub

WOODY PLANTS

Cornus racemosa — Grey dogwood — Medium-sized shrub

Cornus rugosa — Round-leaved dogwood — Medium-sized shrub

WOODY PLANTS

Cornus sericea Red-osier dogwood Medium-sized shrub

WOODY PLANTS

Corylus Hazel   2 species in Ontario
Family: *Betulaceae*

Corylus is composed of shrubs or small trees native to the Northern temperate zone. The nuts of hazels are edible and include the commercially produced hazelnuts and filberts. The nuts are enclosed within a distinctive fringed or leafy husk formed from the floral bracts. The leaves are simple, alternate, deciduous, and ovate to broadly oval. The leaf margins are double-toothed with prominent venation. The pollen flowers occur in pendulous catkins that can be seen throughout the year in clusters of 1–3. The seed flowers appear in early spring before the leaves and are visible as bright red thread-like stigmas emerging from the bud tips.

Winter Key to Corylus
1(a) Bud tips blunt; twigs with glandular hairs; fruit partially enclosed in two spreading leaf-like bracts ....

Corylus americana   American hazel   pg. 190

1(b) Bud tips pointed; twigs more or less hairy but without glandular hairs; fruit enclosed in two fused bracts forming a long tubular beak ....

Corylus cornuta   Beaked hazel   pg. 190

Summer Key to Corylus
1(a) Twigs and petioles with dense hairs and stalked glands; male catkins short-stalked; fruit partially enclosed by two spreading leaf-like bracts ....*Corylus americana*, pg. 190

1(b) Twigs and petioles with hair but lacking stalked glands; male catkins without stalks; fruit enclosed by two fused bracts forming a long tubular beak ....*Corylus cornuta*, pg. 190

WOODY PLANTS

Corylus americana   American hazel   Medium-sized shrub

Corylus cornuta   Beaked hazel   Medium-sized shrub

WOODY PLANTS

Crataegus Hawthorn   ~45 species in Ontario
Family: *Rosaceae*

Crataegus is composed of shrubs and small trees that grow well on open, disturbed sites. The leaves are deciduous, alternately arranged, and simple, but variously lobed and toothed. The fertilized flower develops a fruit resembling a small apple, called a haw. Although the leaves are variable, the hawthorns can be recognized as an identifiable group by twigs with one or two shiny rounded buds at the angles of a winding twig. If two buds are present, one develops into a stiff thorn, which is characteristic of the genus. The bark shreds along the trunk, peeling in narrow strips.

Crataegus spp.   Hawthorn

Crataegus spp.   Hawthorn   Shrubs and small trees

191

WOODY PLANTS

Daphne Daphne
Family: *Thymelaeaceae*

1 species in Ontario

Despite being highly poisonous, with sap that can cause contact dermititis and fruits that can cause kidney failure and death, the species of *Daphne* are often planted as ornamentals for their early-blooming, attractive flowers. One shrubby species, *Daphne mezereum* (February daphne – so named because in its native range in Europe it is known to bloom as early as February), has escaped and become naturalized in eastern Canada. In Ontario, where February daphne is known only from a small number of sites, it doesn't bloom until late April just before the leaves appear. Growing in dense bunches covering the tips of the previous year's woody stems, the flowers are sweet-smelling, pink to light purple, and 4-parted. The fruit is a bright red drupe, 1 cm long, growing in clusters on the stems just below the new year's growth. The leaves are simple, entire, and alternate, though at times they grow densely on the stem and may appear to be whorled.

Daphne mezereum   Daphne

Daphne mezereum   Daphne   Small shrub

WOODY PLANTS

Diervilla Bush honeysuckle   1 species in Ontario
Family: *Caprifoliaceae*

Our only species, *Diervilla lonicera* (northern bush honeysuckle), is a low shrub 0.5–1 m tall, growing in dry, sandy soils in forests, sand dunes, and clearings. Able to regenerate quickly after a fire, *Diervilla lonicera* can become a dominant shrub in dry areas subject to regular burning. The new shoots are green or red, changing over the season to brown or grey, and bear 2 or 4 vertical lines of minute hairs that may be hard to see without a field lens. The flowers are 5-parted and bloom for an extended period during which they change from pale yellow to orange or red. The petals curl backwards, and the lowest petal is more brightly coloured and hairier than the others. The leaves are opposite, simple, deciduous, and toothed. The fruit is a dry brown capsule tapering towards the tip where the five shriveled calyx lobes remain attached. In the winter the persistent distinctive fruits can aid in identification.

Diervilla lonicera   Northern bush honeysuckle

Diervilla lonicera   Northern bush honeysuckle   Low shrub

WOODY PLANTS

Dirca Leatherwood
1 species in Ontario
Family: *Thymelaeaceae*

Dirca is a small genus consisting of three species, all of which are slow growing shrubs native to North America. *Dirca palustris* (eastern leatherwood), the only species that occurs in Ontario, has several unique features. The stems swell at the nodes to form a broad mouth from which the next year's growth emerges. The bud is small, conical, covered in dark silky hairs, and almost completely encircled by the sheathing petiole of the leaf. Perhaps most interesting, the bark of the leatherwood bush is both strong and pliable and can be cut into long strips, which First Nations people used to make ropes, baskets, and thongs. The leaves are simple, alternate, deciduous, and obovate to broadly oval. *Dirca palustris* is one of the first woody plants to bloom in the spring, producing yellow flowers with an elongated tubular corolla and conspicuous stamens.

Dirca palustris   Eastern leatherwood

Dirca palustris   Eastern leatherwood   Medium-sized shrub

WOODY PLANTS

Elaeagnus Oleaster — 3 species in Ontario
Family: *Elaeagnaceae*

Asia is home to the majority of species of *Elaeagnus* while only one, *Elaeagnus commutata* (silverberry), is native to North America. *Elaeagnus commutata* can be found in Ontario along with two introduced species, *Elaeagnus angustifolia* (Russian olive) and *Elaeagnus umbellata* (autumn olive), both of which are cultivated for their fruit in China. From a distance *Elaeagnus* species are easily recognized by the whitish to light grey colouring given to them by the minute silvery or brownish scales covering their leaves, twigs, buds, and fruit. Their leaves are thick, alternate, decidous, entire, and lanceolate to elliptic. Their flowers are white to yellow, 4-parted, have an elongated tubular hypanthium, and occur in axillary clusters. The fruit is silvery, berry-like, ovoid, and has a single stone.

Winter Key to Elaeagnus

1(a) Twigs and buds densely coated in white or silvery hairs ....

Elaeagnus angustifolia — Russian olive — pg. 197

1(b) Twigs and buds covered with rusty scales 2

2(a) Twigs lacking thorns; fruits silvery ....

Elaeagnus commutata — Silverberry — pg. 197

WOODY PLANTS

2(b) Twigs often armed with thorns; fruits scarlet when mature ....

Elaeagnus umbellata   Autumn-olive   pg. 198

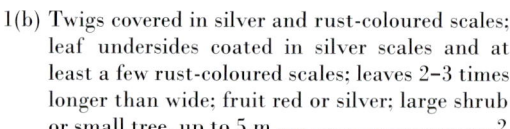

Summer Key to Elaeagnus

1(a) Twigs covered with silvery-white hairs and scales; leaf undersides with silver scales only; leaves 3–8 times longer than wide; fruit yellow with silvery scales; medium-sized tree, to 10 m tall .... *Elaeagnus angustifolia*, pg. 197

1(b) Twigs covered in silver and rust-coloured scales; leaf undersides coated in silver scales and at least a few rust-coloured scales; leaves 2–3 times longer than wide; fruit red or silver; large shrub or small tree, up to 5 m 2

2(a) Mature leaves with silver scales above; branches unarmed; fruit silver and dry .... *Elaeagnus commutata*, pg. 197

2(b) Mature leaves green and smooth above, sometimes with a few scattered scales; branches often armed with thorns; fruit red and juicy .... *Elaeagnus umbellata*, pg. 198

WOODY PLANTS

Elaeagnus angustifolia   Russian olive   Large shrub or small tree

Elaeagnus commutata   Silverberry   Medium-sized shrub

WOODY PLANTS

Elaeagnus umbellata Autumn-olive Large shrub

WOODY PLANTS

Epigaea Trailing arbutus
Family: *Ericaceae*

1 species in Ontario

Ontario's only species, *Epigaea repens* (trailing arbutus), typically grows in dry, sandy, pine forests, but can sometimes be found on moist ground in bogs or deciduous woods. A low, creeping shrub rarely exceeding 15 cm in height, *Epigaea repens* can be easily overlooked as it is often partially concealed by leaf litter. The stems and petioles are densely covered in stiff brown hairs, as are the leaves along their margins and midveins. The leaves are alternate, leathery, evergreen, and have a rounded or heart-shaped base. The richly fragrant flowers appear in small bunches early in the spring before most deciduous plants have their leaves. The flowers are white to pink and 5-parted with the petals fused into a tube along half of their length.

Epigaea repens   Trailing arbutus   Creeping shrub

WOODY PLANTS

Euonymus Spindle-tree   5 species in Ontario
Family: *Celastraceae*

Euonymus is a genus of shrubs and vines. The leaves are opposite, simple, and deciduous. Like *Celastrus*, it produces small inconspicuous flowers in clusters from the axils of the leaves. The fruit is a capsule enclosing seeds set in a red or orange aril.

Winter Key to *Euonymus*
1(a) Stems low, spreading, 10–30 cm tall ....

Euonymus obovatus   Running strawberry-bush   pg. 201

1(b) Stems erect, 1–6 m tall 2

2(a) Buds long, acute; bud scales acute ....

Euonymus atropurpureus   Burning bush   pg. 202

2(b) Buds obtuse; bud scales obtuse ....

Euonymus europaeus   European spindle-tree   pg. 202

WOODY PLANTS

Summer Key to Euonymus

1(a) Low trailing shrub from 10–30 cm tall; flowers 5-merous; fruit warty; petioles 1–5 mm long ....*Euonymus obovatus*, pg. 201

1(b) Erect shrub from 1–6 m tall; flowers 4-merous; fruit smooth; petioles 10–30 mm long 2

2(a) Leaf undersides with small hairs; flowers purple; arils red ....*Euonymus atropurpureus*, pg. 202

2(b) Leaf undersides hairless; flowers greenish-white; arils orange ....*Euonymus europaeus*, pg. 202

Euonymus obovatus — Running strawberry-bush — Trailing shrub

201

WOODY PLANTS

Euonymus atropurpureus — Burning bush — Large shrub

Euonymus europaeus — European spindle-tree — Large shrub

WOODY PLANTS

Fagus Beech

2 species in Ontario

Family: *Fagaceae*

Fagus grandifolia (American beech) is the only beech that is native to Canada and the United States. It is commonly found in association with sugar maple, hemlock, and yellow birch across much of eastern North America. *Fagus grandifolia* is a medium-sized tree with thin, smooth, grey bark that is easily damaged. The leaves are deciduous, alternate, and have straight parallel veins ending in a single tooth. The flowers are unisexual, but both types are found on the same tree. They appear from the axils of new growth at the same time as the leaves. The pollen flowers are in drooping globular clusters at the end of long stalks. The seed flowers produce 4-valved bristly husks each enclosing 2 pyramidal nuts.

Fagus grandifolia   American beech

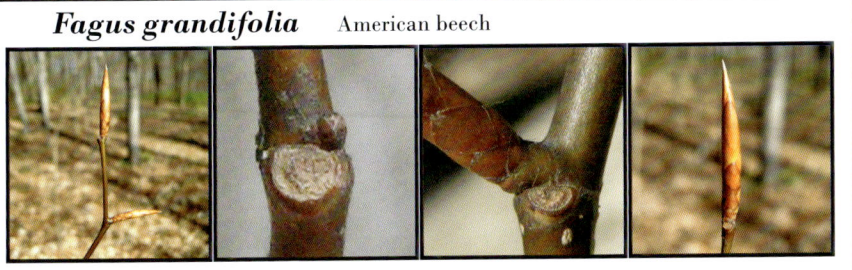

Fagus grandifolia   American beech   Medium-sized tree

WOODY PLANTS

Fraxinus Ash
Family: *Oleaceae*

6 species in Ontario

Three native species of ash commonly occur in southern Ontario: *Fraxinus americana* (white ash), *Fraxinus pennsylvanica* (red ash), and *Fraxinus nigra* (black ash). *Fraxinus pennsylvanica* is often subdivided into three varieties, two of which (var. *pennsylvanica* and var. *subintegerrima*) are found in our area. Two other native species are at the northern end of their range here and much less common. *Fraxinus quadrangulata* (blue ash) is found in only a few scattered locations, and *Fraxinus profunda* (pumpkin ash) only occurs in Essex county. The introduced species *Fraxinus excelsior* (European ash) is sometimes planted as an ornamental and is occasionally found as an escape. The ashes range in size from small to large trees. Ash leaves are deciduous, opposite, and pinnately compound, with 5-11 leaflets per leaf. The small flowers arise from the axils of the previous year's growth in short dense clusters. Pollen flowers and seed flowers are borne on separate trees in white, red, and pumpkin ash. Blue ash has perfect flowers, and black and European ash have variable flowers. The fruit is single-seeded, with the seed in a case enclosed by a narrow wing.

Winter Key to *Fraxinus*
1(a) Twigs with four vertical ridges, square in cross section ....

Fraxinus quadrangulata   Blue ash   pg. 208

1(b) Twigs without ridges, round in cross section 2

2(a) Buds blue-black ....

Fraxinus excelsior   European ash   pg. 210

WOODY PLANTS

2(b) Buds brown ....3

3(a) Buds dark brown; some bark visible between first lateral buds and terminal bud; bark corky, forming scales ....

Fraxinus nigra   Black ash   pg. 210

3(b) Buds reddish-brown; first lateral buds touching, or very nearly touching, the terminal bud; bark firm, forming a diamond pattern 4

4(a) Leaf scars with a deep U- or V-shaped notch at the top 5

4(b) Leaf scars straight or only slightly indented at the top 6

 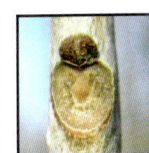

WOODY PLANTS

5(a) Twigs hairy ....

Fraxinus profunda   Pumpkin ash   pg. 209

5(b) Twigs hairless ....

Fraxinus americana   White ash   pg. 211

6(a) Twigs hairy ....

Fraxinus pennsylvanica   Red ash   pg. 209

6(b) Twigs hairless ....

Fraxinus pennsylvanica var. *subintegerrima*   Green ash   pg. 211

WOODY PLANTS

Summer Key to Fraxinus

1(a) Twigs strongly 4-angled ....
...................*Fraxinus quadrangulata*, pg. 208

1(b) Twigs round or oval in cross-section 2

2(a) Twigs and petioles hairy 3

    3(a) Leaflets 8–14 cm long, toothed at least along upper half; leaflet stalks 3–6 mm long, often winged ....*Fraxinus pennsylvanica*, pg. 209

    3(b) Leaflets 10–25 cm long, entire or with irregular rounded teeth; leaflet stalks 5–10 mm long, not winged ....
....................*Fraxinus profunda*, pg. 209

2(b) Twigs and petioles hairless 4

4(a) Lateral leaflets not stalked 5

    5(a) Bark firm, broken into vertical ridges and furrows; leaflet undersides with hairs lining the midvein; fruit with flat wing ....*Fraxinus excelsior*, pg. 210

207

WOODY PLANTS

5(b) Bark soft and corky, usually broken into scales, but occasionally with ridges and furrows; leaflets with a dense tuft of rusty hairs at the point of attachment; fruit often with twisted wing ....*Fraxinus nigra*, pg. 210

4(b) Lateral leaflets stalked (stalk may be winged) 6

6(a) Leaflet undersides pale green and smooth; leaflet stalks 3–6 mm, often winged; fruit wing extending along at least 1/2 the length of the seed ....*Fraxinus pennsylvanica* var. *subintegerrima*, pg. 211

6(b) Leaflet undersides pale whitish-green with minute bumps; leaflet stalks 5–15 mm, not winged; fruit wing extending along no more than 1/3 the length of the seed ....*Fraxinus americana*, pg. 211

Fraxinus quadrangulata Blue ash   Small tree

WOODY PLANTS

Fraxinus pennsylvanica　　Red ash　　　　　　Medium-sized tree

Fraxinus profunda　　Pumpkin ash　　　　　　Medium-sized tree

WOODY PLANTS

Fraxinus excelsior — European ash — Medium-sized tree

Fraxinus nigra — Black ash — Small tree

WOODY PLANTS

Fraxinus pennsylvanica var. *subintegerrima*   Green ash   Tree

Fraxinus americana   White ash   Medium-sized tree

WOODY PLANTS

Gaultheria Wintergreen, Snowberry   2 species in Ontario
Family: *Ericaceae*

Gaultheria is represented by two distinct species in northern Ontario forests. Both species have dark green, shiny, leathery leaves. The first is the trailing vine-like *Gaultheria hispidula* (creeping snowberry) with small, alternate, oval leaves that are green and persistent throughout winter. The flowers are bell-shaped and the fruits are white and smell like citrus and mint. *Linnaea borealis* (twinflower) is a similar species with opposite, toothed leaves that have no odour. The second species is the small (5–15 cm) upright *Gaultheria procumbens* (wintergreen), which has alternate leaves crowded near the tips of the branches. The flowers are tiny and bell-shaped with five lobes at the tip of the fused petals. The fruits are red and mealy and smell like wintergreen.

Summer/Winter Key for Gaultheria

1(a) Leaves less than 1 cm long; stems all trailing; flowers 4-parted; fruit white ....*Gaultheria hispidula*

1(b) Leaves 1–5 cm long; some stems erect; flowers 5-parted; fruit red ....*Gaultheria procumbens*

WOODY PLANTS

Gaultheria hispidula   Creeping snowberry   Trailing shrub

Gaultheria procumbens   Wintergreen   Low shrub

WOODY PLANTS

Gaylussacia Huckleberry   1 species in Ontario
Family: *Ericaceae*

The species of *Gaylussacia* are low (30–100 cm) shrubs that spread underground to form clonal colonies. Ontario's *Gaylussacia baccata* (black huckleberry) thrives in acidic soils, both on wet ground in bogs and fens and on sandy or rocky ground in woods or barrens. The leaves are deciduous, alternate, and simple, with entire margins. The flowers are 5-parted, urn-shaped, yellow-orange to red, and hang in one-sided racemes of 3–7 flowers. In its leaves, growth form, and fruit, *Gaylussacia baccata* closely resembles *Vaccinium myrtilloides* (blueberry; pg. 355), but can be distinguished by the yellow-orange resin dots on its leaves and the ten hard nutlets within its fruits.

Gaylussacia baccata   Black huckleberry   Low shrub

WOODY PLANTS

Gleditsia Honey locust   1 species in Ontario
Family: *Fabaceae*

Gleditsia triacanthos is a thorny tree in the pea family that is found in southwestern Ontario. It has large branched thorns that are found on the trunk and branches and are very sharp. The large (15–20 cm) alternate leaves can be either once or twice compound and have oval leaflets. The zigzagging twigs lack a terminal bud and have very small lateral buds. The flowers are inconspicuous but give rise to distinctive pods that are up to 40 cm long. The pea-like seeds have a very tough seed coat that provides protection and affords viability for many years.

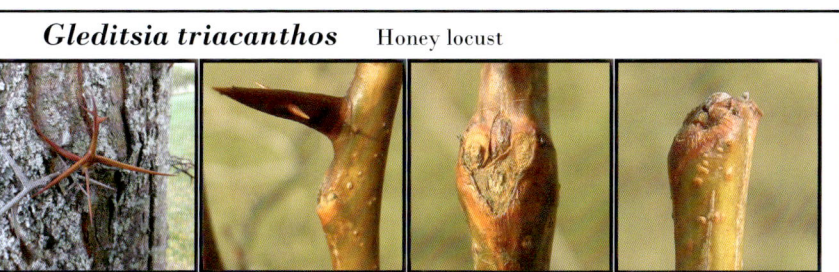

Gleditsia triacanthos   Honey locust

Gleditsia triacanthos   Honey locust   Medium-sized tree

WOODY PLANTS

Gymnocladus Kentucky coffee tree   1 species in Ontario
Family: *Fabaceae*

Gymnocladus dioicus is a large tree occurring in southwestern Ontario. The large (80–100 cm) alternate leaves are doubly compound with up to 70 oval leaflets and do not appear until late in the spring. The twigs are thick and gray/brown with an orange/red pith and no terminal bud. Female trees have clusters of white flowers with five parts (sepal/petals), which give rise to large (15–20 cm), broad, leathery pods that are often covered in a fine powder; male trees have smaller clusters of white flowers that wither after pollen is dispersed.

Gymnocladus dioicus   Kentucky coffee tree

Gymnocladus dioicus   Kentucky coffee tree   Medium-sized tree

WOODY PLANTS

Hamamelis Witch-hazel
1 species in Ontario

Family: *Hamamelidaceae*

Hamamelis virginiana (witch-hazel) is the sole species of this genus native to Canada. It is commonly found in southwestern Ontario and ranges from eastern Wisconsin to Nova Scotia as an understory shrub or small tree. The leaves are deciduous, alternate, and simple, with a wavy or coarsely toothed margin and an asymmetric base. The buds are stalked and lack scales, but are covered with yellowish-brown hairs. Unlike most woody plants in our area, witch-hazel flowers in the autumn as it begins to lose its leaves. The flowers are in groups of three; each flower has four bright yellow twisting petals ~2 cm long. The fertilized flower develops into a hairy two-sided capsule. The capsule matures the following year and then splits explosively apart, forcibly ejecting two shiny black seeds. This species is the source of the astringent witch-hazel that has been used for a variety of medicinal purposes.

Hamamelis virginiana   Witch-hazel

Hamamelis virginiana   Witch-hazel   Large shrub

WOODY PLANTS

Ilex Holly   3 species in Ontario
Family: *Aquifoliaceae*

With its bright red berries and spiny green leaves, *Ilex aquifolium* (English holly) is familar to many as a symbol of Christmas and the source of the traditional holiday colours. Though it can be found in Ontario it is not as common as our native species, *Ilex verticillata* (winterberry) and *Ilex mucronata* (mountain holly). Our native species are erect deciduous shrubs up to 3–4 m tall typically growing in moist conditions near water. The flowers grow singly or in clusters from the leaf axils, and are usually unisexual, the sexes appearing on different plants, but may occasionally be perfect. The fruits are red berry-like drupes that are eaten by birds but are poisonous to humans.

Winter Key to Ilex

1(a) Twigs purple with white lenticels; buds pointed at the tip, purple, lacking persistent stipules; fruit stalks longer than the fruit ....*Ilex mucronata*

1(b) Twigs grey, buff, or reddish-brown; buds blunt, tan-coloured, often with persistent small, spike-like stipules; fruit stalks shorter than the fruit ....

Ilex verticillata   Common winterberry

Summer Key to Ilex

1(a) Leaf margins entire or with a few scattered teeth; petioles red to purple, hairless; twigs purple; petals linear, more than 2 times longer than wide ....*Ilex mucronata*

1(b) Leaf margins with sharp, regular teeth; petioles green, hairy; twigs brown or grey; petals elliptic, less than 2 times longer than wide ....
....*Ilex verticillata*

WOODY PLANTS

Ilex mucronata Mountain holly — Medium-sized shrub

Ilex verticillata Common winterberry — Medium-sized shrub

WOODY PLANTS

Juglans Walnut   3 species in Ontario
Family: *Juglandaceae*

The walnuts are medium-sized to large trees with deciduous, alternate, pinnately compound leaves. The leaflets have very short stalks or are stalkless and range in number from 5 to 23 depending on the species. The pith of the previous year's growth is chambered. The small flowers are green and inconspicuous, appearing with the leaves. They are unisexual with both pollen and seed flowers on the same tree. The pollen flowers are in short catkins of 2–10 that arise from lateral buds. The seed flowers arise in groups of 1–3 closer to the end of the twig. The fruit is a large (5–6 cm) globular husk enclosing a hard shell containing a nut. Two species of *Juglans* are native to southern Ontario: *Juglans nigra* (black walnut), which is uncommon, and *Juglans cinerea* (butternut), which is on Canada's endangered species list.

Winter Key to *Juglans*
1(a) Leaf scar straight along the top, with a downy line ....

Juglans cinerea   Butternut

1(b) Leaf scar notched at the top, without a downy line ....

Juglans nigra   Black walnut

Summer Key to *Juglans*

1(a) Leaves very hairy below, with a large terminal leaflet; fruit elongated ....*Juglans cinerea*

1(b) Leaves slightly hairy below, with a reduced or missing terminal leaflet; fruit globular ....*Juglans nigra*

WOODY PLANTS

Juglans cinerea Butternut — Medium-sized tree

Juglans nigra Black walnut — Medium-sized tree

WOODY PLANTS

Juniperus Juniper   4 species in Ontario
Family: *Cupressaceae*

Junipers are coniferous shrubs or small trees distributed throughout the Northern hemisphere. They are popular as landscape plants, and are therefore quite common in urban settings. In nature, junipers prefer to grow in dry areas and are highly intolerant of shade. Some First Nations people used juniper medicinally to treat diabetes, and recent research has confirmed that juniper may have the ability to mitigate diabetes in mice. The seed cones are comprised of fleshy scales that unite at maturity to produce a dark blue 'berry', often with a dusty white coating. Juniper 'berries' are eaten by birds and small mammals, and are used to provide the primary flavouring for gin. Juniper leaves are evergreen and may be opposite or whorled. The leaves can be of two forms, needle-like (similar to *Taxus*, pg. 340) or scale-like (similar to *Thuja*, pg. 341), both of which can often be found on the same plant.

Summer/Winter Key to Juniperus

1(a) Leaves all needle-like, in whorls of 3, not concealing the twigs; 'berries' borne in leaf axils ....*Juniperus communis*, pg. 223

1(b) Leaves needle-like or scale-like, mostly opposite, concealing the twigs; 'berries' borne on branch tips 2

   2(a) Low, spreading shrub; 'berries' 6–10 mm in diameter ....
....*Juniperus horizontalis*, pg. 223

   2(b) Erect shrub or tree; 'berries' 5–6 mm in diameter ....
....*Juniperus virginiana*, pg. 224

WOODY PLANTS

Juniperus communis Common juniper Low shrub

Juniperus horizontalis Creeping juniper Trailing shrub

WOODY PLANTS

Juniperus virginiana   Eastern red cedar   Very small tree

WOODY PLANTS

Kalmia Laurel

3 species in Ontario

Family: *Ericaceae*

Laurels are evergreen shrubs growing to a height of 60–100 cm in open bogs, swamps, and wet woods. The leaves of laurel species are fatally poisonous to livestock, a fact reflected in some of the common names applied to them: lamb-kill, calf-kill, kill-kid, and sheep-poison. The leathery leaves are simple with entire and slightly inrolled margins, growing in opposite or whorled arrangements. The fruit is a brown globose 5-lobed capsule that often has a persistent style. The flowers are pale to dark pink, hang in clusters of 5–50, and are 5-parted with their petals fused to form a saucer-shaped disk. The flowers utilize an interesting mechanism to ensure the spread of their pollen: the anthers are embedded in depressions in the corolla, causing the filaments to arch backward under tension. When a pollinator lands in the flower the anther is released from its pocket and springs forward to shower the intruder with pollen.

Summer Key to Kalmia

1(a) Stems round; leaves green beneath, with a short stalk, usually more than 1.5 cm wide ....
................................*Kalmia angustifolia*, pg. 225

1(b) Stems flattened; leaves white beneath, stalkless, usually less than 1.5 cm wide ....
................................*Kalmia polifolia*, pg. 226

Kalmia angustifolia   Sheep laurel   Low shrub

WOODY PLANTS

Kalmia polifolia   Bog laurel   Low shrub

WOODY PLANTS

Larix Larch

3 species in Ontario

Family: *Pinaceae*

Of the three species of larch found in Canada, only one, *Larix laricina* (tamarack), is found in southern Ontario. Tamarack ranges from the Rocky Mountains to the Atlantic Ocean as a component of the boreal forest, and extends to the south in our Carolinian range. It is primarily found in swampy areas where other trees cannot establish easily. It grows well on other sites, but is so intolerant of shade that it is usually excluded from them. Larches have soft flexible needles arising singly from the twig. On dwarf shoots that do not extend during the growing season, the needles appear clustered together. Unlike most conifers, the needles of *Larix* are deciduous, becoming yellow and falling from the tree in the autumn. The cones are 1–2 cm long when mature and open in late summer, releasing seeds in the fall. *Larix decidua*, the European larch, is occasionally planted as an ornamental species in southern Ontario.

Winter Key to Larix

1(a) Twigs yellow-buff; seed cones 2–4 cm, with 30 or more scales ....

Larix decidua   European larch   pg. 228

1(b) Twigs orange-brown; seed cones 1–2 cm, with 20 or fewer scales ....

Larix laricina   Tamarack   pg. 228

WOODY PLANTS

Larix decidua European larch   Medium-sized tree

Larix laricina Tamarack   Medium-sized tree

WOODY PLANTS

Ledum   Labrador tea   2 species in Ontario
Family: *Ericaceae*

A spreading or erect evergreen shrub up to 1 m tall growing in bogs, swamps, muskeg, stream edges, and tundra. The leathery leaves are simple, alternate, and densely wrinkled, with entire, strongly inrolled margins. Young twigs and the undersides of leaves are covered in a dense mat of white or rust-coloured hairs. The leaves are strongly fragrant and can be used to make a very flavourful herbal tea, but the tea should only be consumed infrequently in weak concentrations as it contains toxins which can be harmful in large doses. The flowers are white with 5 separate or basally-fused petals, appearing from spring to mid-summer in semi-spherical, in terminal clusters of 10–35.

Ledum groenlandicum   Labrador tea   Low shrub

WOODY PLANTS

Ligustrum Privet
1 species in Ontario
Family: *Oleaceae*

Ligustrum, a close relative to *Syringa* (lilac; pg. 339), is a genus that also has been introduced to North America. It is planted primarily in hedgerows but is commonly found as an escape. The group contains deciduous, semi-evergreen, and evergreen shrubs and trees, but the most commonly planted species in our area is *Ligustrum vulgare*, which is a semi-evergreen shrub. It has opposite, simple, dark green, elliptical leaves that persist through most of the winter, but are eventually deciduous. The flowers are small (~3–4 mm), white, and arranged in small clusters arising from the ends of both main and lateral branches. The fruit is a small black drupe.

Ligustrum vulgare   Common privet

Ligustrum vulgare   Common privet   Shrub

WOODY PLANTS

Lindera   Spicebush

1 species in Ontario

Family: *Lauraceae*

Lindera benzoin is a shrub (up to 3 m tall) commonly found in southern Ontario mostly on the west side of lake Ontario. Its branches and leaves have a spicy aroma when bruised. The alternate leaves are oval to egg-shaped and have smooth margins. The small yellow flowers appear in showy clusters in early spring before the leaves. The fruit is a smooth, bright red berry in clusters of 3–6.

Lindera benzoin   Spicebush

Lindera benzoin   Spicebush   Medium-sized shrub

WOODY PLANTS

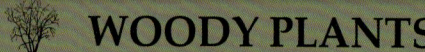

Liriodendron   Tulip tree   1 species in Ontario
Family: *Magnoliaceae*

Liriodendron tulipifera is a large tree up to 35 m tall found in the Carolinian forests along Lake Erie in southern Ontario. The alternate leaves are distinct with 4–6 shallow lobes and a broad notch at the tip. The twigs are each tipped with a large, flat bud with two outer scales; the lateral buds are much smaller. The greenish-yellow, tulip-like flowers are large and showy producing many winged seeds in cone-like clusters on the tips of branches. This is a very fast growing tree.

Liriodendron tulipifera   Tulip tree

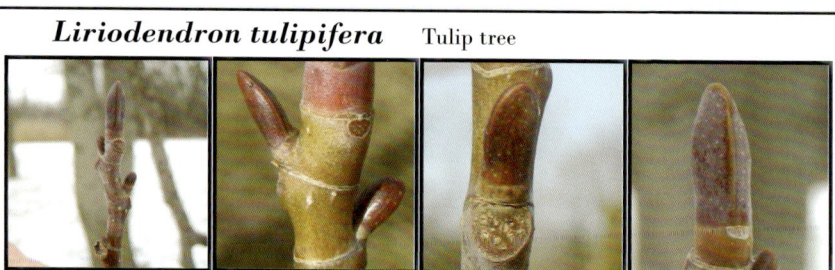

Liriodendron tulipifera   Tulip tree   Large tree

WOODY PLANTS

Lonicera Honeysuckle

16 species in Ontario

Family: *Caprifoliaceae*

The honeysuckles are twining vines or erect shrubs up to 3 m in height. The leaves are deciduous, opposite, simple, and entire. The tubular flowers arise in pairs from the axils of the leaves and are about 1–2 cm long. The fragrant white, yellow, or pink flowers attract several different types of pollinators, depending on length of the flower; hummingbirds and hawkmoths favour species with longer flowers, while shorter flowers are visited primarily by insects. The fruit is a red, blue, or black berry. In southern Ontario there are four native species. Several introduced species and hybrids can also be found.

Winter Key to *Lonicera*

1(a) Branches twining; buds always solitary 2

2(a) Branches smooth, hairless ....

Lonicera dioica   Glaucous honeysuckle   pg. 238

2(b) Branches rough, hairy and glandular ....
......................*Lonicera hirsuta* (no winter profile; for summer profile, see pg. 238)

1(b) Branches erect or arching, but not twining; buds solitary or superposed 3

3(a) Pith of branches white, solid (native species) 4

4(a) Visible bud scales 2, valvate ....
....................*Lonicera villosa* (no winter profile; for summer profile, see pg. 239)

233

WOODY PLANTS

4(b) Visible bud scales 4 or more 5

5(a) Lower bud scales much shorter than the bud ....
..........*Lonicera canadensis* (no winter profile; for summer profile, see pg. 239)

5(b) Lower bud scales roughly as long as the bud ....
....*Lonicera oblongifolia* (not profiled)

3(b) Pith of branches brown, excavated (introduced species) 6

6(a) Twigs hairless 7

Lonicera tatarica   Tatarian honeysuckle   pg. 241

Lonicera × bella   Bell's honeysuckle   pg. 241

WOODY PLANTS

6(b) Young twigs hairy 8

    7(a) Buds pointed, strongly hairy ....

Lonicera maackii   Amur honeysuckle   pg. 240

    7(b) Buds blunt, nearly hairless ....

Lonicera morrowii   Morrow's honeysuckle   pg. 240

235

WOODY PLANTS

Summer Key to Lonicera

1(a) Trailing or twining vine, rarely shrub-like; opposing leaves immediately below flowers more or less joined across the stem; flowers clustered at branch tips above leafy disks 2

2(a) Leaves hairy above; twigs hairy; flowers 2.0–2.5 cm long ....*Lonicera hirsuta*, pg. 238

2(b) Leaves hairless above; twigs hairless; flowers 1.2–1.8 cm long ....*Lonicera dioica*, pg. 238

1(b) Erect shrub; opposing leaves all distinct; flowers in pairs in leaf axils 3

3(a) Stems with solid white pith; styles hairless ...4

4(a) Shrubs less than 1 m high; leaves hairy above; bark peeling in papery layers; inflorescence stalk 1 cm or less; fruit blue, glaucous ....*Lonicera villosa*, pg. 239

4(b) Shrubs more than 1 m high; leaves hairless above; bark shredding; inflorescence stalk longer than 1 cm; fruit red or purple-black not glaucous 5

5(a) Leaves hairless beneath or with scattered, long hairs; flowers with 5 distinct petals; fruits ovoid, axillary pairs distinct ....
....................*Lonicera canadensis*, pg. 239

WOODY PLANTS

5(b) Leaves evenly hairy beneath; flowers with 2 petals; fruits round, axillary pairs more or less fused .... *Lonicera oblongifolia*

3(b) Stems with hollow brown pith; styles hairy 6

6(a) Leaf tips tapering to a long point; flowers and fruit on stalks of 0.5 cm or less ....*Lonicera maackii*, pg. 240

6(b) Leaf tips rounded or with a short point; flowers and fruit on stalks of 0.5 cm or more 7

7(a) Leaves hairy beneath; flowers white, turning yellow; inflorescence stalk 0.5–1.5 cm long ....*Lonicera morrowii*, pg. 240

7(b) Leaves hairless beneath; flowers white or pink; inflorescence stalk 1.5–2.5 cm long ....*Lonicera tatarica*, pg. 241

Note: *Lonicera* × *bella* (pg. 241) is a hybrid between *Lonicera morrowii* and *Lonicera tatarica* and has characteristics intermediate between its two parent species.

WOODY PLANTS

Lonicera hirsuta   Hairy honeysuckle   Woody vine

Lonicera dioica   Glaucous honeysuckle   Woody vine

WOODY PLANTS

Lonicera villosa   Mountain fly honeysuckle   Low shrub

Lonicera canadensis   Fly honeysuckle   Shrub

WOODY PLANTS

Lonicera maackii Amur honeysuckle Shrub

Lonicera morrowii Morrow's honeysuckle Shrub

WOODY PLANTS

Lonicera tatarica Tatarian honeysuckle — Shrub

Lonicera × *bella* Bell's honeysuckle — Shrub

WOODY PLANTS

Magnolia   Magnolia   2 species in Ontario
Family: *Magnoliaceae*

Magnolia acuminata is a rare and endangered medium-sized tree occurring in the Carolinian forests along Lake Erie in southern Ontario. It has large (10–25 cm), alternate, oval leaves with smooth margins and a pointed tip. The twigs are tipped with a large (2 cm long) terminal bud with a single hairy bud scale; the lateral buds are much smaller. The flowers are large, greenish-yellow, and bell-shaped; these are not showy like ornamental species of *Magnolia*. The seeds are found within a cone-like structure on the tip of the branch, which is green and cucumber-like when immature.

Magnolia acuminata   Cucumber tree

Magnolia acuminata   Cucumber tree   Medium-sized tree

WOODY PLANTS

Malus Apple, crabapple   3 species in Ontario
Family: *Rosaceae*

Only two species of apple are native to Canada with only *Malus coronaria* (wild crabapple) in our area. It is a small tree with a crooked trunk and irregular shape. The leaves are deciduous, simple, alternate, and variable in shape, but usually have small lobes and are coarsely toothed. The fruit is a small (4–5 cm) round apple that remains on the tree after the leaves have dropped. The domesticated apple (*Malus domestica*) differs from *Malus coronaria* in having white hairs on the underside of the leaves and leaf stalks and lacks thorns. There are many other introduced species and cultivars planted as ornamentals.

Malus spp.   Apple, Crabapple

Malus spp.   Apple, crabapple   Large shrubs or small trees

WOODY PLANTS

Morus Mulberry   2 species in Ontario
Family: *Moraceae*

Morus rubra (red mulberry) is a rare small tree occurring in southern Ontario. *Morus alba* (white mulberry) is an exotic ornamental that has established in natural ecosystems in southern Ontario. The two species can interbreed to produce hybrids. The alternate leaves are oval to heart-shaped with some leaves that have distinctive asymmetric lobes. Red mulberry tends to have leaves with pointed, long-tapered tips, whereas white mulberry has blunt tipped leaves; hybrids have either form of leaf. The twigs of both species exude a milky sap when cut or broken. The fruits are blackberry-like aggregates that ripen midsummer.

Winter Key to Morus
1(a) Buds 3–4 mm long, closely appressed ....

Morus alba   White mulberry

 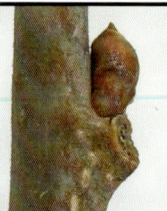  

1(b) Buds 6–8 mm long, slightly diverging from twig ....

Morus rubra   Red mulberry

Summer Key to Morus

1(a) Leaves smooth and shiny above, hairless or with tufts of hair in the vein axils beneath; fruit white to dark purple ....*Morus alba*

1(b) Leaves rough above, hairy beneath; fruit red to dark purple ....*Morus rubra*

WOODY PLANTS

Morus alba White mulberry — Small tree

Morus rubra Red mulberry — Small tree

WOODY PLANTS

Myrica Bayberry, Sweet gale   2 species in Ontario
Family: *Myricaceae*

The species of *Myrica* are shrubs reaching a height of 0.5–2 m that are typically found in moist soils in woods, bogs, swamps, and along shorelines, but some species can also occur in dry, sandy soils in dunes or pine barrens. The twigs are hairy and, along with the leaves, are dotted with yellow glands that produce a pleasant fragrance when bruised. The leaves are alternate, simple, deciduous, and often toothed toward the apex. The plants are dioecious with male flowers in erect scaly catkins and female flowers in inconspicuous cone-like catkins. The fruits of northern bayberry (*Myrica pensylvanica*), also known as wax myrtle, can be boiled in water to release a wax that is used to make naturally fragrant candles.

Winter Key to Myrica
1(a) Buds of two types: upper buds floral, 1–2 cm long, lower buds vegetative, 3–4 mm; bud tips pointed; fruits spiky, not coated in wax ....

Myrica gale   Sweet gale

1(b) Buds all similar; bud tips rounded; fruits warty, covered in white wax ....

Myrica pensylvanica   Northern bayberry

Summer Key to Myrica
1(a) Leaves hairless above; leaf margins toothed toward the apex; fruit brown, spiky, without a waxy coating .... *Myrica gale*

1(b) Leaves hairy above, especially along the midvein; leaf margins entire, occasionally toothed toward the apex; fruit bluish-white, warty, with a waxy coating ....
.... *Myrica pensylvanica*

WOODY PLANTS

Myrica gale Sweet gale — Small shrub

Myrica pensylvanica Northern bayberry — Medium-sized shrub

WOODY PLANTS

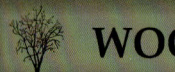

Nyssa Tupelo, Black gum   1 species in Ontario
Family: *Nyssaceae*

Nyssa sylvatica is a rare tree occurring in the Carolinian forests along Lake Erie in southwestern Ontario. The alternate leaves are oval-shaped, shiny, and dark green, found in clusters on dwarf branches along major shoots. The twigs are reddish-brown, containing a distinctive pith with green cross bars. The long (8 mm) terminal bud is covered with 5 scales; the lateral buds diverge from the twig. The flowers are green and inconspicuous, but the fruits are distinctive, dark blue, and plum-like, arranged in clusters on long-stalks that originate in the leaf axils.

Nyssa sylvatica   Black gum

Nyssa sylvatica   Black gum   Small tree

WOODY PLANTS

Ostrya Ironwood
Family: *Betulaceae*

1 species in Ontario

Ostrya is restricted to a single species in Ontario, *Ostrya virginiana* (ironwood or hop-hornbeam). These are medium-sized trees, often found under larger hardwood tree species. As the name implies, ironwood has one of the hardest and heaviest woods of any of the native trees. It has distinctive grayish bark that splits along the trunk into narrow strips that peel slightly at both ends. The leaves are deciduous and alternate in two rows along the twig. The leaves are very similar to *Carpinus caroliniana* (blue beech; pg. 168), with sharp teeth along the margins although the teeth are not as distinctly of two sizes, and some of the veins fork near the ends. The flowers are produced in catkins and develop before the leaves appear. The fruit is a small (~5 mm long) flattened nut that is enclosed in a papery sac.

Ostrya virginiana   Ironwood

Ostrya virginiana   Ironwood   Medium-sized tree

WOODY PLANTS

Parthenocissus Creeper   3 species in Ontario
Family: *Vitaceae*

The two species of creepers native to our region (*Parthenocissus quinquefolia* and *Parthenocissus vitacea*) are readily identified by their five-lobed palmately compound leaves on woody climbing vines. These plants have tendrils that wrap around other plants or objects allowing the vine to climb. *Pathenocissus quinquefolia* also has adhesive discs at the ends of the tendrils, allowing the plant to climb sheer vertical surfaces such as fences or walls. The leaves are alternate and deciduous with coarse teeth along the margins. The flowers are small (~5 mm), greenish, and 5-parted, developing into clusters of dark blue or purple berries. The leaves turn a bright crimson in the fall. *Parthenocissus quinquefolia* is occasionally sold and planted as an ornamental but if left unchecked it can climb over and smother other plants.

Winter Key to Parthenocissus
1(a) Tendrils tipped with adhesive disks, 3–9 branched ....

Parthenocissus quinquefolia   Virginia creeper

1(b) Tendrils not tipped with adhesive disks, 3–5 branched ....

Parthenocissus vitacea   Thicket creeper

Summer Key to Parthenocissus
1(a) Tendrils with 3–5 branches; adhesive discs usually absent; flowers and fruit on dichotomously branching stalks .... *Parthenocissus vitacea*

1(b) Tendrils with 3–9 branches; adhesive discs usually present; flowers and fruit on a central axis .... *Parthenocissus quinquefolia*

WOODY PLANTS

Parthenocissus quinquefolia — Virginia creeper — Climbing vine

Parthenocissus vitacea — Thicket creeper — Climbing vine

251

WOODY PLANTS

Philadelphus Mock-orange — 4 species in Ontario
Family: *Hydrangeaceae*

The mock-oranges are shrubs up to 4 m tall, none of which are native to our province, but due to their use as ornamentals they can occasionally be found as escapees. The ornamental value of the mock-oranges is due to their attractive flowers with a strong orange fragrance. The flowers are showy with four ruffled white petals, four sepals, multiple yellow stamens, and four more-or-less fused styles. The leaves are opposite, simple, oblong to ovate, coarsely toothed to entire, bright green above, and paler beneath. The twigs are tan to dark brown with exfoliating bark. The winter twigs of *Philadelphus* are easily identifiable due to their opposite branching and apparent lack of buds. In fact, the buds are covered by the leaf scar and eventually break through as they begin to swell in the spring.

Philadelphus spp.   Mock-orange

Philadelphus spp.   Mock-orange   Shrubs

WOODY PLANTS

Physocarpus Ninebark   1 species in Ontario
Family: *Rosaceae*

Ontario's only species, common ninebark (*Physocarpus opulifolius*), is a 2–3 m tall shrub with arching stems and peeling, papery bark. It is typically found on sandy, rocky, or gravelly soil along riverbanks and shorelines. Its leaves are simple, alternate, and deciduous. Initially circular, the leaves develop three irregularly toothed lobes as they mature, giving them a maple-like appearance. The white flowers are 5-parted with a hypanthium and numerous stamens, and are borne in flat-topped or rounded terminal clusters. The fruits are small reddish-brown pods that persist in the winter and can aid in identification.

Physocarpus opulifolius   Ninebark

Physocarpus opulifolius   Ninebark   Medium-sized shrub

WOODY PLANTS

Picea Spruce
Family: *Pinaceae*

5 species in Ontario

The spruce trees are an economically important group in Canada, supplying both the pulp and paper industries, and their wood is used extensively in construction. The needles of spruces are evergreen and stiff, arising singly from the twigs. The cones are similar in appearance to the pines with male and female parts in separate structures on the same tree. Three native species, white spruce (*Picea glauca*), red spruce (*Picea rubens*), and black spruce (*Picea mariana*), as well as two introduced species, Norway spruce (*Picea abies*) and Colorado spruce (*Picea pungens*), are often encountered in Ontario. Red spruce is not common in our area but is found in central and eastern Ontario. Black spruce is not a common tree in southern Ontario woodlots, but is sometimes found in wet, swampy areas.

Summer Key to Picea

1(a) Twigs minutely hairy; cones oblong to ovate ...2

2(a) Leaves yellowish-green, shiny, 12–18 mm long; cones 3–5 cm long; cone scales with more or less smooth margins; growing in upland sites, rarely in wet areas ....*Picea rubens*, pg. 255

2(b) Leaves grey-green, glaucous, 6–12 mm long; cones 2–3 cm long; cone scales with wavy margins; growing in wet areas ....*Picea mariana*, pg. 256

1(b) Twigs hairless; cones cylindrical ....3

3(a) Secondary branches drooping; needles dark green; cones 10–18 cm in length ....*Picea abies*, pg. 256

3(b) Secondary branches more or less horizontal; needles green or bluish green; cones 3–12 cm in length...4

WOODY PLANTS

4(a) Needles pointed but not sharp, with an unpleasant odour when broken; cones 3–6 cm long; cone scales fan-shaped, broadest toward apex ....*Picea glauca*, pg. 257

4(b) Needles very sharp-pointed, with odour of resin when crushed; cones 5–12 cm; cone scales roughly diamond shaped, broadest near middle ....*Picea pungens*, pg. 257

Picea rubens   Red spruce   Medium-sized tree

255

WOODY PLANTS

Picea mariana   Black spruce   Small or medium-sized tree

Picea abies   Norway spruce   Large tree

WOODY PLANTS

Picea glauca   White spruce   Medium-sized tree

Picea pungens   Colorado spruce   Medium-sized tree

WOODY PLANTS

Pinus Pine
9 species in Ontario
Family: *Pinaceae*

The pines are a well-known group of trees with several important characteristics. Historically, they provided a considerable economic contribution to early settlements in eastern North America, supplying much of the lumber to Europe's shipyards. Today, pine wood is still a favourite for woodworkers due to its light, uniform texture. Pines are able to grow on nutrient poor sites and where there is little moisture available – places where most other trees cannot establish. The pine is also a symbol of enduring strength and solitude, with a form that has a popular aesthetic quality. The leaves of the pines are modified into needles that are evergreen, and grouped in bundles called fascicles. There are two native species of pine that are found in our area: white pine (*Pinus strobus*) and red pine (*Pinus resinosa*). Scots pine (*Pinus sylvestris*) is a native of Europe, but is commonly planted and has become naturalized.

Summer Key to Pinus

1(a) Needles bundled in groups of 5 ....
....*Pinus strobus*, pg. 259

1(b) Needles bundled in groups of 2 2

2(a) Needles longer than 8 cm, straight; cones ovoid ....*Pinus resinosa*, pg. 259

2(b) Needles 8 cm long or less, twisting or spreading apart; cones conical, often asymmetrical 3

3(a) Needles 2–4 cm long, yellowish-green, spreading apart, sometimes slightly twisted; cones often strongly curving; bark of upper trunk brown ....*Pinus banksiana*, pg. 260

3(b) Needles 4–8 cm long, bluish-green or greyish-green, not spreading apart, spirally twisted; cones straight or slightly curving; bark of upper trunk orange and papery ......
...............................*Pinus sylvestris*, pg. 260

WOODY PLANTS

Pinus strobus   White pine   Large tree

Pinus resinosa   Red pine   Medium-sized tree

WOODY PLANTS

Pinus banksiana   Jack pine   Small tree

Pinus sylvestris   Scots pine   Medium-sized tree

WOODY PLANTS

Platanus   Sycamore   1 species in Ontario
Family: *Platanaceae*

Platanus occidentalis is a very large tree (30–35 m) found in southern Ontario and is often used as an ornamental. The bark is characteristic as it is mottled with green or whitish inner bark and exfoliating brownish outer bark. The alternate maple-like palmate leaves are 3–5 lobed. The twigs zigzag and are marked by stipule scars where leaves from previous years were once attached. The reddish buds are blunt and conical with a single leaf scale. The unisexual flowers are in small circular clusters on different shoots of the same tree. The seeds are achenes clustered into ball-like globular fruits that hang from long (10–15 cm) stalks.

Platanus occidentalis   Sycamore

Platanus occidentalis   Sycamore   Large tree

261

WOODY PLANTS

Populus Poplar, Aspen   6 species in Ontario
Family: *Salicaceae*

The poplars are a difficult group to identify to species, due to many of the same issues as with the willows (see pg. 323). In our region, however, there are only a few native species, which are generally easy to distinguish, but note that hybrids do occur. Poplars have leaves that are deciduous, alternate, simple, and usually broad – about twice as long as wide with a pointed tip. The tiny flowers are borne in catkins, with pollen-producing flowers and seed-producing flowers on different trees. The fruit are small capsules that split open releasing small (1–3 mm) seeds with long white hairs.

Winter Key to Populus

1(a) Terminal buds less than 1 cm long, not resinous; lateral buds with more than 4 scales 2

2(a) Buds dull, with silky whitish hairs; young bark smooth, pale green; older bark dark grey and ridged ....

Populus grandidentata   Largetooth aspen   pg. 266

2(b) Buds glossy, hairless; young and old bark whitish and smooth ....

Populus tremuloides   Trembling aspen   pg. 266

WOODY PLANTS

1(b) Terminal buds 2–2.5 cm long, resinous; lateral buds with 4 or fewer scales 3

3(a) Twigs reddish-brown; buds fragrant ....

Populus balsamifera   Balsam poplar   pg. 265

3(b) Twigs yellowish or greenish; buds not fragrant ....

Populus deltoides spp. *deltoides*   Eastern cottonwood   pg. 265

WOODY PLANTS

Summary Key to Populus

1(a) Leaf stalks rounded or dorsally flattened; leaves egg-shaped ....*Populus balsamifera*, pg. 265

1(b) Leaf stalks laterally flattened; leaves various 2

2(a) Leaves strongly triangular with a long-tapering tip; buds resinous ....*Populus deltoides* ssp. *deltoides*, pg. 265

2(b) Leaves weakly triangular to oval or elliptical with a short-pointed tip; buds not resinous 3

3(a) Leaves with 20–30 teeth per side; leaf stalks longer than leaf blade ....*Populus tremuloides*, pg. 266

3(b) Leaves with 7–15 teeth per side; leaf stalks shorter than leaf blade ....*Populus grandidentata*, pg. 266

WOODY PLANTS

Populus balsamifera — Balsam poplar — Medium-sized tree

Populus deltoides ssp. *deltoides* — Eastern cottonwood — Large tree

WOODY PLANTS

Populus tremuloides   Trembling aspen   Tree

Populus grandidentata   Largetooth aspen   Tree

WOODY PLANTS

Potentilla Cinquefoil

16 species in Ontario

Family: *Rosaceae*

Found throughout Ontario, shrubby cinquefoil (*Potentilla fruticosa*) is a shrub that displays a great deal of variability. It can be found growing in a variety of habitats from wet open ground in fens, on streambanks and on shorelines, to dry, alkaline ground on dunes, rocky slopes, or cliff faces. Its growth form can vary from a low sprawling shrub attaining a height of only 10 cm, to an erect much-branched shrub up to 1 m tall. The shoots and leaves can also vary from glabrous to densely pubescent with long white silky hairs. The leaves are alternate, deciduous, and pinnately compound with 5 (occasionally 3 or 7) lanceolate to elliptic leaflets. As the reddish-brown bark matures it splits into long papery strips or shreds. The flowers are yellow with 5 rounded petals, 5 triangular sepals, 5 floral bracts (alternating with the sepals), 15–25 stamens, and numerous styles.

Potentilla fruticosa   Shrubby cinquefoil

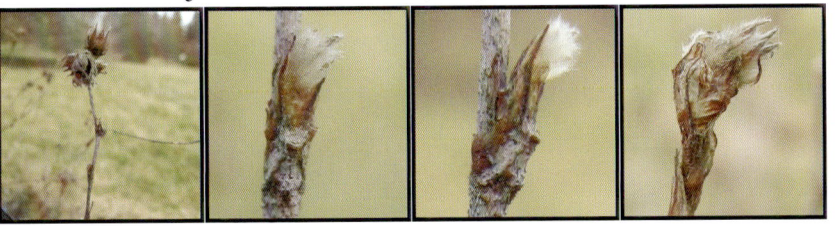

Potentilla fruticosa   Shrubby cinquefoil   Low shrub

267

WOODY PLANTS

Prunus   Cherry, Plum   16 species in Ontario
Family: *Rosaceae*

Unlike most of the other genera in the Rosaceae, the native species of cherry and plum in our area are not difficult to distinguish. Most are small trees, but *Prunus serotina* (black cherry) is a large tree whose wood is valued for furniture and cabinetry. The leaves of *Prunus* are deciduous, alternate, and simple, with sharp single teeth. The flowers usually appear with the leaves. The fruit is a fleshy drupe with a single stone.

Winter Key to Prunus

1(a) Buds reddish, globose, 1–2 mm long, clustered at the tips of twigs; twigs red, up to 2 mm thick ....

Prunus pensylvanica   Pin cherry   pg. 272

1(b) Buds longer than wide, over 2 mm long, solitary at the tips of twigs; twigs thicker than 2 mm ....
....2

2(a) Twigs thorny, terminal bud absent or false ...3

3(a) Bud scales entire or at most notched in the middle; buds reddish-brown, usually under 4 mm long ....

Prunus americana   American plum   pg. 271

WOODY PLANTS

3(b) Bud scales thin, pale, and frayed; buds grey, usually over 4 mm long ....

Prunus nigra Canada plum pg. 271

2(b) Twigs lacking thorns, terminal bud present 4

4(a) Buds glossy, 3–4 mm long; bud scales somewhat fleshy, with orange or green bases and dark brown margins; mature bark black and scaly............................

Prunus serotina Black cherry pg. 273

4(b) Buds dull, sharp-pointed, 4–6 mm long; bud scales thin and dark brown with pale margins; mature bark grey and smooth..

Prunus virginiana Choke cherry pg. 272

WOODY PLANTS

Summer Key to Prunus

1(a) Twigs thorny; fruit 2–3 cm long, with a flattened stone 2

2(a) Leaves often widest below the middle, with sharp teeth; tips of teeth often lacking glands ....*Prunus americana*, pg. 271

2(b) Leaves often widest above the middle, with rounded teeth; tips of teeth bearing glands ....*Prunus nigra*, pg. 271

1(b) Twigs unarmed; fruit 1 cm long or less, with a globose stone 3

3(a) Leaves lanceolate, with irregular rounded teeth; inflorescence axillary, a drooping cluster of 12 or fewer flowers ....*Prunus pensylvanica*, pg. 272

3(b) Leaves oval or elliptical, with regular pointed teeth; inflorescence terminal, an elongated raceme of 20 or more flowers 4

4(a) Shrub to small tree; leaves widest at or above the middle, with 8–11 pairs of lateral veins, undersides hairless or with tufts of hair in vein axils; teeth not curving ....*Prunus virginiana*, pg. 272

4(b) Medium to large tree; leaves widest at or below the middle, with 15 or more pairs of lateral veins, undersides with white or rusty hairs lining the midvein; teeth curving toward leaf tip ....*Prunus serotina*, pg. 273

WOODY PLANTS

Prunus americana American plum — Small tree

Prunus nigra Canada plum — Small tree

WOODY PLANTS

Prunus pensylvanica   Pin cherry   Small tree

Prunus virginiana   Choke cherry   Shrub or small tree

WOODY PLANTS

Prunus serotina Black cherry Up to a medium-sized tree

WOODY PLANTS

Ptelea Hop-tree
1 species in Ontario

Family: *Rutaceae*

Ptelea trifoliata is a rare, small tree occurring in southwestern Ontario. The alternate, compound leaves are composed of three leaflets (trifoliate) joined at the tip of a long (10–15 cm) stalk. The slender twigs have a citrus odour when broken and contain a large, white pith. There is no terminal bud and the lateral buds are very small and sunken into the twig. The inconspicuous flowers are greenish-white with male and female flowers on different trees. The fruits are made up of dense clusters of individually stalked seeds with wings composed of papery tissue with many tiny veins. These fruits were once used to flavour beer!

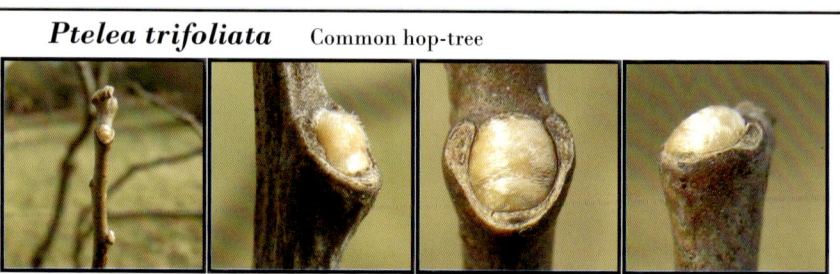

Ptelea trifoliata   Common hop-tree

Ptelea trifoliata   Common hop-tree   Small tree

WOODY PLANTS

Quercus Oak
Family: *Fagaceae*

12 species in Ontario

Like maples or pines, oaks are iconic trees, conveying strength and longevity. The oak has been chosen as the national tree of several countries, including England and the United States, and appears on France's modern coat of arms. Oaks range in size from shrubs to large trees. The leaves on most species are deciduous, but some (the live oaks; none in our area) are evergreen. The leaves are alternately arranged and are usually deeply lobed, each lobe with a prominent vein. Like beeches (pg. 203), the flowers are unisexual with both pollen-producing flowers and seed-producing flowers found on the same tree. The small pollen flowers are clustered on catkins that appear from the previous year's growth or from the bottom of new growth. Seed flowers arise singly or in small clusters from axils on new growth. The fruit is the distinctive acorn, a single nut enclosed in a tough rounded shell with a scaly cup at the base. The oaks can be broadly divided into two groups, the white oaks (most species with rounded lobes) and the red and black oaks (most species with pointed lobes). Hybrids readily occur within these groups, yielding intermediate forms.

Winter Key to Quercus

1(a) Some twigs bearing immature acorns; bark splitting but remaining closely appressed to trunk; buds usually acute (red oaks) 2

2(a) Largest terminal buds 7 mm or longer on average 3

3(a) Buds hairless, dull, straw-coloured; twigs grey or greyish-brown ....

Quercus shumardii   Shumard oak   pg. 287

3(b) Buds hairy, dark red or brown; twigs reddish-brown 4

WOODY PLANTS

4(a) Buds brown with white or grey hairs along their full length; inner bark yellow ....

Quercus velutina   Black oak   pg. 286

4(b) Buds red with hairs only on the upper half of the bud; inner bark red 5

5(a) Buds with a few scattered hairs at the tips, mostly on the scale margins; bud tips sharply pointed ....

Quercus rubra   Red oak   pg. 286

5(b) Buds densely hairy on upper half; bud tips often rounded ....

Quercus coccinea   Scarlet oak   pg. 287

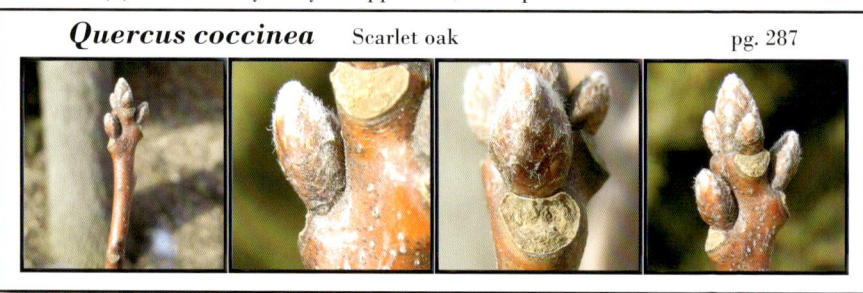

WOODY PLANTS

2(b) Largest terminal buds 6 mm or shorter on average 6

6(a) Terminal buds 2–4 mm long, light brown; acorns 9–13 mm long, cup enclosing 1/4 of the nut ....

Quercus palustris   Pin oak   pg. 288

  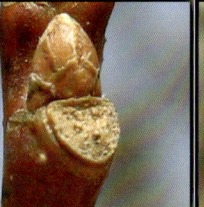 

6(b) Terminal buds 4–6 mm long, reddish brown; acorns 12–20 mm long, cup enclosing 1/3–1/2 of the nut 7

7(a) Upper half of bud hairy ....*Quercus coccinea* (see winter profile pg. 276)

7(b) Buds hairless ....

Quercus ellipsoidalis   Northern pin oak   pg. 288

1(b) Twigs bearing only fully mature acorns; bark splitting and more or less separating from trunk; buds acute or obtuse (white oaks) 8

WOODY PLANTS

8(a) Terminal bud acute, often encircled by long, narrow stipules 9

9(a) Buds hairy; lateral buds appressed; older twigs often developing corky ridges ..

Quercus macrocarpa   Bur oak   pg. 284

9(b) Buds hairless; lateral buds diverging; older twigs not developing corky ridges 10

10(a) Tree, more than 2 m tall ....

Quercus muehlenbergii   Chinquapin oak   pg. 285

10(b) Shrub, less than 2 m tall ....

Quercus prinoides   Dwarf chinquapin oak   pg. 285

WOODY PLANTS

8(b) Terminal bud ovoid to globose, blunt, with or without elongated stipules 11

11(a) Lateral buds appressed; terminal buds often encircled by long, narrow stipules; older twigs often developing corky ridges ....
....*Quercus macrocarpa* (see winter profile pg. 278)

11(b) Lateral buds diverging from twig; terminal buds not encircled by long, narrow stipules; older twigs not developing corky ridges ...12

12(a) Twigs greenish- or yellowish-brown; buds brown; bark on older branches peeling and shaggy ....

Quercus bicolor   Swamp white oak   pg. 284

12(b) Twigs red or reddish-brown; buds reddish-brown; bark on older branches not peeling ....

Quercus alba   White oak   pg. 283

WOODY PLANTS

Summer Key to Quercus

1(a) Leaves with rounded lobes not tipped with bristles, or not lobed but coarsely toothed; mature bark light coloured, splitting and somewhat lifting or peeling from trunk 2

2(a) Leaves lobed, sinuses extending more than 1/3 of the distance between the lobe tip and the midvein 3

3(a) Leaf undersides hairless or with few hairs; leaves lobed along their entire length ....
......................................*Quercus alba*, pg. 283

3(b) Leaf undersides hairy; leaves often coarsely toothed toward the tip and lobed toward the base, but occasionally consistently lobed .....
....4

4(a) Leaves 12–17 cm long, with irregular shallow lobes; leaf undersides with two types of hairs: small stellate hairs and longer erect hairs; branches with peeling bark ....*Quercus bicolor*, pg. 284

4(b) Leaves 15–30 cm long, with deep lobes at least toward the base; leaf undersides with only small stellate hairs; branches often developing corky ridges ....
......................*Quercus macrocarpa*, pg. 284

2(b) Leaves coarsely toothed, sinuses extending less than 1/3 of the distance between the lobe tip and the midvein 5

5(a) Acorn stalks 2–7 cm long; leaf undersides with two types of hairs: small stellate hairs and longer erect hairs ....
..................................*Quercus bicolor*, pg. 284

WOODY PLANTS

5(b) Acorn stalks less than 2 cm long; leaf undersides with only small stellate hairs ...6

6(a) Tree; lateral veins 9 pairs or more ....*Quercus muehlenbergii*, pg. 285

6(b) Shrub; lateral veins 8 pairs or less ....*Quercus prinoides*, pg. 285

1(b) Leaves with pointed lobes tipped by bristles; mature bark dark, splitting but remaining flat against trunk 7

7(a) Leaf undersides white, grey, or light yellowish-brown, and hairy over entire surface ....*Quercus velutina*, pg. 286

7(b) Leaf undersides green and hairless or with tufts of hair in the vein axils 8

8(a) Leaf sinus extending less than 2/3 of the distance between the longest lobe tip and the midvein 9

9(a) Leaves with 7-9 lobes; largest lobes narrowing from their base to their tip; acorn cup covering 1/3 of the nut ....*Quercus rubra*, pg. 286

281

WOODY PLANTS

9(b) Leaves with 5–7 lobes; largest lobes broadening from their base to their tip or with sides parallel; acorn cup covering more than 1/3 of the nut ....*Quercus velutina*, pg. 286

8(b) Leaf sinus extending more than 2/3 of the distance between the longest lobe tip and the midvein 10

10(a) Twigs light brown to light green; buds light yellowish-brown, with a waxy coating, hairless .....*Quercus shumardii*, pg. 287

10(b) Twigs dark brown to reddish-brown; buds dark brown to reddish-brown, without a waxy coating, hairy or hairless 11

11(a) Buds hairy; leaves large, 10–20 cm long; petioles more than 1 mm in diameter 12

12(a) Buds hairy along their entire length; petioles 7–15 cm long; acorn without concentric rings at the tip ....*Quercus velutina*, pg. 286

 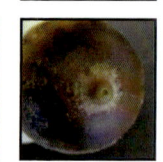

12(b) Buds hairy only on the upper half; petioles 4–6 cm long; acorn with concentric rings at the tip ....*Quercus coccinea*, pg. 287

11(b) Buds hairless; leaves small, averaging 7–10 cm long; petioles 1 mm in diameter or less 13

WOODY PLANTS

13(a) Leaf lobes broadening from their base to their tip; lowest branches ascending or horizontal; acorn cup enclosing 1/3 to 1/2 of the nut; growing in dry upland areas ....
................*Quercus ellipsoidalis*, pg. 288

13(b) Leaf lobes narrowing from their base to their tip; lowest branches descending; acorn cup enclosing 1/4 of the nut; growing in wet areas ....
....................*Quercus palustris*, pg. 288

Quercus alba   White oak   Large tree

WOODY PLANTS

Quercus bicolor   Swamp white oak — Medium-sized tree

Quercus macrocarpa   Bur oak — Small tree

WOODY PLANTS

Quercus muehlenbergii   Chinquapin oak   Medium-sized tree

Quercus prinoides   Dwarf chinquapin oak   Medium-sized shrub

WOODY PLANTS

Quercus velutina Black oak — Medium-sized tree

Quercus rubra Red oak — Medium-sized tree

WOODY PLANTS

Quercus shumardii   Shumard oak   Medium-sized tree

Quercus coccinea   Scarlet oak   Medium-sized tree

WOODY PLANTS

Quercus ellipsoidalis Northern pin oak Medium-sized tree

Quercus palustris Pin oak Medium-sized tree

WOODY PLANTS

Rhamnus Buckthorn   3 species in Ontario
Family: *Rhamnaceae*

Three species of *Rhamnus* are included here: *Rhamnus cathartica* (common buckthorn), *Rhamnus alnifolia* (alder-leaved buckthorn), and *Rhamnus frangula* (glossy buckthorn). The common buckthorn was introduced from Europe and has become naturalized. *Rhamnus alnifolia* is a low shrub (~1 m high), while *Rhamnus cathartica* is larger, occasionally becoming a small tree. *Rhamnus frangula* is sometimes placed in its own genus, *Frangula*. The leaves are deciduous and simple with only a few (3–8) veins per side that curve slightly towards the tip, similar to *Cornus* (see pg. 183). The flowers are inconspicuous and in small clusters in the axils of new growth (as opposed to in heads at the ends of branches in *Cornus*). The fruits are clusters of berries that are black when mature.

Winter Key to Rhamnus
1(a) Buds naked (lacking scales); young twigs with short hairs ....

Rhamnus frangula   Glossy buckthorn   pg. 292

1(b) Buds with scales; twigs hairless ....2

  2(a) Buds alternate; twigs unarmed ....

Rhamnus alnifolia   Alder-leaved buckthorn   pg. 291

WOODY PLANTS

2(b) Buds mostly opposite (occasionally alternate on rapidly growing shoots); some twigs spine tipped ....

Rhamnus cathartica   Common buckthorn   pg. 291

Summer Key to Rhamnus

1(a) Leaves opposite, occasionally alternate on rapidly growing shoots; twigs tipped with thorns ....*Rhamnus cathartica*, pg. 291

1(b) Leaves always alternate; twigs unarmed 2

2(a) Low shrub, 1 m high or less; leaves toothed ....*Rhamnus alnifolia*, pg. 292

2(b) Medium to large shrub, up to 6 m high; leaves entire ....*Rhamnus frangula*, pg. 292

WOODY PLANTS

Rhamnus cathartica   Common buckthorn   Shrub or small tree

Rhamnus alnifolia   Alder-leaved buckthorn   Low shrub

WOODY PLANTS

Rhamnus frangula — Glossy buckthorn — Shrub or small tree

WOODY PLANTS

Rhus Sumac   4 species in Ontario
Family: *Anacardiaceae*

Rhus was at one time considered to be a large genus of over 250 species, but recent molecular data has supported dividing *Rhus* into several genera. The data are still being examined, but this new arrangement currently removes poison ivy and poison sumac from *Rhus* and places them in a new genus, *Toxicodendron* (see pg. 343). Ontario's four species of sumac are shrubs or small trees that are often found growing in colonies along forest edges, along roadsides, and on other exposed sites. Their leaves are alternate, deciduous, and pinnately or ternately compound. The flowers are small, white or yellowish-green, and grow in dense terminal clusters. The fruits are small, red, hairy drupes that are often eaten by birds and small mammals.

Winter Key to Rhus

1(a) Leaf scars round; twigs slender; buds of two types: upper buds floral, 0.5–1 cm long, lower leaf buds small, hidden behind leaf scar ....

Rhus aromatica   Fragrant sumac   pg. 296

1(b) Leaf scars horseshoe shaped, partially encircling bud; twigs stout; buds of one type 2

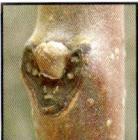

2(a) Twigs hairless, somewhat 3-sided ....

Rhus glabra   Smooth sumac   pg. 297

WOODY PLANTS

2(b) Twigs hairy at least at the tips, round in cross-
section 3

3(a) Twigs with short hair, mostly at the tips; fruit clusters often drooping ....

Rhus copallina   Shining sumac   pg. 296

3(b) Twigs densely covered with long hairs along their full length; fruit clusters
erect ....

Rhus typhina   Staghorn sumac   pg. 297

WOODY PLANTS

Summer Key to Rhus

1(a) Leaflets 3, with a pleasant odour when bruised; inflorescences 0.5–2 cm long; medium shrub up to 2 m tall ....*Rhus aromatica*, pg. 296

1(b) Leaflets 7 or more, without a pleasant odour; inflorescences 5–25 cm long; large shrubs or small trees up to 6 m or more 2

2(a) Leaflets entire; petioles winged between leaflets ....*Rhus copallina*, pg. 296

2(b) Leaflets toothed; petioles not winged 3

3(a) Twigs and petioles hairless or nearly so; fruits with short hairs less than 0.5 mm long ....*Rhus glabra*, pg. 297

3(b) Twigs and petioles hairy; fruits with spiky hairs 1–2 mm long ......*Rhus typhina*, pg. 297

WOODY PLANTS

Rhus aromatica — Fragrant sumac — Low shrub

Rhus copallina — Shining sumac — Shrub or small tree

WOODY PLANTS

Rhus glabra   Smooth sumac   Tall shrub

Rhus typhina   Staghorn sumac   Shrub or small tree

WOODY PLANTS

Ribes Currant, Gooseberry   13 species in Ontario
Family: *Grossulariaceae*

The genus *Ribes* is composed of shrubs with deciduous, alternate, 3- or 5-lobed palmate leaves. There are two groups within the genus: the gooseberries (with prickles or spines), and the currants (without prickles or spines). The flowers are usually 5-parted, in small clusters, and of various colour and shape depending on the species. The fruit is a cluster of small black or red berries. Some of the most common species in our area are included below.

Winter Key to Ribes

1(a) Stems unarmed 2

2(a) Buds and twigs bearing yellow glands ....

Ribes americanum   American black currant   pg. 301

2(b) Buds and twigs without yellow glands 3

3(a) Buds red; buds and twigs with a skunk-like odour when bruised ....

Ribes glandulosum   Skunk currant   pg. 302

WOODY PLANTS

3(b) Buds brown; buds and twigs without skunk-like odour ....

Ribes triste   Wild red currant   pg. 302

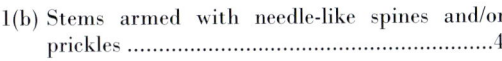

1(b) Stems armed with needle-like spines and/or prickles 4

4(a) Prickles between nodes abundant, roughly the same length as spines at the nodes; buds and twigs glossy, buff ....
Ribes lacustre (no winter profile; for summer profile, see pg. 303)

4(b) Prickles between nodes absent, or if present then significantly shorter than spines at the nodes; young twigs dull grey, bark eventually peeling to reveal reddish-brown to black inner bark 5

5(a) Buds with silky hairs; stems hairless ....
.................*Ribes cynosbati* (no winter profile; for summer profile, see pg. 303)

5(b) Buds hairless; stems hairy ....*Ribes hirtellum* (not profiled)

WOODY PLANTS

Summer Key to Ribes

1(a) Stems unarmed 2

    2(a) Stems erect; fruit black 3

        3(a) Leaf base wedge-shaped; leaves and twigs lacking yellow glands; petioles 1–3 cm long; flowers and fruit 2–5 in a cluster ....*Ribes hirtellum* (not profiled)

        3(b) Leaf base heart-shaped; leaves and twigs with scattered yellow glands; petioles 3–5 cm long; flowers and fruit in racemes of 6 or more ....*Ribes americanum*, pg. 301

    2(b) Stems low, trailing; fruit red 4

        4(a) Twigs and leaves with fetid odour when bruised; terminal leaf lobe ovate; berry with glandular bristles ....*Ribes glandulosum*, pg. 302

        4(b) Twigs and leaves lacking fetid odour; berry smooth; terminal lobe triangular ....*Ribes triste*, pg. 302

1(b) Stems bearing bristles and/or spines, sometimes only on new growth 5

WOODY PLANTS

5(a) Prickles between nodes abundant, roughly the same length as spines at the nodes; flowers and fruit in racemes; fruit with glandular hairs ....*Ribes lacustre*, pg. 303

5(b) Prickles between nodes scattered or absent, if present significantly shorter than spines at the nodes; flower and fruit 2–5 in a cluster; fruit with prickles or smooth 6

6(a) Leaf base heart-shaped; petiole with stalked glands; floral hypanthium bearing stalked glands; fruit spiny, red ....
................................*Ribes cynosbati*, pg. 303

6(b) Leaf base wedge-shaped; petiole with feather-like hairs; floral hypanthium smooth or hairy, but lacking glands; fruit smooth, black..................................
....*Ribes hirtellum* (not profiled)

Ribes americanum   American black currant   Low shrub

301

WOODY PLANTS

Ribes glandulosum   Skunk currant   Low shrub

Ribes triste   Wild red currant   Low shrub

WOODY PLANTS

Ribes lacustre Swamp black currant — Low shrub

Ribes cynosbati Prickly gooseberry — Low shrub

WOODY PLANTS

Robinia Locust
3 species in Ontario
Family: *Fabaceae*

Black locust (*Robinia pseudoacacia*) is native to the eastern United States but it is often planted and has become naturalized in many areas of southern Ontario. It is the only member of the genus found in our area. Black locust is a medium-sized tree with leaves that are deciduous, alternate, pinnately compound, and have 7–19 oval leaflets. The stipules at the base of the leaves are usually modified into stiff spines. The white flowers appear after the leaves and are arranged in long (~15 cm) drooping clusters at the end of new growth. The fruit is a flat pod ~10 cm long that remains on the tree after the leaves have dropped. Like other members of the legume family, the black locust grows in association with nitrogen fixing bacteria – this allows the plant to tolerate poor soil conditions. It can quickly colonize and help reclaim disturbed sites, which is one of the reasons it is widely planted.

Robinia pseudoacacia   Black locust

Robinia pseudoacacia   Black locust   Medium-sized tree

WOODY PLANTS

Rosa Rose
Family: *Rosaceae*   20 species in Ontario

The roses are a group of shrubs and trailing or climbing vines, often with sharp prickles on the stems. The leaves are deciduous, alternate, and pinnately compound with toothed oval leaflets. Roses have been cultivated for centuries and a great variety of floral forms and colours exist. The flowers of our native species are similar in appearance to those of the cherries or apples. The fruit matures in late summer or fall and is called a hip. The hips of some species are high in vitamin C and sometimes used for making jams or brewed for tea.

Winter Key to Rosa

1(a) Trailing or climbing; fruit many per cluster 2

2(a) Most prickles, especially on older stems, straight or nearly so; fruit nearly globose, 8-12 mm ....

Rosa setigera   Prairie rose   pg. 309

2(b) Prickles strongly hooked; fruit ellipsoid, 6-10 mm ....

Rosa multiflora   Multiflora rose   pg. 310

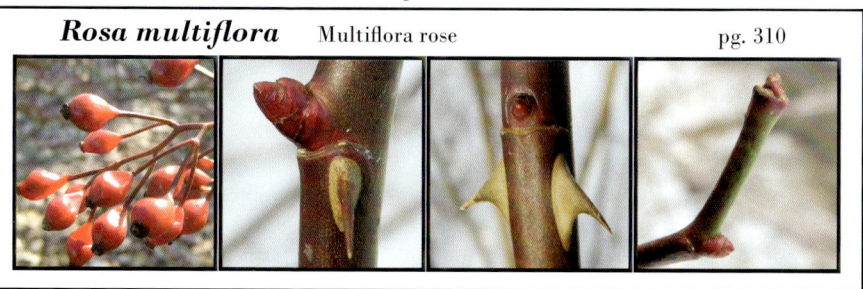

1(b) Stems erect, bushy; fruit solitary or few per cluster 3

305

WOODY PLANTS

3(a) Low shrub, stems less than 1 m tall 4

4(a) Fruit pear-shaped, 1.5–2 cm, smooth; nodes lacking distinct prickles; stems with secondary branches ....

Rosa acicularis   Prickly wild rose   pg. 312

4(b) Fruit globose to ellipsoid, 1.0–1.5 cm, with stalked glands; distinct pairs of prickles at nodes; stems often arising singly, unbranched, from the ground ....*Rosa carolina* (no winter profile; for summer profile, see pg. 311)

3(b) Medium shrub, stems typically 1–3 m tall 5

5(a) Prickles hooked and broad-based, occurring only at nodes; fruit orange-red, 6–10 mm, often with numerous stalked glands ....

Rosa palustris   Swamp rose   pg. 311

5(b) Prickles absent or, if present, some prickles needle-like, occurring at and between nodes at least at the base of stems; fruit 1–1.5 cm, smooth or with few scattered glands 6

WOODY PLANTS

6(a) Prickles, if present, needle-like and occurring only at the base of stems; fruit red, smooth ....

Rosa blanda   Smooth wild rose   pg. 312

6(b) Prickles of two types: hooked broad-based prickles, and shorter needle-like prickles, occurring at and between nodes; fruit orange-red, smooth or with scattered glands ....

Rosa eglanteria   Sweetbrier   pg. 310

307

WOODY PLANTS

Summer Key to Rosa

1(a) Branches reclining, arching, or climbing; flowers several per inflorescence; styles united, protruding 3–4 mm from the hypanthium ....2

2(a) Leaves mostly with 3 leaflets (rarely 5); leaflets 3–9 cm long; stipules entire; flowers 4–8 cm across ....*Rosa setigera*, pg. 309

2(b) Leaves with 7–9 leaflets; leaflets 1–3 cm long; stipules deeply toothed; flowers 2–4 cm across ....*Rosa multiflora*, pg. 310

1(b) Branches erect; flowers solitary to few per inflorescence; styles distinct, barely if at all protruding from hypanthium ....3

3(a) Prickles some or all broad-based, curved; shrubs 2 m or taller at maturity; leaflets mostly 7; pedicels and fruit with stalked glands ....4

4(a) Leaf undersides bearing stalked glands; flowers 2–4 cm across; some prickles broad-based and curved, others needle-like; curved prickles 7–13 mm long, occurring at and between nodes ....*Rosa eglanteria*, pg. 310

4(b) Leaf undersides lacking glands, minutely hairy along veins; flowers 4–6 cm across; all prickles broad-based and curved; prickles 3–6 mm long, occurring only at nodes ....*Rosa palustris*, pg. 311

3(b) Prickles all needle-like; shrubs 2 m or shorter at maturity; leaflets 5–9; pedicels and fruit with or without stalked glands ....5

WOODY PLANTS

5(a) Pedicels and hypanthiums with stalked glands; distinct pairs of prickles at nodes; stems often arising singly, unbranched, from the ground ....*Rosa carolina*, pg. 311

5(b) Pedicels and hypanthiums smooth; nodes lacking distinct prickles; stems with secondary branches 6

6(a) Mature plant 1 m high; prickles dense, along entire stem; stipules broaden towards tip, margins with abundant dark glands; fruit pear-shaped with a distinct neck ....*Rosa acicularis*, pg. 312

6(b) Mature plant 2 m high; prickles sparse or absent, often solely at base of stems; stipule sides parallel or slightly broadening toward tip, margins with few or no dark glands; fruit globose to ellipsoid ....*Rosa blanda*, pg. 312

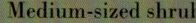

Rosa setigera   **Prairie rose**   **Medium-sized shrub**

WOODY PLANTS

Rosa multiflora — Multiflora rose — Medium-sized shrub

Rosa eglanteria — Sweetbrier — Medium-sized shrub

310

WOODY PLANTS

Rosa palustris Swamp rose Medium-sized shrub

Rosa carolina Pasture rose Low shrub

WOODY PLANTS

Rosa acicularis   Prickly wild rose   Low shrub

Rosa blanda   Smooth wild rose   Low shrub

312

WOODY PLANTS

Rubus Blackberry, Raspberry   17 species in Ontario
Family: *Rosaceae*

The blackberries are a complicated group of plants with a high incidence of hybridization, polyploidy and apomixis. The enormous variation in form has led to the formal description of several hundred species, although some authors consider many forms to be at the subspecies or variety rank. Regardless, it is still a large group. *Rubus* species are perennial but in some species the stems are biennial: one-year-old stems do not produce flowers, and the leaves are sometimes different that those of the second year's growth. Flowers are produced on two-year old stems. Although they are commonly referred to as berries, the fruit is an aggregate of drupes. Numerous species occur in Ontario; some of the most common species are included here.

Winter Key to *Rubus*

1(a) Plants lacking thorns or bristles 2

    2(a) Plant somewhat herbaceous, low and spreading (10–30 cm) ....
        ....*Rubus pubescens* (no winter profile; for summer profile, see pg. 319)

    2(b) Plant woody, taller (0.5–2 m) 3

        3(a) Young bark splitting and peeling, stems terete ....

Rubus odoratus   Purple-flowering raspberry   pg. 318

        -OR-   *Rubus parviflorus* (no winter profile; for summer profile, see pg. 319)

        3(b) Young bark not splitting, stems longitudinally ridged ....

Rubus canadensis   Smooth blackberry   pg. 320

WOODY PLANTS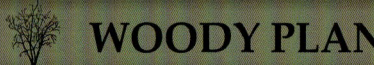

1(b) Plants armed with thorns or bristles 4

4(a) Stems armed only with needle-like bristles ...5

5(a) Stems trailing close to the ground 6

    6 (a) Leaves often persisting into winter; prickles numerous, slender ....
    ....*Rubus hispidus* (no winter profile; for summer profile, see pg. 321)

    6(b) Leaves deciduous; prickles fewer, broad-based ....
    ....*Rubus flagellaris* (no winter profile; for summer profile, see pg. 320)

5(b) Stems erect or arching; leaves entirely deciduous 6

    6(a) Stems round in cross section, bark on older canes peeling ....

Rubus idaeus var. *strigosus*   Wild red raspberry   pg. 321

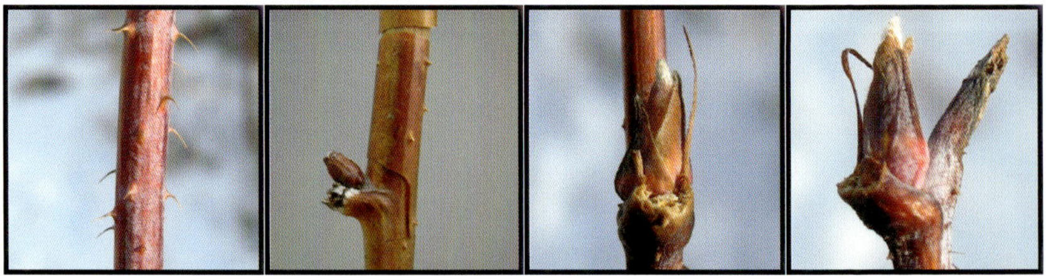

    6(b) Stems longitudinally ridged, bark not peeling; bristles abundant, often curved ....*Rubus setosus* (not profiled)

WOODY PLANTS

4(b) Stems armed with at least some broad-based prickles 7

7(a) Stem longitudinally ridged; canes not rooting at tips 8

8(a) Stems armed with bristles, stalked glands, and broad-based prickles ....

Rubus allegheniensis   Common blackberry   pg. 322

8(b) Stems armed only with scattered, broad-based prickles ....
....*Rubus canadensis* (see winter profile pg. 313)

7(b) Stem roughly round in cross section; canes often rooting at tips ....

Rubus occidentalis   Black raspberry   pg. 322

WOODY PLANTS

Summer Key to Rubus

1(a) Stems unarmed ....2

2(a) Leaves simple; bark vertically splitting and
    peeling in sheets ....3

3(a) Flowers rose to purple (rarely white in an
    albino form); fruit dry, inedible ....
    ....*Rubus odoratus*, pg. 318

3(b) Flowers white; fruit juicy, edible ....
    ....*Rubus parviflorus*, pg. 319

2(b) Leaves compound; bark not peeling in sheets
    ....4

4(a) Stems trailing, 10–40 cm high; leaves all with 3 leaflets;
    petals 5–10 mm long; fruit dark red ....
    ....*Rubus pubescens*, pg. 319

4(b) Stems erect or high-arching, 1–3 m high; leaves of first
    year's growth with 5 leaflets; petals 12–20 mm long; fruit
    purple-black ....*Rubus canadensis*, pg. 320

1(b) Stems armed with bristles or prickles ....5

WOODY PLANTS

5(a) Stems trailing or prostrate, up to 0.5 m high 6

    6(a) Leaves thinner, deciduous, dull; terminal leaflet broadest near the middle with a slender tip; stems armed with broad-based, slightly curved prickles ....*Rubus flagellaris*, pg. 320

    6(b) Leaves thick, evergreen, often glossy; terminal leaflet often broadest above the middle with a more rounded tip; stems armed with dense needle-like bristles ..*Rubus hispidus*, pg. 321

5(b) Stems erect, up to 3 m high 7

    7(a) Leaf undersides grey or whitish; stems round in cross-section, glaucous; inflorescence of 10 or fewer flowers 8

        8(a) Prickles straight, needle-like; canes not rooting at tips; fruit red ....*Rubus idaeus*, pg. 321

        8(b) Prickles broad-based, curved; canes rooting at tips; fruit purple-black ....*Rubus occidentalis*, pg. 322

    7(b) Leaf undersides green; stems longitudinally ridged, not glaucous; inflorescence of more than 10 flowers 9

317

WOODY PLANTS

9(a) Stems densely covered with soft, needle-like bristles only; low shrub, rarely more than 1 m high; petals 7–10 mm long .... *Rubus setosus* (not profiled)

9(b) Stem with at least some rigid, broad-based prickles; medium shrub, 2 m or higher at maturity; petals 10–20 mm long 10

10(a) Stems with stalked glands, bristles, and scattered broad-based prickles; leaf undersides hairy, at least in vein axils; fruit up to 20 mm long .... *Rubus allegheniensis*, pg. 322

10(b) Stems with few, scattered, broad-based prickles only; leaf undersides hairless; fruit up to 12 mm long .... *Rubus canadensis*, pg. 320

Rubus odoratus   Purple-flowering raspberry   Medium-sized shrub

WOODY PLANTS

Rubus parviflorus Thimbleberry Medium-sized shrub

Rubus pubescens Dwarf raspberry Creeping shrub

WOODY PLANTS

Rubus canadensis Smooth blackberry — Medium-sized shrub

Rubus flagellaris Northern dewberry — Trailing shrub

WOODY PLANTS

Rubus hispidus — Swamp dewberry — Trailing shrub

Rubus idaeus — Wild red raspberry — Medium-sized shrub

WOODY PLANTS

Rubus occidentalis — Black raspberry — Medium-sized shrub

Rubus allegheniensis — Common blackberry — Medium-sized shrub

WOODY PLANTS

Salix Willow
Family: *Salicaceae*
43 species in Ontario

The willows as a whole are a difficult group to identify to species with certainty. The key characteristics often appear at different times, in association with different parts of the plant. Making a positive identification may require following an individual throughout the growing season in order to observe flower, fruit, and leaf morphology, and in some cases may only be possible for female individuals if the definitive features are only observable on the fruit. These issues are further complicated by extensive hybridization between species. The leaves of *Salix* are deciduous, mostly alternate, simple, and usually long and narrow. The flowers appear before the leaves and are very small (~1 mm or less), occuring in catkins that last for only a week or two. Male and female flowers are produced on different individuals. The fruit are small capsules that split open releasing small (1-2 mm) seeds with many long white hairs.

Salix spp.   Willow

Salix spp.   Willow   Shrubs and trees

WOODY PLANTS

Sambucus   Elderberry
2 species in Ontario

Family: *Adoxaceae*

The elderberries, or elders, are erect shrubs or small trees mostly occuring in the Northern Hemisphere. The twigs and stems are unique due to their conspicuously warty surface and the presence of a large, spongy pith. The branching is opposite. The leaves are pinnately compound with 5–11 serrate leaflets. When bruised, the leaves and twigs emit a rank odour. Theflowers are small and white with 5 petals and occur in large, many-branched terminal cymes. The fruit is a red or black berry-like drupe that contains 3–5 small stones and is a favourite food of many birds and mammals. The blossoms of some species have been used to produce a flavouring for liquors and syrups, while the berries of some species have been used to make wines and jams.

Winter Key to Sambucus

1(a) Pith of older twigs white; buds sessile, widest at the base, 3–5 mm long ....

Sambucus canadensis   American elderberry

1(b) Pith of older twigs brown; buds somewhat stalked, widest at the middle, 10 mm long ....

Sambucus pubens   Eastern red elderberry

Summer Key to Sambucus

1(a) Leaflets 5–11 (most often 7); pith of older branches white; inflorescence wider than long; fruit purple-black .... *Sambucus canadensis*

1(b) Leaflets 5–7 (most often 5); pith orange to brown; inflorescence as long or longer than wide; fruit red .... *Sambucus pubens*

WOODY PLANTS

Sambucus canadensis American elderberry   Shrub

Sambucus pubens Eastern red elderberry   Shrub

WOODY PLANTS

Sassafras   Sassafras — 1 species in Ontario
Family: *Lauraceae*

Sassafras albidum is a rare, small tree occurring in southwestern Ontario. The alternate leaves are distinctively shaped with three veins and up to three unique lobes. The greenish twigs are smooth and brittle with oval terminal and lateral buds. All of the vegetation has a spicy fragrance when bruised or broken. The flowers are small, greenish-yellow, and in clusters with male and female flowers on different trees. The blue berry-like fruits are perched on a reddish cup with a long stalk; the seeds are large and stone-like.

Sassafras albidum   Sassafras

Sassafras albidum   Sassafras — Small tree

WOODY PLANTS

Shepherdia Buffalo berry — 1 species in Ontario
Family: *Elaeagnaceae*

Our only species of *Shepherdia*, buffalo berry (*Shepherdia canadensis*), is a moderate shrub of 2 m found in coarse gravelly or rocky soils in open woods or along shorelines. The twigs, buds, and leaves of buffalo berry are covered in shiny, rusty-brown scales, much like in *Elaeagnus* (see pg. 195), but buffalo berry can easily be distinguished by its opposite branching. The somewhat leathery leaves are simple, opposite, entire, and oval to elliptic. The upper leaf surface is dark green with sparse stellate (star-shaped) hairs while the lower surface is silvery-white, densely hairy, and dotted with rust-coloured scales. The flowers are unisexual, small, greenish-yellow, and generally inconspicuous. The fruit is red, berry-like, and edible, though extremely bitter. Some First Nations people collected the fruits, mixed them with sweeter berries such as raspberry, and whipped them into a tasty, salmon-coloured froth.

Shepherdia canadensis Buffalo berry — Low shrub

WOODY PLANTS

Smilax Greenbrier   6 species in Ontario
Family: *Smilacaceae*

Smilax is composed of species that are either woody or herbaceous vines. Herbaceous species such as *Smilax herbacea* (herbaceous carrion flower) and woody vines such as *Smilax tamnoides* (bristly greenbrier) are both common in southern Ontario. The genus is defined by alternate, simple leaves that are broadly oval with prominent longitudinal nerves interspersed with net-like veins. The stipules at the base of the leaves are prolonged as tendrils that aid in climbing as these species can overgrow small shrubs, fencerows, and boulders. The flowers are unisexual with six yellow or greenish-yellow tepals (petals/sepals) and are arranged in umbels on long stalks. The berry-like fruits are dark blue to black with six seeds. This is a monocot genus formerly included in family Liliaceae.

Winter Key to Smilax

1(a) Older twigs with scattered prickles; prickles broad-based, green with black tips, up to 12 mm long ....
............*Smilax rotundifolia* (no winter profile; for summer profile, see next page)

1(b) Older twigs with dense prickles; prickles needle-like, blackish, 3–10 mm long ....

Smilax tamnoides   Prickly greenbrier

Summer Key to Smilax

1(a) Stems 4-sided; prickles needle-like, black; leaves 5–10 cm long, with 3–5 primary veins ....*Smilax rotundifolia*

1(b) Stems round or with more than 4 sides; prickles large and broad-based, green with black tips; leaves 8–12 cm long, with 5–7 primary veins ....*Smilax tamnoides*

WOODY PLANTS

Smilax rotundifolia — Round-leaved greenbrier — Woody vine

Smilax tamnoides — Prickly greenbrier — Woody vine

WOODY PLANTS

Solanum Nightshade
Family: *Solanaceae*

10 species in Ontario

In Ontario, *Solanum* is composed mostly of introduced herbaceous species, including potato (*Solanum tuberosum*) and tomato (*Solanum lycopersicum*). *Solanum dulcamara* (bittersweet nightshade) is a semi-woody perennial vine that was introduced from Europe and Asia and is a common weed in and around southern Ontario woodlots. *Solanum dulcamara* has alternate, deciduous, simple, entire leaves. Both lobed and unlobed leaves usually occur on the same plant; the lower stem leaves are unlobed and the upper stem leaves are commonly lobed near the base. The leaves and stems can be glabrous or downy. The flowers are distinctive, with 5 purple petals that are often reflexed and stamens that unite in a distinctive yellow cone around the style. The fruits are poisonous red berries that hang in small branched clusters.

Solanum dulcamara — Bittersweet nightshade

Solanum dulcamara — Bittersweet nightshade

Semi-woody vine

WOODY PLANTS

Sorbus Mountain-ash
3 species in Ontario
Family: *Rosaceae*

The mountain-ashes are, in fact, unrelated to the true ashes from which their name is derived based on the superficial similarity of their leaves. In Ontario there are three species of mountain-ash all of which are large shrubs or small trees. Our species have alternate, deciduous, pinnately compound leaves composed of 9–17 sharply toothed leaflets. Individual flowers are small, but they form large, showy, white clusters that are wider than they are long. The fruits are orange or red pomes and can usually be seen on the tree in large clusters long after the leaves have fallen. The fruits are often eaten by birds, especially waxwings and thrushes, and spread in their droppings. Folklore in both Europe and eastern Canada holds that a particularly heavy fruitset in nountain-ash indicates the coming of a harsh winter.

Winter Key to Sorbus
1(a) Buds covered in dense white woolly hairs ....

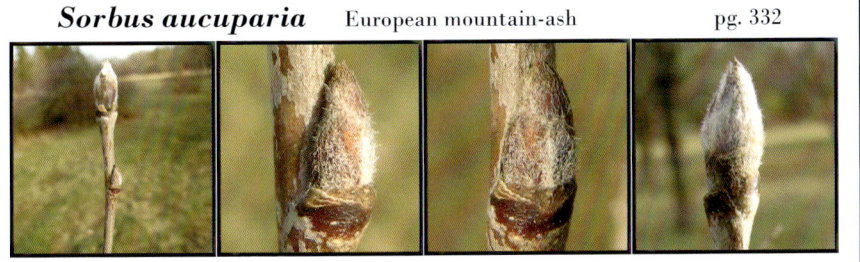

Sorbus aucuparia   European mountain-ash   pg. 332

1(b) Buds hairless or with a few white hairs at the tip ....

Sorbus americana   American mountain-ash   pg. 333

Sorbus decora   Showy mountain-ash   pg. 333

WOODY PLANTS

Summer Key to Sorbus

1(a) Twigs, leaf undersides, and hypanthium densely hairy ....
....*Sorbus aucuparia*

1(b) Twigs, leaf undersides, and hypanthium hairless or nearly so...2

2(a) Leaves 3–5 times longer than wide; flowers 5–6 mm in diameter; fruit 4–6 mm in diameter ....*Sorbus americana*

2(b) Leaves 2–3 times longer than wide; flowers 10 mm in diameter; fruit 7–10 mm in diameter ....*Sorbus decora*

Sorbus aucuparia   European mountain-ash   Small tree

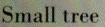

WOODY PLANTS

Sorbus americana American mountain-ash Shrub or small tree

Sorbus decora Showy mountain-ash Shrub or small tree

WOODY PLANTS

Spiraea Spiraea

8 species in Ontario

Family: *Rosaceae*

The species of *Spiraea* are shrubs up to 1.5 m tall with many stiff, erect branches and are a source of many prized ornamentals. Garden catalogues are full of dozens of cultivars from North America, Europe, and Asia, renowned for their hardiness and beautiful inflorescences. The dense inflorescences can be flat-topped or upright clusters of small white, pink, or purple flowers. Like other members of the rose family (Rosaceae), the flowers possess a hypanthium to which their five petals, five sepals, and 10–50 stamens are attached. The leaves are simple, alternate, deciduous, and usually finely toothed, but may also be lobed or entire. The shrubs of *Spiraea* produce salicylates, chemical compounds from which the drug aspirin was derived. The First Nations people of southern Ontario brewed a tea from the leaves of *Spiraea* which they drank to relieve pain and ease fevers.

Winter Key to Spiraea
1(a) Stems densely woolly ....

Spiraea tomentosa   Steeple-bush   pg. 336

1(b) Stems hairless or with scattered hairs 2

2(a) Inflorescence broadly pyramidal, lower branches elongated; inflorescence branches hairless or slightly hairy; stems usually reddish- or purplish-brown .....

Spiraea alba var. *latifolia*   Broad-leaved meadowsweet   pg. 336

WOODY PLANTS

2(b) Inflorescence narrowly pyramidal; inflorescence branches often densely hairy; stems usually yellowish-brown ....

Spiraea alba var. ***alba***   Narrow-leaved meadowsweet   pg. 337

Summer Key to Spiraea

1(a) Leaf undersides with a dense covering of white or tan woolly hairs; flowers bright pink (rarely white) ....
....*Spiraea tomentosa*, pg. 336

1(b) Leaf undersides hairless or nearly so; flowers white (rarely pale pink) 2

2(a) Leaves coarsely toothed, 1.5–4 cm wide; stems reddish-brown; inflorescence a broad pyramid; branches of inflorescence hairless; flowers white or pale pink ....
....*Spiraea alba* var. *latifolia*, pg. 336

2(b) Leaves finely toothed, 1–1.8 cm wide; stems yellowish-brown; inflorescence a narrow pyramid; branches of inflorescence hairy; flowers white ....*Spiraea alba*, pg. 337

335

WOODY PLANTS

Spiraea tomentosa — Steeple-bush — Small shrub

Spiraea alba var. *latifolia* — Broad-leaved meadowsweet — Small shrub

WOODY PLANTS

Spiraea alba   Narrow-leaved meadowsweet   Small shrub

WOODY PLANTS

Staphylea Bladdernut

1 species in Ontario

Family: *Staphyleaceae*

Staphylea trifolia is a shrub commonly found in southern Ontario. The opposite, compound leaves are trifoliate (3 leaflets) and the terminal leaflet has a long stalk while the two lateral leaflets are sessile (not stalked); the leaflet margins are finely toothed. The twigs are green and mottled with gray lines or stripes. The flowers appear before the leaves have fully expanded and are white, cylinder-like, and in drooping clusters. The fruit is a distinctive triangular capsule (5–8 cm long) that is inflated at maturity and is tipped with a persistent thread-like style; the seeds are loose and rattle within the capsule.

Staphylea trifolia Bladdernut

Staphylea trifolia Bladdernut — Large shrub

WOODY PLANTS

Syringa Lilac

1 species in Ontario

Family: *Oleaceae*

The lilacs are native to southeastern Europe and Asia, but have been cultivated and planted as ornamentals in gardens worldwide. They are valued for their showy, fragrant flowers and are generally hardy, requiring little care. Several hundred cultivars exist, most of which can be seen at the Royal Botanical Gardens in Hamilton, which has the world's largest collection. In our area, lilac has naturalized and can be seen along roadsides and abandoned fields. The lilacs range in size from shrubs to small trees. The leaves are usually heart-shaped, deciduous, simple, opposite, and entire. The flowers are tubular, ~1 cm long with 4 petals and can be white or various shades of purple and pink. The flowers appear at the same time as the leaves, in large clusters at the ends of the previous years growth. The fruit is a flat, 2-sided capsule that splits when mature releasing 2 winged seeds. *Syringa vulgaris* is the most commonly cultivated species.

Syringa spp.   Lilac

Syringa spp.   Lilac   Shrubs or small trees

339

WOODY PLANTS

Taxus Yew
1 species in Ontario
Family: *Taxaceae*

Only two species of *Taxus* are native to Canada, only one of which, *Taxus canadensis* (Canada yew), occurs in our region. It is a spreading shrub of the forest understory. Several cultivars with varying growth forms are also planted as ornamentals. The leaves of Canada yew are modified into flat evergreen needles 1.5–3 cm long. Pollen-producing and seed-bearing structures are found on separate trees. The pollen cones are ~0.5 cm long and arise from the underside of the twig on the previous year's growth. The ovules are born singly on short lateral stalks also from the previous year's growth. The fertilized seed is set within a red cup-shaped aril, about 1 cm across, superficially resembling a berry. The yew is a common food source for wild game such as deer and moose, but is poisonous to livestock.

Taxus canadensis Canada yew — Spreading shrub

WOODY PLANTS

Thuja Cedar
1 species in Ontario
Family: *Cupressaceae*

Eastern white cedar (*Thuja occidentalis*) is a small tree that is primarily found in swamps and bogs or on dry sites with shallow soils over limestone bedrock. Several varieties have been cultivated and it is often planted for use in hedges due to its full evergreen form that is very tolerant to being trimmed. The wood is often used for fence posts and rails as it is light, splits easily, and is decay resistant. The leaves of the eastern white cedar are modified into small (3–4 mm long) scale-like structures that are arranged in four rows, in overlapping pairs flattened along the twig. Both the pollen-producing cones and seed-producing cones are small (~0.5 cm) and globular, appearing at the ends of the twigs. Although *Thuja* species are commonly called cedars, they are in a different genus than the true cedars (*Cedrus*).

Thuja occidentalis   Eastern white cedar   Small tree

341

WOODY PLANTS

Tilia Basswood, Linden   5 species in Ontario
Family: *Malvaceae*

Tilia americana is the only species of this genus whose range extends north into Canada. It is a medium-sized to large tree, with deciduous, simple, alternate leaves that are heart-shaped and often have an asymmetric base. The leaf margins have coarse single teeth that are tipped with glands. The flowers appear from the leaf axils of new growth after the leaves have fully developed. They are arranged in small clusters at the end of a stalk that arises from the midrib of a long tongue-like leafy bract. The flowers are 5-parted and 1–2 cm across with a fragrant odour. The fruits are round, woody capsules ~0.5 cm long, covered with short brown hairs. Basswood has very light soft wood which makes it a popular choice for carvers.

Tilia americana   American basswood

Tilia americana   American basswood   Large tree

WOODY PLANTS

Toxicodendron   Poison ivy, Poison sumac   1 species in Ontario
Family: *Anacardiaceae*

Toxicodendron used to be included within the closely related genus, *Rhus* (see pg. 293). Being able to distinguish between the two genera is a useful skill because the species of *Toxicodendron* all produce the oil urushiol, which can cause a severe allergic reaction upon contact with the skin. Reliable distinctions are found in their fruits and flowers: the fruits of *Toxicodenderon* are hairless and white while those of *Rhus* are hairy and red; the inflorescences of *Toxicodendron* grow from the leaf axils while the inflorescences of *Rhus* grow from the branch tips. Only two species of *Toxicodendron* grow in Ontario: poison ivy (*Toxicodendron radicans*) and poison sumac (*Toxicodendron vernix*). Poison ivy is a highly variable plant. The leaves are always alternate and deciduous, but may be glossy or matte, green or purple, toothed or entire, and, though they are usually ternately compound, have been known to bear five leaflets. The growth form varies from a small shrub (ssp. *rydbergii*) to a trailing or climbing vine (ssp. *radicans*) and it can be found in a wide range of habitats from forest understory to open areas, in wet, dry or rocky soil. Poison sumac is a shrub or small tree that grows in wet or swampy areas. Its leaves are alternate, deciduous, and pinnately compound with 7–13 entire leaflets.

Winter Key to Toxicodendron

1(a) Erect shrub, 3–5 m high; twigs stout; buds sessile ..... *Toxicodendron vernix*, pg. 345

1(b) Climbing vine or low shrub less than 1 m high;
       twigs slender; buds stalked 2

   2(a) Low shrub ....

Toxicodendron radicans ssp. *rydbergii*   Poison ivy   pg. 345

WOODY PLANTS

2(b) Climbing vine ....

Toxicodendron radicans ssp. ***radicans***   Poison ivy   pg. 346

Summer Key to Toxicodendron

1(a) Leaflets 7–13 (rarely 5), margins entire ....
....*Toxicodendron vernix*, pg. 345

1(b) Leaflets 3, margins entire, toothed, or lobed 2

2(a) Low shrub rarely more than 1 m tall; petioles hairless ....
........................*Toxicodendron radicans* ssp. *rydbergii*, pg. 345

2(b) Climbing or scrambling vine; petioles hairy ....
........................*Toxicodendron radicans* ssp. *radicans*, pg. 346

WOODY PLANTS

Toxicodendron vernix   Poison sumac   Large shrub

Toxicodendron radicans ssp. *rydbergii*   Poison ivy   Low shrub

WOODY PLANTS

Toxicodendron radicans ssp. *radicans*   Poison ivy   Woody vine

WOODY PLANTS

Tsuga Hemlock
Family: *Pinaceae*

1 species in Ontario

The single species of hemlock native to eastern North America is eastern hemlock (*Tsuga canadensis*). It is a large tree normally found on moist sites in cooler locations as pure stands, or in association with sugar maple, American beech, yellow birch, white spruce, and white pine. It is quite shade tolerant and able to persist in the understory. The evergreen needles of eastern hemlock are flat and 1–2 cm long with thin white stripes (composed of individual stomata) on either side of the midrib beneath. The needles are arranged spirally on flexible twigs, but are curved at the base to give the appearance of being in two ranks. The seed-producing cones are small (~2 cm long), and mature in the fall, releasing seeds throughout the winter.

Tsuga canadensis   Eastern hemlock   Large tree

WOODY PLANTS

Ulmus Elm
8 species in Ontario
Family: *Ulmaceae*

Mature elm trees were once much more common on the southern Ontario landscape than they are today. Elms were once widely planted in urban areas, as they are easy to germinate and transplant, and grow quickly. Dutch elm disease (a fungus) has, however, killed most of the mature trees, and continues to afflict trees as they reach a certain size. Mature elms are medium-sized to large trees. They have deciduous, alternate, simple leaves with toothed margins and asymmetrical leaf bases. They have small inconspicuous flowers that appear before the leaves. The fruit is a small, winged, flat seed.

Winter Key to Ulmus

1(a) Buds light brown or orange, with dark margins
....2

2(a) Branches and older twigs with corky ridges ....

Ulmus thomasii Rock elm   pg. 353

2(b) Branches and older twigs without corky ridges 3

3(a) Twigs hairless or slightly hairy; flower and leaf buds of equal size ....

Ulmus americana White elm   pg. 352

WOODY PLANTS

3(b) Twigs densely hairy; flower buds distinctly plumper than leaf buds ....

Ulmus procera   English elm   pg. 351

1(b) Buds dark brown to blackish; older twigs grey
    and stout 4

5(a) Buds with brown hairs; twigs without prominent lenticels; inner bark somewhat
    mucilaginous ....

Ulmus glabra   Scotch elm   pg. 351

5(b) Buds tipped with rusty hairs; twigs with prominent lenticels; inner bark
    strongly mucilaginous ....

Ulmus rubra   Slippery elm   pg. 352

WOODY PLANTS

Summer Key to Ulmus

1(a) Leaves small, 5–8 cm long, with an average of 12 pairs of lateral veins; fruits 10–15 mm long, entirely hairless .... *Ulmus procera*, pg. 351

1(b) Leaves larger, 8–20 cm long, with an average of 15 or more pairs of lateral veins; fruits either hairy or, if hairless, 20–25 mm long 2

2(a) Leaves very rough above, with an average of 15 pairs of lateral veins; often 4 or more lateral veins forking; flowers stalkless or nearly so 3

3(a) Leaves 8–16 cm long, margins hairless; leaves often with 3 points at the tip; fruit 20–25 mm long, entirely hairless, oval .... *Ulmus glabra*, pg. 351

3(b) Leaves 15–20 cm long, margins with a fringe of hairs; leaves with only one point at the tip; fruit 10–15 mm long, hairy over seed, almost circular .... *Ulmus rubra*, pg. 352

2(b) Leaves smooth or sometimes slightly rough above, with an average of more than 15 pairs of lateral veins; often 3 or fewer lateral veins forking; flowers on long stalks, pendulous 4

4(a) Leaves 10–15 cm long, smooth to somewhat rough above; leaf bases asymmetrical; branches without corky wings, remaining smooth; fruit 8–10 mm long, hairy only on the margins .... *Ulmus americana*, pg. 352

4(b) Leaves 5–10 cm long, smooth above; leaf bases symmetrical or nearly so; branches developing corky wings; fruit 10–15 mm, hairy over entire surface .... *Ulmus thomasii*, pg. 353

WOODY PLANTS

Ulmus procera   English elm   Large tree

Ulmus glabra   Scotch elm   Large tree

WOODY PLANTS

Ulmus rubra   Slippery elm   Medium-sized tree

Ulmus americana   White elm   Large tree

WOODY PLANTS

Ulmus thomasii   Rock elm   Medium-sized tree

WOODY PLANTS

Vaccinium Blueberry — 12 species in Ontario
Family: *Ericaceae*

Most of the species of *Vaccinium* produce edible berries, and two groups in particular are commercially produced on a large scale: blueberries and cranberries. In addition to humans, many birds and animals feed on the berries of *Vaccinium*. Black bears are known to gorge themselves on blueberries during the season. Despite a few tropical exceptions, *Vaccinium* is predominantly a Northern temperate genus of sandy, rocky, or acidic soils. Ontario has 11–12 species of *Vaccinium* ranging from creeping vines to small or medium sized shrubs typically found in sphagnum bogs, lakesides, barrens, or dry rocky woods. The leaves are small (in most species, from 0.5–5 cm), alternate, simple, deciduous or evergreen, and oval to elliptic. The flowers are 4- or 5-parted, bell- or -urn-shaped, and white, pink, or green.

Winter Key to Vaccinium

1(a) Twigs more or less angled, hairless or with hair running in thin vertical lines ....*Vaccinium angustifolium*

1(b) Twigs round, densely hairy with stiff white hairs ....
....*Vaccinium myrtilloides*

Summer Key to Vaccinium

1(a) Leaf margins finely toothed; leaf undersides hairless or with sparse hair along the veins; berries sweet, 6–12 mm in diameter ....
...................................*Vaccinium angustifolium*

1(b) Leaf margins entire; leaf undersides hairy; berries sour, 6–9 mm in diameter ....
......................................*Vaccinium myrtilloides*

WOODY PLANTS

Vaccinium angustifolium   Lowbush blueberry   Low shrub

Vaccinium myrtilloides   Velvet-leaf blueberry   Low shrub

WOODY PLANTS

Viburnum Viburnum, Arrowwood   9 species in Ontario
Family: *Adoxaceae*

Viburnum species are shrubs or small trees mostly occuring in the Northern Hemisphere. Their leaves are opposite, simple, and may be entire, toothed, or lobed, often with warty glands on the petiole just below the leaf blade. The leaves of some species strongly resemble, and can be easily confused with, those of the maples. The flowers are 5-parted, small (3–5 mm), white or rarely pink, and occur in rounded or flat-topped terminal clusters. Some species bear a ring of large, showy, sterile flowers around the perimeter of the cluster to attract pollinators. The fruits are red or bluish-black berry-like drupes with a single flattened stone. Many species are valued as ornamentals for their attractive flowers and fruits.

Winter Key to Viburnum

1(a) Terminal bud naked, outer 'scales' (actually immature leaves) with finger-like ridges along the margins; twigs with a dense covering of stellate hairs ....2

2(a) Twigs reddish or purplish; leaf scars broad ....*Viburnum lantanoides* (no winter profile; for summer profile, see pg. 364)

2(b) Twigs greyish-brown; leaf scars narrow ....

Viburnum lantana   Wayfaring tree   pg. 365

1(b) Terminal bud with scales, scale margins smooth; twigs hairless or with unbranched hairs ....3

3(a) Buds with 2 valvate scales ....4

WOODY PLANTS

4(a) Buds brown or grey, long and slender, pointed 5

5(a) Buds smooth, purplish-grey; 2 terminal bud scales often completely covering the bud ....

Viburnum lentago   Nannyberry   pg. 365

5(b) Buds rough, brown; 2 terminal bud scales sometimes not touching at the base ....

Viburnum nudum var. *cassinoides*   Wild raisin   pg. 366

4(b) Buds bright red, oblong or ovoid, blunt or with a short point 6

357

WOODY PLANTS

    6(a) Young twigs red, slender ....

Viburnum opulus   European cranberry bush   pg. 362

    6(b) Young twigs yellow or buff, slightly thicker ....

Viburnum trilobum   Highbush cranberry   pg. 363

    3(b) Buds with more than 2 visible scales 7

    7(a) Bud scales 6; buds diverging slightly from twig ....

Viburnum rafinesquianum   Downy arrow-wood   pg. 363

358

WOODY PLANTS

7(b) Bud scales 4; buds appressed .... 8

8(a) Lower pair of bud scales short, not reaching middle of bud; twigs hairy ......

Viburnum acerifolium Maple-leaved viburnum pg. 362

8(b) Lower pair of bud scales usually reaching the middle of the bud or higher; twigs hairless ....

Viburnum recognitum Southern arrow-wood pg. 364

WOODY PLANTS

Summer Key to Viburnum

1(a) Leaves maple-like with 3 lobes and palmate veins 2

2(a) Petioles lacking glands; inflorescence of small, uniform flowers; fruit purple-black ....*Viburnum acerifolium*, pg. 362

2(b) Petioles with wart-like glands near the blade; inflorescence of small flowers ringed by large, showy, neutral flowers; fruit red 3

3(a) Leaves hairy beneath; petiole glands concave, 0.9–1.5 mm in diameter; stipules with sharp, bristle-like tips ....*Viburnum opulus*, pg. 362

3(b) Leaves hairless or with scattered hairs lining the veins; petiole glands with round or flat tops, 0.4–0.8 mm in diameter; stipules with thick, rounded tips ....*Viburnum trilobum*, pg. 363

1(b) Leaves lacking lobes; venation pinnate 4

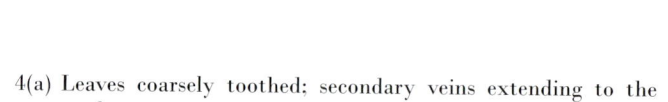

4(a) Leaves coarsely toothed; secondary veins extending to the teeth 5

5(a) Leaves with 4–11 teeth per side; petioles less than 1 cm long, hairy, usually with linear stipules; stone of fruit flattened ....*Viburnum rafinesquianum*, pg. 363

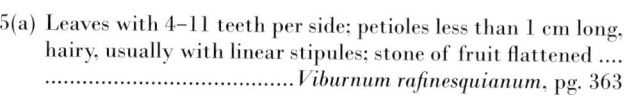

WOODY PLANTS

5(b) Leaves with 9–21 teeth per side; petioles 1–3 cm long, hairless or nearly so, usually lacking stipules; stone of fruit ellipsoid .... *Viburnum recognitum*, pg. 364

4(b) Leaves entire or finely toothed; leaf veins branching and winding before reaching the margin 6

6(a) Leaves and twigs with stellate hairs; buds naked 7

7(a) Sprawling or loosely ascending shrub; leaves broad-ovate to almost rounded, 10–21 cm long; petioles 1–6 cm long; inflorescence stalkless, ringed by large, showy, neutral flowers ....
.... *Viburnum lantanoides*, pg. 364

7(b) Erect shrub; leaves narrow-ovate, 5–12 cm long; petioles 0.7–2.5 cm long; inflorescence stalked, of uniform flowers ....
.... *Viburnum lantana*, pg. 365

6(b) Leaves and twigs hairless or with simple hairs; buds with scales 8

8(a) Leaves with regular, sharp teeth; petioles with wavy margins; inflorescence stalkless or with a short stalk of less than 0.5 cm; fruit 12–15 mm long ....
.... *Viburnum lentago*, pg. 365

8(b) Leaves entire or with irregular, rounded or pointed teeth; petioles with straight margins; inflorescence on a stalk 0.5–5 cm long; fruit 6–9 mm long ....
Viburnum nudum var. *cassinoides*, pg. 366

WOODY PLANTS

Viburnum acerifolium   Maple-leaved viburnum   Small shrub

Viburnum opulus   European cranberry bush   Shrub or small tree

WOODY PLANTS

Viburnum trilobum Highbush cranberry — Large shrub

Viburnum rafinesquianum Downy arrow-wood — Small shrub

WOODY PLANTS

Viburnum recognitum — Southern arrow-wood — Medium-sized shrub

Viburnum lantanoides — Hobblebush — Small shrub

WOODY PLANTS

Viburnum lantana Wayfaring tree — Shrub or small tree

Viburnum lentago Nannyberry — Large shrub

WOODY PLANTS

Viburnum nudum var. *cassinoides* Wild raisin Large shrub

WOODY PLANTS

Vitis Grape
Family: *Vitaceae*

4 species in Ontario

Grapevines are woody climbing vines with tendrils. Their deciduous leaves are simple, lobed, and are often rounded or heart-shaped at the base. The bark is loose and sheds in strips along the stem. They also have small 5-parted flowers that fruit into a cluster of reddish-purple or blue berries, 1–2 cm across. *Vitis vinifera* is the species that all grape cultivars used for food or wine were originally derived from.

Vitis spp.   Grape

Vitis spp.   Grape   Woody vine

WOODY PLANTS

Zanthoxylum Prickly-ash   1 species in Ontario
Family: *Rutaceae*

The only species of this southern genus to reach our province is the common prickly-ash (*Zanthoxylum americanum*), a prickly shrub or small tree found growing in thickets along roadsides, forest edges, and fencerows mainly in the Carolinian zone of southern Ontario. It is said that chewing the bark causes a numbing effect in the mouth, hence an alternative common name, the toothache tree. The leaves are alternate, deciduous, and pinnately compound with 5–11 leaflets. The leaflets are entire or crenate, ovate to elliptic, dark green on top, and much paler beneath. As with many other members of the citrus family (Rutaceae), the leaves are dotted with translucent glands that produce a spicy citrus fragrance when bruised. A pair of wide-based spines are found at the leaf axil, and pairs of smaller prickles appear along the petiole. The flowers are small, greenish and inconspicuous, and appear in the early spring. The fruit is an aromatic red pod that eventually splits in two to reveal one or two shiny black seeds.

Zanthoxylum americanum   Prickly-ash

Zanthoxylum americanum   Prickly-ash   Shrub or small tree

WOODY PLANTS

Woodlot Herbs

Carole Ann Lacroix, Chris Earley, Brian Lacey & Steven Newmaster

Introduction

Woodlots are habitat for many species of herbs. An herb (short for herbaceous plant) is a plant with a fleshy stem that dies down to the ground at the end of the growing season, although the roots or underground stems may still be alive. They can live for more than one year (annual), two years (biennial), or more than two years (perennial). Annuals must produce flowers and set fruit during the growing season. Some can do this within a short amount of time, such as the spring ephemerals. A biennial plant will typically grow vegetatively for the first year (as a basal rosette of leaves) and then produce a flowering stem and set seed in the second year, after which the plant dies. Although the stems of herbs are not made up of woody tissue, these plants do have vascular tissue to transport water. In Ontario there are 1900 species of herbs of which 1000 species are common terrestrial herbs. Our research indicates that there are approximately 200 species of herbs commonly found in woodlots and you might expect to find 20–40 species in any particular woodlot depending on how nutrient-rich the site is or if there have been plenty of disturbances that may support a rich flora of colonizing species.

Identifying herbs requires some basic knowledge of floral structure and the use of a herbarium. The chapter 'The Key to Floral Diversity' (pg. 101) provides a good introduction to diversity in floral structure, which is used to classify plants. Floral characters can be used to quickly relate any herbaceous plant in flower to a family concept. Once you know what family a plant belongs to you can explore the diversity of genera and species at your local herbarium. A herbarium is a natural history museum/collection of pressed, preserved plants that are stored, catalogued, and arranged systematically in cabinets. It serves as a vital reference when you need to identify a plant and provides a source of information about plants – what they look like, when they flower, where they are found, what chemicals they have in them, and a DNA record. These records are voucher specimens for research including some of the original type specimens that were used in the naming of a species. The University of Guelph Herbarium has over 100,000 samples. These specimens are cared for over time (some samples date back to the 1500s; e.g. Fig. 1) so that current and future generations can explore botanical diversity and use the collection in support of conservation, ecology, and sustainable development.

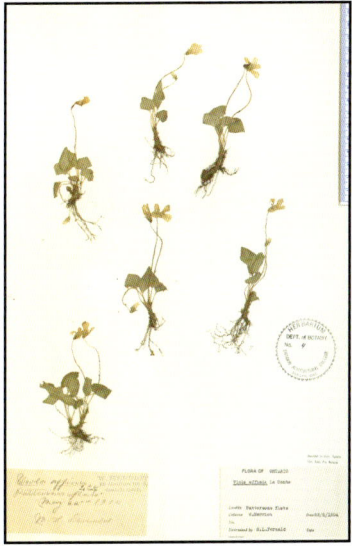

Figure 1. *University of Guelph OAC Herbarium voucher dated 1904*

WOODLOT HERBS

An Identification Key to Woodlot Herbs

1. Flowers regular 2
   2. Flowers with four or five regular parts 3
      3. Flowers with four regular parts 4
         4. Leaves opposite 5
            5. Leaves simple .... *Cornus canadensis*, pg. 386
            5. Leaves divided 6
               6. Leaves ternately compound .... *Cardamine diphylla*, pg. 383
               6. Leaves deeply divided .... *Cardamine concatenata*, pg. 382
         4. Leaves alternate 7
            7. Leaves serrate .... *Alliaria petiolata*, pg. 377
            7. Leaves entire .... *Maianthemum canadense*, pg. 395
      3. Flowers with five regular parts 8
         8. Leaves alternate 9
            9. Flowers yellow 10
               10. Leaves simple .... *Caltha palustris*, pg. 381
               10. Leaves divided 11
                   11. Leaves glabrous; basal leaves simple, cauline leaves divided ....
                        .... *Ranunculus abortivus*, pg. 403
                   11. Leaves hairy, all leaves divided 12
                       12. Sepals recurved; leaflets of a similar size ....
                            .... *Ranunculus recurvatus*, pg. 405
                       12. Sepals ascending; leaflets of 2 different sizes ....
                            .... *Geum aleppicum*, pg. 391
            9. Flowers white or purple 13
               13. Flowers white; leaves with irregularly divided segments; fruits with
                   long, persistent styles .... *Geum canadense*, pg. 392
               13. Flowers light lilac to purple; leaves with regularly divided segments
                   and white mottled blotches ...... *Hydrophyllum virginianum*, pg. 394
        8. Leaves basal, opposite, or whorled 14
           14. Leaves basal 15
              15. Leaves basal and simple 16
(*Hepatica acutiloba* may key here and has 4–8 white to purple petals - see pg. 393)
                 16. Flowers white or yellow 17
                     17. Flowers white; leaves serrate .... *Tiarella cordifolia*, pg. 410
                     17. Flowers yellow; leaves crenate ..... *Ranunculus ficaria*, pg. 404
                 16. Flowers pink with dark pink lines; leaves entire ....
                         .... *Claytonia caroliniana*, pg. 385
              15. Leaves basal and divided 18
                 18. Flowers yellow; leaflets irregularly toothed ....
                        .... *Waldsteinia fragarioides*, pg. 425
                 18. Flowers white; leaflets regularly toothed ....
                        .... *Fragaria virginiana*, pg. 389
           14. Leaves opposite or whorled 19
              19. Leaves whorled; flowers white .... *Trientalis borealis*, pg. 411
              19. Leaves opposite 20
                 20. Leaves divided .... *Geranium robertianum*, pg. 390
                 20. Leaves simple 21
                     21. Leaves entire .... *Phlox divaricata*, pg. 398

WOODLOT HERBS

   21. Leaves serrate .... *Mitella diphylla* (not profiled)
 2. Flowers with three or six regular parts 22
  22. Flowers with three regular parts 23
   23. Leaves all basal .... *Asarum canadense*, pg. 380
   23. Leaves cauline, in a single whorl 24
    24. Flowers white .... *Trillium grandiflorum*, pg. 413
    24. Flowers red or wine-coloured .... *Trillium erectum*, pg. 412
  22. Flowers with six regular parts 25
   25. Leaves basal or opposite 26
    26. Leaves basal; flowers blue .... *Scilla siberica*, pg. 407
    26. Leaves opposite 27
     27. Leaves simple .... *Podophyllum peltatum*, pg. 399
     27. Leaves divided .... *Caulophyllum thalictroides*, pg. 384
   25. Leaves alternate 28
    28. Flowers white 29
     29. Flowers in a raceme at the top of an erect stem ....
      .... *Maianthemum stellatum*, pg. 397
     29. Flowers in a panicle at the end of a curved ascending stem ....
      .... *Maianthemum racemosum*, pg. 396
    28. Flowers yellow or greenish-white 30
     30. Leaves pierced by the stem; flowers yellow 31
      31. Leaves green beneath; flowers 2–4 cm long ....
       .... *Uvularia perfoliata*, pg. 416
      31. Leavss whitish beneath; flowers 2.5–5 cm long ....
       .... *Uvularia grandiflora*, pg. 415
     30. Leaves sessile or with a short petiole; flowers yellowish-green or greenish-white 32
      32. Leaves glabrous .... *Polygonatum biflorum*, pg. 401
      32. Leaves hairy on smaller veins beneath ....
       .... *Polygonatum pubescens*, pg. 402
1. Flowers irregular or parts numerous/indistinguishable 33
 33. Flower parts irregular 34
  34. Leaves basal 35
   35. Leaf trifoliate .... *Arisaema triphyllum*, pg. 379
   35. Leaves simple 36
    36. Flowers white 37
     37. Flowers white with brown lines on lower 3 petals, beardless or slightly bearded .... *Viola macloskeyi*, pg. 420
     37. Flowers white (to deep violet) and bearded ....
      .... *Viola odorata* (not profiled)
    36. Flowers blue, pink, purple, or violet 38
     38. Flowers blue, beard long and without knobs ....
      .... *Viola septentrionalis*, pg. 424
     38. Flowers pink, purple, or violet 39
      39. Flowers pinkish-purple, tips of beard hairs with knobs ....
       .... *Viola cucullata*, pg. 419
      39. Flowers pale violet, beardless .... *Viola selkirkii*, pg. 423
  34. Leaves alternate 40
   40. Leaves entire 41
    41. Flowers yellow; plant 20–80 cm tall .. *Cypripedium calceolus*, pg. 387

WOODLOT HERBS

    41. Flowers pink-purple (sometimes white) with a fringed lower lip; plant 8–15 cm tall ....*Polygala paucifolia*, pg. 400
  40. Leaves divided or crenate **42**
    42. Leaves divided; sepals red, petals yellow with long spurs (1.5–2.5 cm) .. ..*Aquilegia canadensis*, pg. 378
    42. Leaves crenate **43**
      43. Flowers white or yellow **44**
        44. Flowers white and petals tinted purple on the outside with dark lines in the throat ....*Viola canadensis*, pg. 417
        44. Flowers yellow with purplish-brown veins in the throat ....*Viola pubescens*, pg. 421
      43. Flowers light blue or violet **45**
        45. Flowers light blue, spur 4–5 mm long ....*Viola conspersa*, pg. 418
        45. Flowers violet, spur 7–12 mm long ....*Viola rostrata*, pg. 422
33. Flower parts numerous/indistinguishable **46**
  46. Leaves basal **47**
    47. Flowers yellow or white **48**
      48. Flowers white; leaf lobed, with red latex ....*Sanguinaria canadensis*, pg. 406
      48. Flowers yellow **49**
        49. Leaves deeply pinnately lobed ....*Taraxacum officinale*, pg. 408
        49. Leaves roundly heart-shaped, with teeth and shallow lobes, appearing after flowers set seed ....*Tussilago farfara*, pg. 414
    47. Flowers whitish to deep purple, petals 4–8; leaves 3-lobed ....*Hepatica acutiloba*, pg. 393
  46. Leaves cauline, 2–3 times ternately compound **50**
    50. Flowers greenish, with no petals and 4–5 sepals, sepals with a purplish blush; leaflets crenately toothed ....*Thalictrum dioicum*, pg. 409
    50. Flowers white; leaflets sharply toothed **51**
      51. Fruits white; stigma wider than ovary .....*Actaea pachypoda*, pg. 375
      51. Fruits red; ovary wider than stigma ....*Actaea rubra*, pg. 376

WOODLOT HERBS

Doll's eyes
Actaea pachypoda

Family: *Ranunculaceae*

Habitat: Rich deciduous woods.

Description: Leaves two to three times ternately compound. Flowers white, in a raceme. Fruit white.

Notes: The leaves superficially resemble *Caulophyllum thalictroides* (pg. 384) and *Thalictrum dioicum* (pg. 409). The species is named for the fruit, which is white with a black dot in the centre, resembling an eye. Also known as white baneberry.

Poisonous Compounds: All parts are poisonous. Can blister the skin.

WOODLOT HERBS

Red baneberry
Actaea rubra

Family: *Ranunculaceae*

Habitat: Rich deciduous woods.

Description: Leaves two to three times ternately compound. Flowers white, in a raceme. Fruit red.

Notes: The leaves superficially resemble *Caulophyllum thalictroides* (pg. 384) and *Thalictrum dioicum* (pg. 409). It is very difficult to tell *Actaea rubra* and *Actaea pachypoda* (previous page) apart when they are flowering but extremely easy when they are in fruit. *Actaea rubra* usually has red fruit with slender pedicels and *Actaea pachypoda* has white fruit with stout pedicels. The two species may occasionally hybridize.

Poisonous Compounds: All parts are poisonous. Can blister the skin.

WOODLOT HERBS

Garlic mustard
Alliaria petiolata

Family: *Brassicaceae*

Habitat: Weedy colonizer, can grow in full sun or shade. Spread by humans.

Description: Flowers white, small with four petals. Basal leaves reniform with crenate teeth. Cauline leaves alternate and deltoid.

Notes: Has a strong odour of garlic when the leaves are crushed.

Poisonous Compounds: None known. The leaves, flowers, and fruit are edible. Can be used in salads or made into pesto. Leaves at all times of the year have a higher value of vitamin A than spinach. The leaves and tops have a higher value of vitamin C, by weight, than oranges.

WOODLOT HERBS

Columbine
Aquilegia canadensis

Family: *Ranunculaceae*

Habitat: Rich deciduous woods.

Description: Flowers drooping with long red spurs that attach to the yellow tubular portion of the petals. Leaves ternately decompound. Leaves resemble those of *Caulophyllum thalictroides* and *Thalictrum dioicum*.

Notes: In the same family as *Aconitum* and *Actaea*, which are poisonous. The seeds and roots of *Aquilegia canadensis* are highly poisonous and contain cardiogenic toxins which cause heart palpitations and severe gastroenteritis when consumed.

Poisonous Compounds: Highly poisonous; effects similar to the extremely poisonous alkaloid aconite.

WOODLOT HERBS

Jack-in-the-pulpit
Arisaema triphyllum

Family: *Araceae*

Habitat: Rich and moist temperate deciduous forests.

Description: Leaves usually two (depending on the age of the plant), trifoliate and basal. Fruit red when mature.

Notes: The name 'jack-in-the-pulpit' describes its unusual flowers. The flowers are the on the stalk ('jack') located under the colourful hood ('the pulpit'). This plant will remain vegetative for five to seven years after germination, and when the plant flowers for the first time, it only produces male flowers. This happens for the next few years until the roots have stored enough photosynthates and have enough energy to produce female flowers and seeds.

Poisonous Compounds: Contains calcium oxalate crystals, with highest concentrations found in the leaves. Small doses of oxalate toxin can cause swelling, choking, and sensations of burning in the mouth and throat. If larger doses are consumed, severe digestive upset, breathing difficulties, and possibly convulsions, coma, and death, may occur. Recovery is possible, but permanent liver and kidney damage is likely.

WOODLOT HERBS

Wild ginger
Asarum canadense

Family: *Aristolochiaceae*

Habitat: Rich deciduous woods.

Description: Leaves growing from a trailing rhizome. Flowers purple, found below the leaves, usually resting on the ground.

Notes: The flowers are pollinated by beetles, which have easy access due to the position of the flowers. The roots taste like ginger and can be turned into a candy, considered quite a treat by early pioneers. Used by many First Nations for colds, indigestion, and as a general tonic.

Poisonous Compounds: None known.

WOODLOT HERBS

Marsh marigold
Caltha palustris

Family: *Ranunculaceae*

Habitat: Wet woods, meadows, swamps, bogs, and shallow water.

Description: Basal leaves round with a deep sinus and crenate teeth. Stems hollow. Flowers bright yellow, up to 4 cm wide.

Notes: A yellow dye can be obtained from the petals. Has been used to treats warts.

Poisonous Compounds: Poisonous. Contains protoanemonin and helleborine; toxic to the heart and causes inflammation of the stomach. Violent purgative.

WOODLOT HERBS

Five-parted toothwort
Cardamine concatenata
syn. *Dentaria laciniata*

Family: *Brassicaceae*

Habitat: Rich deciduous woods.

Description: The flowers have four white petals and are indistinguishable from the flowers of *Cardamine diphylla* (next page). The three whorled leaves are so deeply lobed that they appear to be 5-parted.

Notes: The leaves can be used in a salad to add a peppery taste.

Poisonous Compounds: None recorded.

WOODLOT HERBS

Broadleaf toothwort
Cardamine diphylla
syn. *Dentaria diphylla*

Family: *Brassicaceae*

Habitat: Rich deciduous woods.

Description: Flowers 4-merous, white. Leaves opposite and ternately divided, with coarse crenate teeth.

Notes: The flowers are indistinguishable from the flowers of *Cardamine concatenata* (previous page). These two species interbreed and so you may find leaves ranging in form between the typical *C. diphylla* leaves and the deeply divided leaves of *C. concatenata*.

Poisonous Compounds: None recorded.

WOODLOT HERBS

Blue cohosh
Caulophyllum thalictroides

Family: *Berberidaceae*

Habitat: Rich deciduous woods.

Description: Flowers green or purple. Leaves resembling those of *Thalictrum* (pg. 409) and *Aquilegia* (pg. 378).

Notes: The pea-sized seeds have been used as a substitute for coffee (see next section about poisons!).

Poisonous Compounds: Leaves and seeds contain the alkaloid methylcytisine, glycosides, and saponins that cause gastroenteritis. There have been reports of poisoning of children who ate the blue fruits.

WOODLOT HERBS

Spring beauty
Claytonia caroliniana

Family: *Montiaceae*

Habitat: Rich deciduous woods.

Description: Flowers pink with dark pink nectar guides on the inside. Plants with one or two basal leaves.

Notes: Commonly found with *Claytonia virginiana*, which has identical flowers but much narrower leaves.

Poisonous Compounds: None recorded.

WOODLOT HERBS

Bunchberry
Cornus canadensis

Family: *Cornaceae*

Habitat: Moist acid woods and bogs.

Description: Stems growing from a woody rhizome with clusters of four to five leaves at the top. Leaf veins curving toward the leaf tips. Flower with four showy white bracts that resemble petals. Flowers inconspicuous, greenish-purple, tubular, in the middle of the large white bracts.

Notes: The fruits are edible with a mild apple flavour. Birds are the main dispersal agents of the seeds, consuming the fruit during their fall migration. In Alaska, bunchberry is an important forage plant for mule deer, black-tailed deer and moose.

Poisonous Compounds: None recorded.

WOODLOT HERBS

Lady's slipper orchid
Cypripedium calceolus

Family: *Orchidaceae*

Habitat: Wet woods.

Description: Perennial herbaceous herb. Leaves two to several with parallel main veins. Flower with a yellow pouch.

Notes: The name 'lady's slipper orchid' describes the lower pouch, which resembles a a shoe but is actually two fused petals. Has been used medicinally as a sedative and antispasmodic; it can be substituted for the European valerian, which is used for its sedative qualities.

Poisonous Compounds: Various species of *Cypripedium* are known to cause contact dermatitis resembling a poison ivy rash.

WOODLOT HERBS

Trout lily
Erythronium americanum

Family: *Liliaceae*

Habitat: Rich deciduous woods.

Description: Flowers yellow, drooping, with reflexed petals. Leaves one or two, basal, mottled.

Notes: The spots on the leaves are said to resemble the markings on a trout.

Poisonous Compounds: All parts of the plant, but especially the bulb and the fresh leaves, are strongly emetic and should not be used internally.

WOODLOT HERBS

Wild strawberry
Fragaria virginiana

Family: *Rosaceae*

Habitat: Open woodlands and clearings, often in disturbed areas.

Description: Leaves basal, trifoliate, toothed. Flowers with five round white petals. Fruit resembles a small strawberry.

Notes: Wild strawberries are used as food plants by the larvae of a number of moth and butterfly species. The fruit are eaten by many game birds (e.g. grouse, pheasants) and song birds (e.g. crows, catbirds, finches, pine grosbeaks, sparrows, veeries, cedar waxwings). The fruits and leaves are also eaten by many mammals (e.g. opossums, rabbits, skunks, squirrels, chipmunks, mice, deer).

Poisonous Compounds: Edible.

WOODLOT HERBS

Herb-robert
Geranium robertianum

Family: *Geraniaceae*

Habitat: Disturbed forests.

Description: Leaves opposite, divided into three to five leaflets. Terminal leaflet stalked.

Notes: Introduced from Eurasia. Used in traditional medicine for the treatment of inflammatory diseases and cancer. The seeds are eaten by mourning doves and the entire plant is eaten by deer.

Poisonous Compounds: Not recommended for prolonged or excessive use because of high tannin content that may damage the liver over time.

WOODLOT HERBS

Yellow avens
Geum aleppicum

Family: *Rosaceae*

Habitat: Meadows, thickets, and deciduous woods.

Description: Flowers with five yellow petals. Terminal leaflet of basal leaves wedge-shaped at the base and deeply cleft. Leaves with small leaflets alternating with larger ones.

Notes: Used by the Cree, Iroquois, Malecite, Micmac and Ojibwa First Nations for various aliments.

Poisonous Compounds: None known.

WOODLOT HERBS

White avens
Geum canadense

Family: *Rosaceae*

Habitat: Dry or moist woods.

Description: Flowers with five white petals. Uppermost leaves simple and sessile, lower cauline leaves with three obovate leaflets.

Notes: Roots used by the Chippewa and Iroquois First Nations for gynecological aid; a decoction of the whole plant was used as a love medicine.

Poisonous Compounds: None known.

WOODLOT HERBS

Hepatica
Hepatica acutiloba

Family: *Ranunculaceae*

Habitat: Rich deciduous woods.

Description: Flowers range in colour from white to a dark purple. Basal leaves shallowly 3-lobed.

Notes: The 'doctrine of signatures', popular in medieval times, used parts of plants that resembled parts of the body. Because the leaves of hepatica resemble a liver they were used for treatments of liver ailments.

Poisonous Compounds: Poisonous in large doses.

393

WOODLOT HERBS

Waterleaf
Hydrophyllum virginianum

Family: *Boraginaceae*

Habitat: Rich deciduous woods.

Description: Flowers whitish to blue/purple, arranged in a inflorescence resembling a scorpion's tail.

Notes: Named after the white patches found on mature leaves that resemble water marks. The Iroquois used a decoction or chewed the roots to treat cracked lips and mouth sores.

Poisonous Compounds: None known.

WOODLOT HERBS

Canada mayflower
Maianthemum canadense

Family: *Asparagaceae*

Habitat: Coniferous and deciduous woods.

Description: Plants 10–25 cm tall, with 1–3 leaves and clusters of 12–25 star-shaped white flowers. Berries round, green mottled with red when young, maturing to dark red.

Notes: The fruit are eaten by grouse, mice, and chipmunks.

Poisonous Compounds: The berries may be poisonous.

WOODLOT HERBS

False Solomon's seal
Maianthemum racemosum
syn. *Smilacina racemosa*

Family: *Asparagaceae*

Habitat: Rich deciduous woods.

Description: Flowers white, in a panicle (cluster) at the end of the stem. Leaves alternate, finely hairy beneath, in one plane arising from a curved, ascending stem.

Notes: The Algonquin used an infusion of the plant as a tea for sore backs. The fruits are eaten by grouse, wood thrushes, veeries, and deer mice.

Poisonous Compounds: None recorded.

WOODLOT HERBS

Starry false Solomon's seal
Maianthemum stellatum
syn. *Smilacina stellata*

Family: *Asparagaceae*

Habitat: Rich deciduous woods.

Description: Flowers white, in a raceme at the apex of the erect stem. Leaves alternate, sessile and finely hairy.

Notes: The Delaware First Nation used the roots to stimulate the stomach and cleanse the system.

Poisonous Compounds: None recorded.

WOODLOT HERBS

Wild blue phlox
Phlox divaricata

Family: *Polemoniaceae*

Habitat: Rich deciduous forests.

Description: Flowers blue to purple and showy, with five petals, clustered at the apex of the plant. Grows to 25–50 cm with opposite, entire, sessile leaves.

Notes: Blooms in early spring. Early medical practitioners made a tea from the leaves of *Phlox* species to treat eczema and to 'purify the blood'. A tea made from the boiled roots was once thought to cure venereal diseases.

Poisonous Compounds: None known.

WOODLOT HERBS

Mayapple
Podophyllum peltatum

Family: *Berberidaceae*

Habitat: Rich deciduous woods.

Description: Leaf single, irregularly lobed, peltate. Flowers solitary, white, arising from below the leaf.

Notes: Many First Nations used it as a purgative.

Poisonous Compounds: The whole plant except the fruit contains the active poisonous compound podophyllotoxin.

WOODLOT HERBS

Gaywings
Polygala paucifolia

Family: *Polygalaceae*

Habitat: Moist rich woods.

Description: Plants perennial, with rhizomes emerging from small tubers and stems 8–15 cm tall. Leaves several, entire, near the top of the plant. Flowers rose-purple to white.

Notes: This plant is often misidentified as an orchid because of its showy irregular flowers that have two lower petals fused into a boat-shaped structure. The genus *Polygala* has the common name 'milkwort', derived from Latin for many or much (*poly*) and milk (*gala*). It was believed to increase milk production in nursing mammals.

Poisonous Compounds: None recorded.

WOODLOT HERBS

Smooth Solomon's seal
Polygonatum biflorum

Family: *Asparagaceae*

Habitat: Rich deciduous woods.

Description: Leaves alternate, entire, sessile, and glabrous. Can be confused with *Polygonatum pubescens*, but the flowers are smaller in this species, 14–22 mm.

Notes: The Chippewa used a decoction of the roots, steamed and inhaled, for headaches.

Poisonous Compounds: Known to have anthraquinone in the berries, which causes vomiting and diarrhea.

WOODLOT HERBS

Hairy Solomon's seal
Polygonatum pubescens

Family: *Asparagaceae*

Habitat: Rich deciduous woods.

Description: Flowers greenish-yellow, growing singly or in pairs from nodes under the leaves. Leaves alternate, entire, and almost sessile on the curved, ascending stem. Hairs present on the underside of the leaves (only visible with magnification).

Notes: Women of the Abnaki First Nation (division of the Algonquin) used it as a treatment for spitting up blood.

Poisonous Compounds: Known to have anthraquinone in the berries, which causes vomiting and diarrhea.

WOODLOT HERBS

Small-flowered crowfoot
Ranunculus abortivus

Family: *Ranunculaceae*

Habitat: Moist or dry rich woods.

Description: Petals five, yellow, shorter than the sepals. Leaves dimorphic. Basal leaves round, irregularly toothed, with heart-shaped bases. Basal leaves round and irregularly toothed with heart-shaped bases. Stem leaves divided into narrow segments.

Notes: One of the few Ontario *Ranunculus* sp. that only grows in hardwood forests. All the buttercups have nutritious seeds that are eaten by many birds (e.g. wood ducks, grouse, wild turkeys, sparrows) and mammals (e.g. muskrats, rabbits, skunks, squirrels, chipmunks, mice) that sometimes also eat the whole plant.

Poisonous Compounds: When *Ranunculus* plants are handled, naturally occurring ranunculin is broken down to form protoanemonin, which is known to cause contact dermititis in humans. All *Ranunculus* species are poisonous when eaten fresh by cattle, horses, and other livestock, however their acrid taste and the blistering of the mouth means they are usually left uneaten.

403

WOODLOT HERBS

Lesser celandine
Ranunculus ficaria

Family: *Ranunculaceae*

Habitat: Escaped from cultivation. Prefers damp ground.

Description: Petals eight to twelve, yellow, shiny. Basal leaves round with a heart-shaped base, irregularly toothed.

Notes: Has been used for thousands of years in the treatment of haemorrhoids and ulcers. It is not recommended for internal use because it contains several toxic components.

Poisonous Compounds: When *Ranunculus* plants are handled, naturally occurring ranunculin is broken down to form protoanemonin, which is known to cause contact dermititis in humans. All *Ranunculus* species are poisonous when eaten fresh by cattle, horses, and other livestock, however their acrid taste and the blistering of the mouth means they are usually left uneaten.

WOODLOT HERBS

Hooked crowfoot
Ranunculus recurvatus

Family: *Ranunculaceae*

General: Named after the hooked beak on the achenes.

Habitat: Moist or dry woods and along streams.

Description: Petals five, pale yellow, no longer than the sepals. Both upper and lower leaves deeply cleft into 3 egg shaped segments.

Notes: The Cherokee used *Ranunculus recurvatus* as a sedative and to treat dermitis. The Iroquois used this plant as a laxative, to treat venereal diseases, and as a toothache remedy. This species has many poisonous compounds, however (see below).

Poisonous Compounds: When *Ranunculus* plants are handled, naturally occurring ranunculin is broken down to form protoanemonin, which is known to cause contact dermititis in humans. All *Ranunculus* species are poisonous when eaten fresh by cattle, horses and other livestock, however their acrid taste and the blistering of the mouth means they are usually left uneaten.

WOODLOT HERBS

Bloodroot
Sanguinaria canadensis

Family: *Papaveraceae*

Habitat: Moist to dry woods.

Description: Leaf solitary, basal, irregularly lobed. Flower solitary, white, with 8–12 petals.

Notes: Named after the red latex that is released by wounding any part of the plant. Deer feed on the plants in early spring.

Poisonous Compounds: Bloodroot produces morphine-like alkaloids, primarily the toxin sanguinarine. The alkaloids are transported to and stored in the rhizome. Salves containing bloodroot have been used in the past but cause extreme burns to contacted tissue. Although folklore suggests that natives have used it internally as an emetic, it is not advised.

WOODLOT HERBS

Siberian squill
Scilla siberica

Family: *Asparagaceae*

Habitat: Rich deciduous woods.

Description: Flower deep blue. Leaves linear, in a basal rosette. Plants 10–30 cm tall.

Notes: Native to Europe and Asia. Escaped from cultivation in North America.

Poisonous Compounds: Entire plant contains cardiac glycosides, which can potentially cause poisoning if ingested.

WOODLOT HERBS

Dandelion
Taraxacum officinale

Family: *Asteraceae*

Habitat: Lawns and disturbed sites.

Description: Leaves basal, in a rosette, irregularly lobed and toothed. Inflorescence resembles one flower, but it is actually made up of a hundred or more small yellow tubular flowers.

Notes: The leaves can be used in salads. Young flower buds can also be eaten or made into wine. Numerous First Nations use this plant medicinally. The leaves and seed heads are eaten by many birds (e.g. grouse, ring-necked pheasants, wild turkeys, blackbirds, goldfinches, sparrows, towhees) and mammals (porcupines, rabbits, chipmunks, deer mice, deer).

Poisonous Compounds: None recorded.

408

WOODLOT HERBS

Early meadow-rue
Thalictrum dioicum

Male flower

Family: *Ranunculaceae*

Habitat: Rich, moist, deciduous woods.

Description: Stem single, arising from the ground and dividing into three leaves, which are then divided twice more (ternately decompound).

Notes: The leaves closely resemble those of *Caulophyllum thalictroides* except that *Thalictrum* sp. has flowers that appear later and have no petals, whereas *Caulophyllum* flowers have five sepals and petals. *Thalictrum* has separate male and female plants. The Cherokee used the roots for diarrhea and vomiting and the Iroquois used it as a wash for sore eyes and for heart palpitations. A suspicious record exists of the plant being used 'to make you crazy', which may be an indication of toxins present.

Poisonous Compounds: Although there are none recorded, because this species is found in the family *Ranunculaceae*, which has other genera that have poisonous substances, one should be very hesitant to use it.

Fruit

409

WOODLOT HERBS

Foamflower
Tiarella cordifolia

Family: *Saxifragaceae*

Habitat: Rich woods.

Description: Flowers with five delicate petals. Leaves heart-shaped, basal, shallowly lobed with serrate teeth.

Notes: The grouping of the flowers, which have a combination of white petals and white filaments, produces a 'foam' apppearance. The Cherokee used an infusion of roots to stimulate appetite or to treat sore mouths.

Poisonous Compounds: None recorded.

WOODLOT HERBS

Starflower
Trientalis borealis

Family: *Primulaceae* (syn. *Myrsinaceae*)

Habitat: Rich woods and bogs.

Description: Plants with a single stalk, arising from ground with a whorl of 5–9 leaves near the top of the plant, below the often single flower (sometimes 2–3 flowers).

Notes: Named after its rotate white-petalled flower. The First Nations Montagnais (related to Algonquin) used an infusion for general sickness and tuberculosis.

Poisonous Compounds: None recorded.

WOODLOT HERBS

Red trillium
Trillium erectum

Family: *Melanthiaceae*

Habitat: Rich, moist, deciduous woods.

Description: Flower stalk 1–5 cm long and reflexed or curved below the three whorled leaflets. Petals usually maroon-coloured.

Notes: Known as a emmenagogue (promotes menstruation) and uterine stimulant by the Cherokee. Hence, large doses may bring on miscarriages.

Poisonous Compounds: None recorded.

WOODLOT HERBS

White trillium
Trillium grandiflorum

Family: *Melanthiaceae*

Habitat: Rich deciduous woods.

Description: Leaves three, sessile, whorled, surpassed by one large white flower with three petals.

Notes: If you come across a white trillium with green stripes do not think you have discovered a new species. This is only *T. grandiflorum* that has been infected with a virus that causes the green stripes.

Poisonous Compounds: None recorded.

WOODLOT HERBS

Coltsfoot
Tussilago farfara

Family: *Asteraceae*

Habitat: Well drained soil. Usually found along road shoulders. One of the first plants to bloom in the spring. Looks superficially like dandelion.

Description: Coltsfoot flowers before any leaves appear. The peduncle has scales and is topped with a bright yellow flower. The mature leaves are green above and white below (tomentose underside).

Notes: Named for the shape of the leaf which was thought to resemble a horse's hoof. Used as a salt substitute. Bake coltsfoot in the oven until it's slightly blackened and then crumble over food to add a salty taste. Not strong enough to cook with.

Poisonous Compounds: The discovery of toxic pyrrolizidine alkaloids in the plant has resulted in liver health concerns. In response, the German government banned the sale of coltsfoot.

WOODLOT HERBS

Large-flowered bellwort
Uvularia grandiflora

Family: *Colchicaceae*

Habitat: Rich deciduous woods especially in calcareous areas.

Description: Flowers yellow, drooping. Tepals glabrous. Leaves alternate, simple, hairy beneath, pierced by the stem.

Notes: Bumblebees and other species of bees feed from the nectar and collect pollen from the flowers. Deer are known to eat these plants.

Poisonous Compounds: None known.

415

WOODLOT HERBS

Perfoliate bellwort
Uvularia perfoliata

Family: *Colchicaceae*

Habitat: Rich deciduous woods especially in calcareous areas.

Description: Flowers yellow, drooping. Tepals glandular, papillose. Leaves alternate, simple, glabrous beneath, pierced by the stem.

Notes: The Iroquois used an infusion of the roots as a cough medicine. Bumblebees and other species of bees feed from the nectar and collect pollen from the flowers. Deer are known to eat these plants.

Poisonous Compounds: None recorded.

WOODLOT HERBS

Canada violet
Viola canadensis

Family: *Violaceae*

Habitat: Rich deciduous woods.

Description: Flowers mauve-purple on the back of the petals with dark brown-purple nectar guides on the inside throat. Lateral petals bearded. Leaves glabrous, found on the stem.

Notes: One of the earliest flowering violets. Some species of *Viola* are used in perfumes. The Flambeau Ojibwas (Wisconsin) and Potawatomis (Michigan) made a tea of the plant to treat heart pains. *Viola canadensis* leaves added to a soup will thicken it like okra. All *Viola* seeds are consumed by birds (e.g. mourning doves, grouse, juncos) and mice. Rabbits also eat the leaves.

Poisonous Compounds: None known.

WOODLOT HERBS

Dog violet
Viola conspersa

Family: *Violaceae*

Habitat: Woods and meadows.

Description: Leaves glabrous, found on the stem. Flowers light blue-violet, with a 4–5 mm long spur. Lateral petals bearded.

Notes: Can be confused with *Viola rostrata* (pg. 422), which has a much longer spur. The Ojibwa used an infusion of the whole plant for heart trouble. Most violet species flowers contain rutin, which maintains the strength and intergrity of capillary walls and are therefore good for varicose veins.

Poisonous Compounds: None recorded.

WOODLOT HERBS

Marsh blue violet
Viola cucullata

Family: *Violaceae*

Habitat: Bogs, swamps, and other wet places. Full sun or shade.

Description: Leaves all basal. Petals light violet to sometimes white.

Notes: Looks very similar to *Viola septentrionalis* (pg. 424). The only difference is that the hairs in the beard on the lateral petals are shorter than 1 mm in *Viola cucullata* and have an expanded knob at the end. *Viola cucullata* is the provincial flower of New Brunswick. The Cherokee used a poultice of leaves to treat a headache or boils.

Poisonous Compounds: None recorded.

WOODLOT HERBS

Northern white violet
Viola macloskeyi

Family: *Violaceae*

Habitat: Beside streams or very shallow water.

Description: Leaves all basal with rounded teeth, ranging from glabrous to hairy. Flowers white with brown-purple veins on the inside throat of the three lowest petals. Lateral petals lightly bearded or glabrous.

Notes: Can be distinguished from *Viola canadensis* (pg. 417) by the leaves, which are only basal. Rich in vitamins A and C. The flowers and leaves can be used in salads.

Poisonous Compounds: None recorded.

WOODLOT HERBS

Downy yellow violet
Viola pubescens

Family: *Violaceae*

Habitat: Rich deciduous woods.

Description: One of the only yellow violets found in southern Ontario. Leaves found on the stem. All the vegetative parts are hairy.

Notes: One of the earliest flowering violets. The Cherokee used an infusion for dysentery, colds, and coughs. The roots were also used as a poultice for boils and as a wash for face eruptions.

Poisonous Compounds: None known.

421

WOODLOT HERBS

Long-spurred violet
Viola rostrata

Family: *Violaceae*

Habitat: Shady slopes and woodlands.

Description: Leaves glabrous, found on the stem. Flowers light violet, with dark nectar guides and a spur 7–12 mm long. Lateral petals beardless.

Notes: *Viola rostrata* is known to hybidize with *Viola conspersa* (dog violet; pg. 418). Violets are known to have salicylic acid (aspirin contains acetylsalicylic acid) and large amounts of some vitamins.

Poisonous Compounds: None recorded.

WOODLOT HERBS

Great-spurred violet
Viola selkirkii

Family: *Violaceae*

Habitat: Woods and shady ravines.

Description: Leaves basal, with crenate teeth and rounded bases that close together and are often overlapping. Flowers pale violet, beardless, with a spur 4–7 mm long.

Notes: Named for Thomas Douglas, Earl of Selkirk (1771–1820), by Frederick Pursh. Douglas was a Scottish philanthropist that was responsible for initiating a large number of settlements in Canada (Prince Edward Island, Lake St. Clair, and Manitoba) with immigrants from Scotland. Douglas hired the botanist Pursh to do a plant survey of the new colony at Red River in Manitoba.

Poisonous Compounds: None recorded.

WOODLOT HERBS

Northern blue violet
Viola septentrionalis

Family: *Violaceae*

Habitat: Moist or wet disturbed areas. Can be weedy.

Description: Leaves all basal. Petals deep violet to white or lavender.

Notes: Looks very similar to *Viola cucullata* (pg. 419). The only difference is that the hairs in the beard on the lateral petals are longer than 1 mm in *Viola septentrionalis* and the hairs are not expanded into a knob at the end. The Cherokee used an infusion to treat colds.

Poisonous Compounds: None recorded.

WOODLOT HERBS

Barren strawberry
Waldsteinia fragarioides

Family: *Rosaceae*

Habitat: Dry or moist deciduous woods.

Description: Leaves basal, trifoliate, toothed. Leaflets with a V-shaped base. Flowers with five round petals. Fruit looks like a small strawberry.

Notes: The Iroquois used it as a blood remedy and as a poultice applied to snakebites.

Poisonous Compounds: Edible, but not tasty.

425

WOODLOT LICHENS

Woodlot Lichens

Troy McMullin & Steven Newmaster

Introduction

Lichen diversity is considerable within woodlot ecosystems. Currently, there are approximately 17,500 species of lichens described by science and many are still to be discovered, perhaps some right here in our understudied woodlots. In Ontario, there are over 1000 lichens species, 65% of which are found within forested ecosystems. Woodlots of central and southern Ontario comprise over 250 species of lichens, including 100 uncommon or rare species. Most of the common woodlot lichens (~130 species) are found throughout central and southern Ontario, with only a few (~30 species) that are more common in the Carolinian forests of southwestern Ontario. Not all of the common woodlot lichens can be found in any particular woodlot. Our preliminary studies of woodlot floristic diversity recorded 167 species of plants including trees, shrubs, herbs, ferns, bryophytes, and lichens (see Fig. 2, pg. 10). Lichens comprised 20% of the total species diversity. When exploring woodlots in southern Ontario you should expect to find 15–50 species.

Lichens are particularly sensitive to air pollution, which is perhaps the most influential environmental variable affecting lichen diversity. Unlike vascular plants, lichens lack roots; they obtain their nutrients from the atmosphere, precipitation, and the water that washes over them as moisture percolates through the forest canopy (e.g., dripping from tree branches). Lichens also lack a protective waxy cuticle making them more absorbent for nutrients and also for atmospheric pollutants, which can be fatal to many species. Some lichen species are tolerant to air pollution and others are very sensitive; there is a gradient of pollution tolerance making lichens excellent bioindicators. The fallout from a single emission point is easy to track as less tolerant species will occur progressively farther away.

There are several ecological factors that contribute to patterns in lichen diversity. These were discovered by botanists who explored lichen diversity in many different types of forests. You can do this yourself! Go for a meander through your local woodlot hunting for lichens. A keen observer will notice the splash of yellow on the trunks of mature trees, the leafy greenish-grey cover on large dead branches, or the minute gardens of colour growing on old fence posts. These are all lichens and they are generally quite small, but once you get an eye for them you will start to see them regularly on trees, rocks and soil (microhabitats). Without names, you may recognize the same community of lichens growing on specific microhabitats. This is called habitat specificity. It is well documented for lichens and there are names for those species that grow on rocks ('saxicolous'), logs/wood ('lignicolous'), or bark ('corticolous'). It gets even more specific as corticolous lichens can be categorized as those that are found on conifer or deciduous bark. The deciduous cortiphiles can be further divided into those that are found only on particular species of trees; pH (bark acidity) and the presence of cations such as magnesium are important in explaining this fine-scaled habitat specificity. Although trees are an important microhabitat (Fig. 1) it has been documented that the variety of microhabitats in a forest is critical for maintaining lichen diversity. The age of a woodlot or forest is a critical factor in explaining lichen diversity, which is generally higher in old-growth forests.

WOODLOT LICHENS

It is important to understand that there is a temporal gradient associated with disturbance and the development of microhabitats. That is, with time new habitats develop, such as stumps and logs, or unique microclimates such as areas of high humidity within dense canopies. This temporal gradient is important at many scales; at small scales (decades) the time since gap formation or when a tree falls in the forest provides a succession of unique microhabitats for different lichens that have specific light, humidity and nutrient requirements. At a larger scale (centuries), the time since a stand disturbance such as a fire or logging is important in determining the lichens that will be present; different communities of lichens are found in mature white pine forests at 100 years vs. 250 years. A greater temporal scale (millennia) would include the time since a glacial event, after which species migration is dependent on time-dependent factors such as the ability of lichens to disperse; some species disperse spores into the upper atmosphere circumnavigating the globe and other species disperse via vegetative structures that are carried by other organisms, such as on the slimy backs of gastropods (slugs)! These are only a few of the ecological factors that contribute to patterns in lichen diversity.

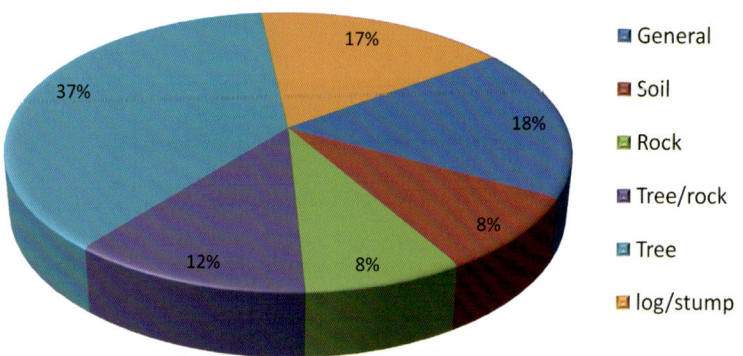

Figure 1. Lichen habitat specificity

Lichens provide ecosystem services and functions. They are important contributors to forest nutrient cycling. Some lichens fix nitrogen that eventually ends up in the soil providing essential nutrients for plants. They play a crucial role in the establishment of new communities on bare rock or soil. Many vertebrates and invertebrates use lichens for food and shelter. Lichens are the primary winter diet of the rare and endangered woodland caribou of Ontario and the arctic range caribou. Other animals, such as bats, birds, and squirrels, utilize lichens for nesting material.

People use lichens, too. Perhaps the most common use was for making dyes prior to the manufacturing of synthetic dyes. Lichens are also used by indigenous cultures around the world for medicine, perfume, toothpaste, clothing, teas, food, and decorations. Lichens have unique acids and other uncommon chemicals that have many utilities. They are, however, slow growing and sensitive to environmental change, making them difficult to cultivate in large agricultural facilities. Some of the slowest growing species develop at a rate of less than 1 mm a year, while faster growing ones can average up to a few millimetres a year.

WOODLOT LICHENS

What is a Lichen?

A lichen is composed of a mycobiont (fungi), and a photobiont (algae, cyanobacteria, or both). When they are lichenized, they symbiotically live together in a form that does not resemble either of them living independently. Typically, the mycobiont will produce an outer cortex, which makes up the body of the lichen (thallus) with webby hyphae (medulla) inside, within which the photobiont occurs (Fig. 2). The photobiont is generally clustered towards the upper part of the medulla, right below the upper cortex, where it photosynthesises more easily and produces mobile carbohydrates shared with the mycobiont.

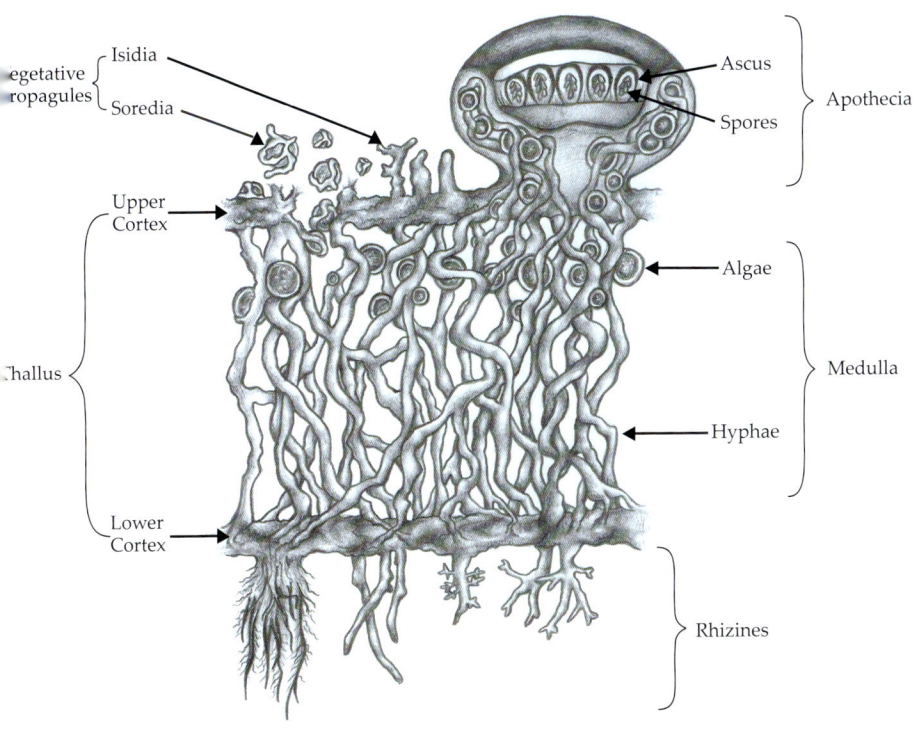

Figure 2. The basic structure of a foliose lichen

WOODLOT LICHENS

Lichen Structure

Lichens occur in a variety of forms and colours. Species are typically divided into three forms: foliose, fruticose, and crustose. Foliose species are leaf-like with lobes; they usually have a distinct upper and lower surface (cortex) (Fig. 3). Fruticose species are branch-like and typically have the same type of cortex throughout (Fig. 4). Crustose species grow within their substrate and appear crust-like on the surface (Fig. 5). To remove a crustose species the substrate upon which it grows must be removed as well. On the bark and branches of some trees, small areas of discolouration can occasionally be found. These are crustose lichens that are either immature or perhaps suppressed by air pollution.

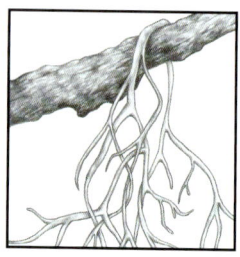
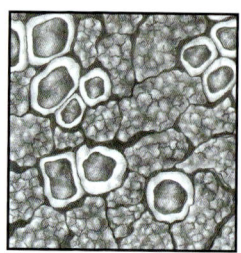

Figure 3. An example of the basic foliose lichen form.

Figure 4. An example of the basic fruticose lichen form.

Figure 5. An example of the basic crustose lichen form.

Lichen Reproduction

The mycobiont and photobiont within a lichen reproduce sexually (independently) and vegetatively (jointly). Sexually, the photobiont reproduces by cellular division, but it is generally confined within the lichen. The mycobiont produces spores in fruiting bodies within or on the thallus of a lichen; the spores are released into the air and when they encounter an appropriate microhabitat the fungus will grow, unless a photobiont is present, then a lichen may grow. Vegetatively, some lichens improve their chance of reproductive success by producing minute asexual propagules that are composed of cells from both the mycobiont and photobiont. These propagules have a range of forms, but they are broadly divided into soredia and isidia. Soredia are rounded structures emerging from within the thallus, and they lack any cortical tissue and appear powdery as a result. Isidia are larger finger-like projections produced on the upper cortex of the thallus and have cortical tissue around them. Unlike soredia, isidia have to be broken off of the thallus. Both vegetative propagules are transported by wind, water, and animals, and when they land in a suitable microhabitat a new lichen thallus will grow. Broken pieces of a thallus (fragmentation) can also form new thalli similar to the way that isidia and soredia do.

WOODLOT LICHENS

Lichen Language
Like any science, lichenology has its own language. Technical terms have been avoided in this chapter, but there are a few new words that you will need to become familiar with to use the identification key and fully understand the diagnostic descriptions on each of the species profile pages. These terms are presented in the illustrated glossary at the end of the book and are also described throughout the key each time a new term is presented.

Go Explore
It is now your time to explore. Once you get up close to lichens, ideally with a hand lens, you will see that they are charismatic and photogenic. The following section will help you to get acquainted with some of the lichen species common to woodlots through photos, descriptions and an identification key. Remember to appreciate the small things in life and watch your step!

Lichens and Allied Fungi of the Arboretum
To learn more about the lichen diversity in woodlots of southern Ontario, the Arboretum (408 acres) at the University of Guelph in Guelph, Ontario, was used as a case study. Thirty-five of the most common field-identifiable lichens and allied fungi in the Arboretum were selected. A dichotomous key and profiles with photos and diagnostic characters of the 35 species are presented in this chapter. Twenty-two additional species are discussed in the comments section of the species profiles. Woodlots throughout southern Ontario were examined and these 57 species were regularly present, particularly the 'Big Six' (see next page).

WOODLOT LICHENS

The Big Six
When it comes to lichens, a small number of species make-up the majority of what is seen. The lichen community in the Arboretum and in woodlots of southern Ontario are no exception. Of the 35 species profiled in this chapter, the following 'Big Six' are very common in the region; they regularly colonize woodlots in high abundance:

Candelaria concolor

Parmelia sulcata

Physcia adscendens

Physcia millegrana

Physcia stellaris

Xanthomendoza fallax

WOODLOT LICHENS

An Identification Key to the Lichens of the Arboretum

Before using the key, read the following:
* All species descriptions are for dry specimens; colours and texture may change when they are wet.
* Species discussed in the comments section on the profile pages are **bolded** if they have a profile page.
* Some diagnostic characteristics will require a hand lens or a stereo microscope to properly see.
* The measurement for lobe sizes is done across the width of the last major lobe (Fig. 6).
* A rating of commonness was given to each species based on the number of times it was encountered in the Arboretum: very uncommon, 1–2 times; uncommon, 3–5 times; common, >5 times, but not ubiquitous; very common, ubiquitous.
* Any collecting in the Arboretum requires permission.

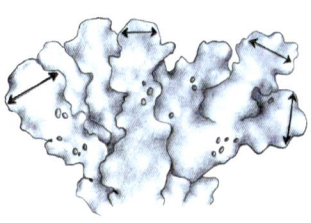

Figure 6. Where to measure the width of foliose lichen lobes.

1. Body of the lichen (thallus) crustose 2
    2. Thallus pink, growing on dead or dying species of *Parmelia* or *Physcia* ....
    .... ***Marchandiomyces corallinus***, pg. 452
    2. Thallus not pink, not growing on dead or dying lichens 3
        3. Minute black stalks (stubble) growing on polypores, <1.0 mm tall; uncommon ...
        .... ***Phaeocalicium polyporaeum***, pg. 455
        3. Minute black stalks absent; various substrates 4
            4. Apothecia elongated (lirellae), black, script-like; thallus white; on trees; very uncommon .... ***Graphis scripta***, pg. 449
            4. Apothecia round or absent; thallus not distinctly white 5
                5. Apothecia present 6
                    6. Apothecia chestnut brown, disk densely covered in pruina and appearing smoky grey with dark margins, <2.0(–2.5) mm in diameter; thallus thin, pale grey to pale beige; on rocks; uncommon ....
                    .... ***Sarcogyne regularis***, pg. 467
                    6. Apotheca orange, yellow or appearing grey-blue, <1.5 mm in diameter..7
                        7. Apothecial disks orange 8
                            8. Apothecial margins grey; thallus grey or imperceptible; on wood and bark; very common .... ***Caloplaca cerina***, pg. 439
                            8. Apothecial margins orange to pale orange; thallus yellow; on rock; very uncommon .... ***Caloplaca flavovirescens***, pg. 440
                        7. Apothecial disks yellow or appearing grey-blue 9
                            9. Apothecia yellow; thallus immersed or with scattered yellow areoles; on calcareous rock; very common .. ***Candelariella aurella***, pg. 442
                            9. Apothecia appearing grey-blue; thallus yellow green and sorediate throughout; on bark; very common .... ***Arthonia caesia***, pg. 438
                5. Apothecia absent 10
                    10. Thallus discontinuous, comprised of pale green and white powdery soredia, distinct fibrous margin lacking, has the appearance of icing sugar splashed on moist tree bases and woody debris; uncommon ....
                    .... ***Lepraria finkii***, pg. 451

WOODLOT LICHENS

  10. Thallus continuous, comprised of fine yellow-green soredia, often with a distinct fibrous margin; on bark, usually above the base of the tree; uncommon ....*Lecanora thysanophora*, pg. 450
1. Thallus fruticose or foliose **11**
 11. Thallus fruticose **12**
  12. Thallus branched, tufted, bush-like, yellow-green, longitudinally ridged; on bark and wood; uncommon ....*Evernia mesomorpha*, pg. 447
  12. Primary thallus foliose, basal leaflets (squamules); secondary thallus fruticose, erect stalks (podetia) with no branching or sparsely branched, not longitudinally ridged or bush-like; on wood, soil, or tree bases **13**
   13. Tips of the podetia with irregular bright red fruiting bodies (apothecia), cups and soredia absent ....*Cladonia cristatella*, pg. 444
   13. Tips of the podetia with pale to dark brown apothecia if present, cups present or absent, soredia present **14**
    14. Podetia virtually always without cups at the tips, if cups are present they are <1.0 mm wide, base of podetia often with a cortex that is <1.0(–2.0) mm high; uncommon ....*Cladonia coniocraea*, pg. 443
    14. Podetia with cups at the tip >1.0 mm wide **15**
     15. Podetia <2.0 cm tall, deep cups at the tips, soredia throughout; very uncommon ....*Cladonia fimbriata*, pg. 445
     15. Podetia <9.0 cm tall, shallow cups at the tips, small stalks on the margin of the cups with brown apothecia at the tips; very common ....*Cladonia rei*, pg. 446
 11. Thallus foliose **16**
  16. Thallus a distinct shade of yellow or orange **17**
   17. Thallus with vegetative propagules resembling a fine powder (soredia) ....
....**18**
    18. Thallus shades of yellow with a green hue; lobes <0.4 mm wide; very common ....*Candelaria concolor*, pg. 441
    18. Thallus shades of orange, mature lobes >1.0 mm **19**
     19. Soredia formed in crescent shaped bird's nests along margins between the upper and lower cortices, mostly yellow to green-yellow; very common ....*Xanthomendoza fallax*, pg. 468
     19. Soredia formed on the surface, marginal, mostly orange to yellow-orange, very uncommon ....
....*Xanthomendoza ulophyllodes*, pg. 469
   17. Soredia absent **20**
    20. Lobes leafy, concave, <3.5 mm wide; thallus in circular to oval patches; bright orange apothecia almost always present; common .....
....*Xanthoria parietina*, pg. 471
    20. Thallus becoming cushion-like in appearance; lobes <0.7(–1.0) mm wide at base, <0.3(–0.5) mm wide at tip, finely divided and overlapping, flat to convex; apothecia crowded; very uncommon ....
....*Xanthoria polycarpa*, pg. 472
  16. Thallus shades of grey-green, green, or brown **21**
   21. Thallus shades of brown **22**
    22. Minute finger-shaped vegetative propagules (isidia) on the upper surface; pale to white patches in areas where the isidia have broken off; lobes <6.0(–7.0) mm wide; very common ....
....*Melanelixia subaurifera*, pg. 453
    22. Isidia absent **23**

WOODLOT LICHENS

23. Interior layer (medulla) red-orange; soredia granular, mostly marginal including the lobe tips; common ....***Phaeophyscia rubropulchra***, pg. 460
23. Medulla white or pale yellow **24**
   24. Upper surface with frosty white appearance (pruina), often restricted to the lobe tips **25**
      25. Medulla white; soredia white to blue-grey; common ....***Physconia detersa***, pg. 464
      25. Medulla pale yellow; soredia pale yellow to white; very uncommon ....***Physconia enteroxantha***, pg. 465
   24. Pruina absent **26**
      26. Small colourless spike-like hairs on the lobe tips; soredia marginal; very uncommon ....***Phaeophyscia hirsuta***, pg. 457
      26. Colourless spike-like hairs on the lobe tips absent **27**
         27. Soredia only on upturned lobe tips, hemispherical, appearing pompon like; very common ....***Phaeophyscia pusilloides***, pg. 459
         27. Soredia marginal or on the upper surface (laminal), not hemispherical **28**
            28. Soredia primarily on the lobe tips or marginal, occasionally somewhat laminal, coarse; lobes <2.0(–3.0) mm wide, lobules may occur at the tips; common ...***Phaeophyscia adiastola***, pg. 456
            28. Soredia primarily laminal, coarse or powdery (farinose); lobules absent **29**
               29. Soredia farinose, in delimited soralia; lobes <1.5(–2.0) mm wide; common ....***Phaeophyscia orbicularis***, pg. 458
               29. Soredia coarse, not in delimited soralia, distinct pale to white patches in areas where the soredia have been abraded; lobes <6.0(–7.0) mm wide; very common ....***Melanelixia subaurifera***, pg. 453
21. Thallus shades of grey-green, green, or yellow-green **30**
   30. Minute finger-shaped vegetative propagules (isidia) on the upper surface, sparse near the lobe tips; pseudocyphellae white, conspicuous on the upper surface near the lobe tips; very uncommon ....***Punctelia rudecta***, pg. 466
   30. Isidia absent **31**
      31. Soredia absent **32**
         32. Thallus yellow-green; on siliceous rocks; uncommon ....***Xanthoparmelia cumberlandia***, pg. 470
         32. Thallus shades of grey to green-grey; on bark, rarely on rock; very common ....***Physcia stellaris***, pg. 463
      31. Soredia present **33**
         33. Lower surface white or pale; upper surface with distinct irregular white mottling (maculae) **34**
            34. Soredia inside inflated hood-shaped lobe tips, long hairs on the margins (cilia); very common ....***Physcia adscendens***, pg. 461

34. Soredia on the lobe margins, cilia absent; very common ... .*Physcia millegrana*, pg. 462
33. Lower surface black or brown, marginal areas may be lighter; maculae absent **35**
   35. Thallus distinctly yellow-green; soredia irregular, in clusters scattered across the upper surface; common ....*Flavoparmelia caperata*, pg. 448
   35. Thallus shades of grey, grey-green, or green **36**
     36. Interior layer (medulla) red-orange; soredia granular, mostly marginal including the lobe tips; common ....*Phaeophyscia rubropulchra*, pg. 460
     36. Medulla white or pale yellow **37**
       37. Upper surface with frosty white appearance (pruina), often restricted to the lobe tips............**38**
         38. Medulla white; soredia white to blue-grey; common ....*Physconia detersa*, pg. 464
         38. Medulla pale yellow; soredia pale yellow to white; very uncommon ....*Physconia enteroxantha*, pg. 465
       37. Pruina absent...**39**
         39. Small colourless spike-like hairs on the lobe tips; soredia marginal; very uncommon ....*Phaeophyscia hirsuta*, pg. 457
         39. Colourless spike-like hairs on the lobe tips absent **40**
           40. Upper surface with pseudocyphellae and subtle ridges and depressions; rhizines heavily branched (squarrose); very common ....*Parmelia sulcata*, pg. 454
           40. Upper surface without pseudocyphellae or ridges and depressions; rhizines unbranched **41**
             41. Soredia only on upturned lobe tips, hemispherical, appearing pompon like; very common...................................... ....*Phaeophyscia pusilloides*, pg. 459
             41. Soredia marginal or on the upper surface (laminal), not hemispherical **42**
               42. Soredia primarily on the lobe tips or marginal, occasionally somewhat laminal, coarse; lobes <2.0(–3.0) mm wide, lobules may occur at the tips; common ....*Phaeophyscia adiastola*, pg. 456
               42. Soredia primarily laminal, coarse or powdery (farinose); lobes <1.5(–2.0) mm wide; lobules absent; common .. *Phaeophyscia orbicularis*, pg. 458

WOODLOT LICHENS

Arthonia caesia

Vernacular Name: Frosted comma lichen

Synonyms: *Chrysothrix caesia*; *Allarthonia caesia*

Form: Crustose

Photobiont: Green-algae, chlorococcoid

Upper Cortex (thallus): Shades of pale yellow-green, finely granular

Lower Cortex: Not applicable

Medulla: White

Lobes: Not applicable

Soredia: Very common, shades of pale yellow-green, occurring throughout, continuous

Isidia: None

Rhizines: Not applicable

Apothecia: Common, appearing blue to grey-blue because of a pruinose covering – dark brown otherwise, <0.3(–0.4) mm in diameter, without margins, inconspicuous without a hand lens

Spores: Colourless, 4-celled, two cells at one end can be larger than the other two, 15.0–24.0 x 4.0–6.0 μm

Pseudocyphellae: None

Chemistry: Thallus with usnic acid

Substrates: Very common in the Arboretum, occurring on the bark and wood of many tree species

Comments: Often over-looked because of its size, but this species is extremely common in southern Ontario. It is usually fertile, but when it is sterile it can appear similar to **Lecanora thysanophora** (pg. 450), which can be distinguished by its larger thalli, larger granules of soredia, and a fibrous margin. Recent molecular studies suggest that this species may belong in the genus *Chrysothrix*, but further analysis is required to confirm the name change.

WOODLOT LICHENS

Caloplaca cerina

On the smooth bark of Populus tremuloides

Vernacular Name: Grey-rimmed firedot lichen

Synonyms: *Caloplaca stillicidiora*; *Caloplaca gilva*; *Caloplaca ulmorum*

Form: Crustose

Photobiont: Green algae, chlorococcoid

Upper Cortex (thallus): Shades of grey to blue-grey, immersed, often difficult to see

Lower Cortex: Not applicable

Medulla: White

Lobes: Not applicable

Soredia: None

Isidia: None

Rhizines: Not applicable

Apothecia: Common, lecanorine, <1.5(–2) mm in diameter, disks orange with grey margins, disks concave when young and flat when mature, with or without pruina

Spores: Colourless, thick septum dividing two cell cavities connected by a narrow channel (polarilocular), 2-celled, septum 5–8 μm wide

Pseudocyphellae: None

Chemistry: Orange disks with anthraquinones

Substrates: Commonly occurring in the Arboretum on wooden benches and the bark of deciduous trees, particularly *Populus* spp.

Comments: A small species, but once you get an eye for the tiny bright orange disks it is hard to miss. There are no other *Caloplaca* species in southern Ontario that have grey margins around orange disks, making this species easy to identify in the field.

On a wooden bench

WOODLOT LICHENS

Caloplaca flavovirescens

Vernacular Name: Sulphur-firedot lichen
Synonym: *Caloplaca erythrella*
Form: Crustose
Photobiont: Green-algae, chlorococcoid
Upper Cortex (thallus): Shades of yellow, cracked throughout
Lower Cortex: Not applicable
Medulla: White

Lobes: Not applicable
Soredia: None
Isidia: None
Rhizines: Not applicable

Apothecia: Common, <0.8(–1.2) mm in diameter, disks orange with orange margins that are the same shade as the disks or paler

Spores: Colourless, thick septum dividing two cell cavities connected by a narrow channel (polarilocular), 2-celled, septum 4–6 μm wide

Pseudocyphellae: None

Chemistry: Thallus and apothecia with anthraquinones

Substrates: Found on limestone rocks in the Japanese Garden and the Gosling Wildlife Gardens in the Arboretum. Only known to occur on base-rich rocks or cement.

Comments: The cracked yellow thallus without soredia is distinctive for rock inhabiting *Caloplaca* species in southern Ontario. *Caloplaca flavorubescens* is very similar in appearance, but it only occurs on bark.

WOODLOT LICHENS

Candelaria concolor

Vernacular Name: Candleflame lichen (lemon lichen)

Synonym: *Lichen concolor*

Form: Foliose, but can appear crustose from a distance

Photobiont: Green-algae, chlorococcoid

Upper Cortex (thallus): Lemon-yellow, greening in the shade, smooth to coarse, matte

Lower Cortex: White, shiny

Medulla: White, thin

Lobes: Overlapping rosettes, <0.5(–1.0) mm wide, <1.0(–1.5) mm long, continuous center and divided tips, rough edges

Soredia: Common, yellow, marginal to terminal, granular

Isidia: None

Rhizines: Common, white, simple

Apothecia: Uncommon, disks orange-brown, laminal to terminal, lecanorine, <1.5(–2.0) mm in diameter, dull, smooth

Spores: Colourless, oblong, 6.0–7.0 x 3.0–6.0 μm, 2-celled, >30 per ascus

Pseudocyphellae: None

Chemistry: Thallus with calycin and compounds related to pulvinic acid

Substrates: Found in large colonies throughout the Arboretum, typically on old deciduous trees, known to occur on rocks as well

Comments: Can appear similar to *Xanthomendoza fallax* (pg. 468). These two species are often found growing together in abundant colonies in the Arboretum; depending on the amount of shade or sun exposure their colours can be similar. *Xanthomendoza fallax* is typically a darker shade of orange-yellow and has larger lobes. They are both relatively small, but easy to distinguish in the field without a hand lens.

The larger more orange coloured thalli are Xanthomendoza fallax. *These two species are often found growing together.*

WOODLOT LICHENS

Candelariella aurella

Vernacular Name: Hidden goldspeck lichen

Synonyms: *Candelariella cerinella*; *Candelariella epixantha*; *Candelariella deflexa*

Form: Crustose

Photobiont: Green-algae, chlorococcoid

Upper Cortex (thallus): Absent or scattered yellow granules, granules 0.5–1.5 mm in diameter

Lower Cortex: Not applicable

Medulla: White, indistinct

Lobes: Not applicable

Soredia: None

Isidia: None

Rhizines: Not applicable

Apothecia: Very common, yellow, lecanorine, <1.0(–1.2) mm in diameter, flat to slightly convex, usually scattered

Spores: Colourless, 1-celled, 10.0–18.0 x 5.0–6.0 μm, ellipsoid, straight or curved, 8 per ascus

Pseudocyphellae: None

Chemistry: Pulvinic acid derivatives

Substrates: Limestone. In the Arboretum, it was commonly found on limestone, but it is also known to occur on other types of rock, trees, and wood.

Comments: A small species that can be tough to spot at first, but it is hard to miss once you get familiar with it. *Candelariella vitellina* is very similar in appearance, but it grows on noncalcareous rocks, particularly granite. On rocks, these two species can usually be separated by substrate, but if they are on bark or trees, counting the number of spores in the asci is the only way to distinguish them with certainty. *Candelariella vitellina* has 16–32 spores per ascus.

WOODLOT LICHENS

Cladonia coniocraea

Vernacular Names: Lesser powderhorn lichen (powderhorn lichen, lesser toothpick Cladonia, tiny toothpick Cladonia, awl-shaped stump lichen, common powderhorn)

Synonym: *Cenomyce coniocraea*

Form: Primary thallus (basal leaflets) foliose, secondary thallus (podetia) fruticose

Photobiont: Green-algae, Trebouxioid

Primary Thallus (basal leaflets) Upper Cortex: Grey-green to olive-brown, smooth, rounded lobes <6.0(−7.0) mm wide, concave to convex

Primary Thallus (basal leaflets) Lower Cortex: White, occasionally becoming sorediate along the margins

Secondary Thallus (podetia): Grey-green, sorediate, occasionally with basal leaflets on the lower half, <4.0(−5.0) mm tall, <2.0(−2.5) mm in diameter, tapered, straight, rising from centre of basal leaflets, typically unbranched and cupless, occasionally branched or cup-bearing, cups <1.0(−1.5) mm wide

Medulla: White

Lobes: See 'Primary Thallus'

Soredia: Common, pale green, powdery, covering the podetia and on the margins of the lower surface of the basal leaflets

Isidia: None

Rhizines: None

Apothecia: Very uncommon, brown, at the tips of the podetia, biatorine, <1.3(−1.8) mm in diameter, convex

Spores: Colourless, ellipsoid, 10.0–16.0 x 3.0–5.0 μm, 1-celled, 8 per ascus

Pseudocyphellae: None

Chemistry: Primary and secondary thallus with fumarprotocentraric and protocetraric acids

Substrates: Commonly found on cedar fences throughout the Arboretum, also known to occur on tree bases and woody debris

Comments: *Cladonia ochrochlora* is often similar in appearance. It is distinguished by patchy and granular soredia instead of continuous and powdery, larger basal leaflets (<7.5 mm wide), and it produces cups more frequently instead of points at the tips of the podetia. Unfortunately, intermediary species are common and the two species are often considered together for ecological studies.

WOODLOT LICHENS

Cladonia cristatella

Vernacular Name: British soldiers

Synonyms: None

Form: Primary thallus (basal leaflets) foliose, secondary thallus (podetia) fruticose

Photobiont: Green-algae, Trebouxioid

Primary Thallus (basal leaflets) Upper Cortex: Shades of of grey-green to yellow-green, 2.0–3.0 mm long

Primary Thallus (basal leaflets) Lower Cortex: White

Secondary Thallus (podetia): Shades of yellow-green, <2.0(–3.0) cm tall, unbranched or with a few branches near the tips that can resemble a cup, no true cups formed, usually with a continuous cortex or occasionally slightly areolate

Medulla: White

Lobes: See 'Primary Thallus'

Soredia: None

Isidia: None

Rhizines: None

Apothecia: Very common, usually red, rarely pale yellow to orange, at the tips of the podetia, biatorine, convex

Spores: Colourless, 1-celled, 8 per ascus

Pseudocyphellae: None

Chemistry: Primary and secondary thallus with barbatic, didymic, and usually usnic acid

Substrates: Found on wooden fences and archways in the Arboretum. May also occur on soil, tree bases, stumps, or fallen logs

Comments: This is the only species of *Cladonia* in the region with red apothecia and no soredia or cups, which makes it easy to identify in the field. It usually lacks basal leaflets on the podetia, or only has a sparse covering near the base, but occasionally the podetia are densely covered in basal leaflets. Pollution tolerant.

WOODLOT LICHENS

Cladonia fimbriata

Vernacular Name: Trumpet lichen

Synonym: *Cladonia major*

Form: Primary thallus (basal leaflets) foliose, secondary thallus (podetia) fruticose

Photobiont: Green-algae, Trebouxioid

Primary Thallus (basal leaflets) Upper Cortex: Grey-green, <10.0(–11.0) mm long, <4.0(–4.5) mm wide, persistent, ascending, scalloped margin, flat to concave, margins turned up

Primary Thallus (basal leaflets) Lower Cortex: White, darkening towards base

Secondary Thallus (podetia): Shades of green-grey to green-brown, <4.0(–4.5) cm tall, <2.0(–2.5) mm wide, unbranched, closed trumpet-shaped cups <5.0 mm in diameter, cups with even margins, on the upper surface of the basal leaflets

Medulla: White

Lobes: See 'Primary Thallus'

Soredia: Common, powdery, covering the entire surface of the podetia including the interior of the cup, also on the upper cortex of the basal leaflets

Isidia: None

Rhizines: None

Apothecia: Uncommon, brown, on cup margin, biatorine, <1.5(–2.0) mm in diameter, flat to convex, occasionally stalked

Spores: Colourless, oblong, 8–14 x 3.0–4.4 μm, 1-celled, 8 per ascus

Pseudocyphellae: None

Chemistry: Primary and secondary thallus with fumarprotocetaric acid

Substrates: Found at one location in the Arboretum, on a cedar fence, also known to occur on tree bases and woody debris

Comments: Can appear similar to a number of *Cladonia* species with goblet shaped cups, particularly those in the *Cladonia chlorophaea* group. ***Cladonia fimbriata*** is distinguished by its soredia, which are relatively small and powdery in appearance instead of larger granules with a rougher appearance, and by its even cup margin. The width of the cup is also narrower giving it more of a trombone shape than a goblet.

WOODLOT LICHENS

Cladonia rei

Vernacular Name: Wand lichen
Synonym: *Cladonia nemoxyna*
Form: Primary thallus (basal leaflets) foliose, secondary thallus (podetia) fruticose
Photobiont: Green-algae, Trebouxioid
Primary Thallus (basal leaflets) Upper Cortex: Pale green, smooth, <3.0(-3.5) mm long, <2.0(-2.5) mm wide, persistent to disappearing, ascending, wavy or scalloped margins
Primary Thallus (basal leaflets) Lower Cortex: White; smooth
Secondary Thallus (podetia): Shades of grey-green to brown, rough, basal leaflets at the base, <9.0(-10.0) cm tall, <3.5(-4.0) mm thick, occasionally branching up to 3 times, shallow cups closed and lopsided, with or without a cortex, typically leaning, rising from the upper surface of basal leaflets
Medulla: White
Lobes: See 'Primary Thallus'
Soredia: Common on the podetia, uncommon on the margins of the lower surface of the basal leaflets, granular, sparse in areas
Isidia: None
Rhizines: None
Apothecia: Common, brown, on the tips of proliferations from the cup margins, biatorine
Spores: Colourless, 1-celled, 8 per ascus
Pseudocyphellae: None
Chemistry: Secondary thallus with homosekikaic acid and occasionally fumarprotocetaric and/or sekikaic acids
Substrates: Found on old cedar fences throughout the Arboretum, also known occur on soil, tree bases and woody debris
Comments: The slender leaning podetia, sparse soredia in some areas, and lopsided shallow cups distinguish this species. Its chemistry is also diagnostic, but a chemical analysis such as thin layer chromatography is required to detect it.

WOODLOT LICHENS

Evernia mesomorpha

Vernacular Name: Boreal oakmoss lichen

Synonym: *Evernia thamnodes*

Form: Fruticose

Photobiont: Green-algae, chlorococcoid

Upper Cortex (thallus): Shades of yellow-green, wrinkled with coarse soredia

Lower Cortex: Not applicable

Medulla: White

Branches: Forming tufts <10.0(–15.0) cm long, longitudinally ridged, main branches <1(–2) mm wide

Soredia: Common, yellow-green, coarsely granular, throughout, not continuous

Isidia: None

Rhizines: Not applicable

Apothecia: Very uncommon, lecanorine, marginal, <6.0 mm in diameter, on short stalks, disks red-brown and concave

Spores: Colourless, ellipsoid, 1-celled, 8 per ascus

Pseudocyphellae: Common, white

Chemistry: Thallus with divaricatic and usnic acid

Substrates: In the Nature Reserve of the Arboretum it was found on conifer and deciduous trees. Outside of the Nature Reserve it was only found on wooden benches. Typically occurs on trees, occasionally on rocks.

Comments: *Evernia prunastri* is similar in appearance, but can be distinguished by its flattened branches with distinct upper and lower cortices. The upper cortex is yellow-green, while the lower cortex is white. *Evernia prunastri* was once thought to be extirpated in southern Ontario because it had not been reported in over 70 years, but it was recently discovered in 2008 on the Bruce Peninsula and in 2009 in the Arboretum. In contrast, **Evernia mesomorpha** is very common in southern Ontario woodlots. It is also pollution tolerant.

WOODLOT LICHENS

Flavoparmelia caperata

Vernacular Name: Common greenshield lichen

Synonyms: *Pseudoparmelia caperata; Parmelia caperata; Parmelia cylisphora; Parmelia flavicans*

Form: Foliose

Photobiont: Green-algae, *Trebouxia*

Upper Cortex (thallus): Shades of yellow-green, bright, smooth, wrinkled, dull to shiny

Lower Cortex: Black, browning toward margins

Medulla: White

Lobes: Rounded, wavy, elongated, <13.0(−15.0) mm wide, radiating, overlapping, scalloped and ascending margins, widening at tips

Soredia: Common, yellow-green, laminal, granular to pustular, in shallow regions of the lobe surface or on apothecial margins

Isidia: None

Rhizines: Common, black, white tipped, browning toward lobe margins, simple

Apothecia: Uncommon, disks brown, laminal, lecanorine, <12.0(−14.0) mm in diameter, concave

Spores: Colourless, ellipsoid, 15.0–24.0 x 8.0–13.0 μm, 1-celled, 8 per ascus

Pseudocyphellae: None

Chemistry: Thallus with usnic acid and atranorin; medulla with protocetraric and caperatic acid

Substrates: Found on exposed trees throughout the Arboretum, but known to occasionally occur on rocks as well

Comments: The large, wrinkled and yellow-green thallus with laminal soredia and no pseudocyphellae make this a distinctive species in the field. It has been referred to as 'the 40 mile an hour lichen' because its colour and relatively large size make it distinguishable from a car while driving by.

WOODLOT LICHENS

Graphis scripta

Vernacular Name: Common script lichen

Synonym: *Lichen scriptus*

Form: Crustose

Photobiont: Green-algae, *Trentepohlia*

Upper Cortex (thallus): White to shades of grey-green and grey-brown, smooth to wrinkled, dull

Lower Cortex: Not applicable

Medulla: White

Lobes: Not applicable

Soredia: None

Isidia: None

Rhizines: Not applicable

Apothecia: Very common, black, laminal, also called lirellae, 1.0–5.0 x 0.2–0.4 mm, tapered ends, straight and curved, raised, occasionally branching, longitudinal slit-like disk light brown to white

Spores: Colourless, cylindrical, 25.0–70.0 x 6.0–10.0 µm, 5–15-celled, 8 per ascus

Pseudocyphellae: None

Chemistry: No known lichen substances

Substrates: Found on *Fagus grandifolia* (American beech) in the Nature Reserve of the Arboretum, also known to colonize a wide variety of other tree types

Comments: The elongated apothecia are known as lirellae. On birch trees the lirellae sometimes follow the grain of bark and can appear similar to lenticels. Most species in the genus *Graphis* occur in more tropical climates. *Graphis scripta* is the only species from this genus in southern Ontario, which makes it distinctive in the field.

WOODLOT LICHENS

Lecanora thysanophora

Vernacular Name: Mapledust
Synonyms: None
Form: Crustose
Photobiont: Chlorococcoid
Upper Cortex (thallus): Composed primarily of granular soredia that are yellow-green, margins often white and fibrous
Lower Cortex: Not applicable
Medulla: White
Lobes: Not applicable
Soredia: Very common, yellow-green, occurring throughout except at the margins
Isidia: None
Rhizines: Not applicable
Apothecia: Absent
Spores: Not applicable
Pseudocyphellae: None
Chemistry: Thallus with atranorin, usnic acid, and zeorin, occasionally with porphyrilic acid
Substrates: Most common on maples, but frequent on other deciduous trees, and occasionally on rock. In the Arboretum, it was found on a variety of deciduous trees in the Nature Reserve.
Comments: Similar in appearance to sterile specimens of *Arthonia caesia* (pg. 438), which are usually fertile. When sterile, *A. caesia* has finer soredia, smaller thalli (covering less area), and it lacks a fibrous margin. *Lepraria finkii* (next page) can also be similar in appearance, but it tends to grow closer to tree bases, its soredia are larger and less continuous, and lacks a fibrous margin.

WOODLOT LICHENS

Lepraria finkii

Vernacular Name: Fluffy dust lichen

Synonyms: *Lepraria lobificans*; *Crocynia americana*; *Crocynia aliciae*

Form: Crustose

Photobiont: Green-algae, unicellular

Upper Cortex (thallus): Shades of light green to white to yellow green, dull, loose

Lower Cortex: Not applicable

Medulla: White

Lobes: Not applicable

Soredia: Very common, shades of light green to white, covering thallus and extending beyond, powdery, appearing fluffy

Isidia: None

Rhizines: None

Apothecia: None

Spores: None

Pseudocyphellae: None

Chemistry: Thallus with atranorin, constictic and stictic acids and zeorin

Substrates: Found at three locations in the Wild Goose Woods of the Arboretum, all on moist woody debris or tree bases. Also known to occur on rocks and mosses

Comments: There are a number of similar species of *Lepraria* in southern Ontario that require chemical analyses to distinguish. *Lepraria finkii* is the most common. In the Arboretum, it is the only species that occurs outside of the Nature Reserve. *Lecanora thysanophora* (previous page) may appear similar, but it tends to grow higher on trees, its soredia are finer and more continuous, and it has a distinctive fibrous margin.

451

WOODLOT LICHENS

Marchandiomyces corallinus

Vernacular Name: None known

Synonym: *Illosporium corallinum*

Form: Lichenicolous fungi

Substrates: First found in the Arboretum by University of Guelph student Nicole Daoust on **Physcia stellaris** (photo). It has since been found throughout the Arboretum on a variety of *Physcia* species. Also know to occur on species of *Parmelia*.

Comments: A bright pink lichenicolous fungi (a fungi that is only known to grow on lichens). It is a coralloid sclerotia that occurs on dead or dying areas of a variety lichen species in the genera *Parmelia* and *Physca*.

WOODLOT LICHENS

Melanelixia subaurifera

Vernacular Names: Abraded camouflage lichen (abraded brown shield)

Synonyms: *Parmelia subaurifera*; *Melanelia subaurifera*

Form: Foliose

Photobiont: Green-algae, *Trebouxia*

Upper Cortex (thallus): Shades of brown, often red-brown in the center to olive brown towards the margins, smooth, wavy to wrinkled, dull centre, shiny near the edge

Lower Cortex: Black to brown, smooth to wrinkled, dull to shiny

Medulla: White

Lobes: Rounded, <6.0(–7.0) mm wide, appressed

Soredia: Common, brown, laminal, granular, becoming isidioid, when abraded a white to yellow area remains

Isidia: Common, laminal, cylindrical, <0.25(–0.50) mm long, <0.1 mm wide, when broken off a white to yellow area remains

Rhizines: Common, black to brown, simple

Apothecia: Very uncommon, laminal, lecanorine

Spores: Colourless, ellipsoid, 10.0–13.0 x 5.5–7.0 µm, 1-celled, 8 per ascus

Pseudocyphellae: Uncommon to common, inconspicuous, on the upper surface

Chemistry: Medulla with lecanoric acid

Substrates: Found on the branches of several types of deciduous trees throughout the Arboretum. Also known to occasionally occur on rocks

Comments: The only lichen known from the Arboretum that is brown with soredia and/or isidia that when rubbed off leave white to yellow patches. It can appear similar to two other species in southern Ontario that are also brown, *M. subargentifera* and *M. fuliginosa*. *Melanelixia subargentifera* is distinguished by its marginal soredia, and *M. fuliginosa* by its lack of soredia and long (<1.5(–2.0) µm) branching isidia.

WOODLOT LICHENS

Parmelia sulcata

Vernacular Names: Hammered shield lichen (powdered shield, furrowed shield, waxpaper lichen)
Synonym: *Parmelia rosiformis*
Form: Foliose
Photobiont: Green-algae, Trebouxioid
Upper Cortex (thallus): Grey to grey-green or blue-green, subtle ridges and depressions, reticulate pattern of white pseudocyphellae, adnate, shiny
Lower Cortex: Black towards the center, becoming brown towards the lobe margins
Medulla: White, with a continuous algal layer
Lobes: Elongate, wavy, round or angular margins, <5.0(–5.5) mm wide
Soredia: Common, white to grey, laminal and marginal, along cracks in the upper cortex, granular
Isidia: None
Rhizines: Very common, black, more squarrose near the center of the thallus, simple towards the lobe margins
Apothecia: Very uncommon, disks brown to red-brown, laminal, lecanorine, <12.0(–13.0) mm wide, concave, margins occasionally sorediate
Spores: Colourless, ellipsoid, 1-celled, 10.0–18.0 x 5.0–12.0 μm, 8 per ascus
Pseudocyphellae: Very common, white, irregular, marginal and laminal
Chemistry: Medulla with salazinic acid and occasionally lobaric acid
Substrates: Found throughout the Arboretum on trees of all types, wood, and less often on rock
Comments: This is one of the larger species in the Arboretum. ***Physcia stellaris*** (pg. 463) and *Physcia aipolia* are the only species in the Arboretum that are similar in size and colour, but they lack soredia, the upper surface is covered with maculae, and they typically have an abundance of apothecia. Species in southern Ontario that may appear similar are *Parmelia squarrosa*, which has isidia instead of soredia, and *Parmelia saxatilis*, which also has isidia instead of soredia and its rhizines are simple instead of squarrosely branched. Tolerant to most air pollution and nitrophilous.

WOODLOT LICHENS

Phaeocalicium polyporaeum

Vernacular Name: Polypore stubble

Synonyms: *Calicium polyporaeum*; *Mycocalicium polyporaeum*; *Calicium fungivorum*

Form: Crustose

Photobiont: Absent

Upper Cortex (thallus): Immersed

Lower Cortex: Not applicable

Medulla: White?, indistinct

Lobes: Not applicable

Soredia: None

Isidia: None

Rhizines: Not applicable

Apothecia: Common, black, at the tip of the stalk, <0.16(–0.20) mm in diameter, variable in shape, typically convex

Stalk: Black, dull, <0.8(–1.3) mm tall, <0.07(–0.10) mm thick, unbranched

Spores: Pale brown, shape varies, commonly ellipsoid, 7.8–19.5 x 2.0–4.0 µm, (1–)2(–4)-celled, thin septum

Pseudocyphellae: None

Chemistry: No known lichen substances

Substrates: Found on several polypores throughout the Arboretum. Known to colonise on the fruiting-bodies of *Hirschioporus pargamenus*, *Trametes versicolor* and *Trichaptum biformis*

Comments: This is the only stubble species found growing on polypores in the Arboretum. Species of *Chaenotheca* will also grow on polypores in southern Ontario, but they have spherical spores. The pale brown spores of *P. polyporaeum* distinguish it from other species of *Phaeocalicium*. The thin septum in the spore is faint and can be difficult to see. This species will only colonise old or dead tissue.

WOODLOT LICHENS

Phaeophyscia adiastola

Vernacular Names: Powder-tipped shadow lichen (granulated shadow)
Synonym: *Physcia adiastola*
Form: Foliose
Photobiont: Green-algae, *Trebouxia*
Upper Cortex (thallus): Shades of green-grey to brown
Lower Cortex: Black, often becoming lighter at the lobe margins, dull to shiny
Medulla: White

Lobes: Elongated, <2.0(–3.0) mm wide, occasionally overlapping, flat to concave, rounded tips, adnate, lobules may occur at the tips

Soredia: Common, shades of green to brown, marginal, rarely laminal, granular, becoming isidioid, not confined to soralia

Isidia: None

Rhizines: Common, black, white tipped, simple, conspicuous

Apothecia: Uncommon, disks dark brown to black, marginal, lecanorine, <3.0(–3.5) mm in diameter

Spores: Dark brown, ellipsoid, 2-celled, 17.0–25.0 x 6.0–10.0 μm, 8 per ascus

Pseudocyphellae: None

Chemistry: No known lichen substances

Substrates: Found on mossy rocks and deciduous trees throughout the Arboretum. Also known to occasionally grow directly on rocks

Comments: Can appear similar to four other species of *Phaeophyscia* that occur in southern Ontario. *Phaeophyscia orbicularis* (pg. 458) can be distinguished by its finer and more laminal soredia. *Phaeophyscia rubropulchra* (pg. 460) has a red medulla instead of white. *Phaeophyscia pusilloides* (pg. 459) has soredia on upturned lobe tips appearing hemispherical or pompon-like. *Phaeophyscia hispidula* has rhizines that project from the margins, a larger thallus (11.0 cm in diameter compared to 6.0 cm), wider lobes (6.0 mm compared to 3.0 mm) and strongly upturned lobe tips.

WOODLOT LICHENS

Phaeophyscia hirsuta

Vernacular Name: Hairy shadow lichen

Synonyms: *Phaeophyscia cernohorskyi*; *Physcia hirsuta*; *Physcia cernohorskyi*

Form: Foliose

Photobiont: Green-algae, *Trebouxia*

Upper Cortex (Thallus): Grey to brown, colourless cortical hairs on margins, maculate

Lower Cortex: Black

Medulla: White

Lobes: Rounded, <1.0(–1.5) mm wide

Soredia: Common, green to grey, marginal, powdery

Isidia: None

Rhizines: Common, black, simple, not extending beyond lobe margins

Apothecia: Uncommon, disks dark brown to black, lecanorine, colourless hairs on margin

Spores: Brown, elongated, 2-celled

Pseudocyphellae: None

Chemistry: No known lichen substances

Substrates: Found on a mature and exposed *Populus × canescens* (grey poplar) in the Arboretum. Also known to occur on other deciduous trees

Comments: Can appear similar to two other species in southern Ontario, *Phaeophyscia kairamoi* and *P. hirtella*, which also produce colourless cortical hairs. *Phaeophyscia kairamoi* is distinguished by its granular soredia becoming isidioid and cortical hairs growing with the soredia. *Phaeophyscia hirtella* is distinguished by its lack of soredia.

WOODLOT LICHENS

With abraded soredia

Phaeophyscia orbicularis

Vernacular Names: Mealy shadow lichen (granulated shadow, wreath lichen)
Synonym: *Physcia orbicularis*
Form: Foliose
Photobiont: Green-algae, *Trebouxia*
Upper Cortex (thallus): Grey to brown, smooth
Lower Cortex: Black, lightening at margin, dull to shiny
Medulla: White, occasionally with a yellowish upper layer
Lobes: Elongated, <1.5(−2.0) mm wide, radiating, flat to convex, divided tips, overlapping
Soredia: Common, black, grey to white, occasionally green, laminal, occasionally marginal, powdery
Isidia: None
Rhizines: Common, black, often white tipped, simple, commonly extending beyond the lobe margins, may occur on the margins of the apothecia

Apothecia: Uncommon to common, disks dark brown to black, lecanorine, <2.0(−2.5) mm in diameter
Spores: Brown, ellipsoid, 19.0–38.0 x 7.5–11.0 µm, 2-celled, 8 per ascus
Pseudocyphellae: None
Chemistry: Medulla occasionally with zeorin
Substrates: Found on rocks, trees and benches throughout the Arboretum
Comments: Can appear similar to *Physciella melanchra* and *Phaeophyscia hispidula* in southern Ontario. *Physciella melanchra* is distinguished by its pale lower surface. *Phaeophyscia hispidula* is distinguished by its larger lobes (<6.0[−7.0] mm wide instead of <1.5[−2.0] mm) and more marginal soredia. ***Phaeophyscia adiastola*** (pg. 456) can also appear similar, but it also has more marginal soredia, which are granular instead of powdery.

WOODLOT LICHENS

Phaeophyscia pusilloides

Vernacular Names: Pompon-tipped shadow lichen (pompon shadow)

Synonym: *Physcia pusilloides*

Form: Foliose

Photobiont: Green-algae, *Trebouxia*

Upper Cortex (thallus): Shades of green-grey to brown

Lower Cortex: Black

Medulla: White

Exposed specimen becoming brown

Lobes: Often overlapping, <1.0(−1.2) mm wide, upturned where soredia occur

Soredia: Common, yellow-green, fine, in soralia, on upturned lobe tips, soralia hemispherical appearing pompon-like, shades of green to brown

Isidia: None

Rhizines: Common, black, occasionally white tipped, dense, simple, often extending beyond the lobe margins

Apothecia: Uncommon, dark brown to black, laminal, lecanorine

Spores: Dark brown, 2-celled, laminal

Pseudocyphellae: None

Chemistry: No known lichen substances

Substrates: Found on the bark of a variety of deciduous trees throughout the Arboretum. Also known to occasionally occur on calcareous rocks

Shaded specimen

Comments: Can appear similar to three other species of *Phaeophyscia* in southern Ontario: ***Phaeophyscia rubropulchra*** (pg. 460) is distinguished by its red medulla; ***P. orbicularis*** (previous page) by its laminal soredia; and ***P. adiastola*** (pg. 456) by its larger more granular soredia that are not on raised lobe tips and do not form pompon-like mounds.

WOODLOT LICHENS

Phaeophyscia rubropulchra

Vernacular Name: Orange-cored shadow lichen

Synonyms: *Physcia endochrysea; Physcia rubropulchra; Physcia orbicularis* f. *rubropulchra*

Form: Foliose

Photobiont: Green-algae, *Trebouxia*

Upper Cortex (Thallus): Green-grey to brown, smooth

Lower Cortex: Black

Medulla: Red-orange, often exposed by insects

Lobes: Narrow, ascending tips, <1.2(–1.7) mm wide

Soredia: Common, often darker than the thallus, on lobe tips, occasionally laminal, granular

Isidia: None

Rhizines: Common, black, occasionally with white tips, simple

Apothecia: Uncommon to common, disks dark brown to black, lecanorine, <1.0(–1.5) mm in diameter

Spores: Brown, round to ellipsoid, 2-celled

Pseudocyphellae: None

Chemistry: Medulla with anthraquinones and unidentified fatty acids

Substrates: Found on a variety of tree types throughout the Arboretum. Also known to grow on rocks

Comments: Can appear similar to *P. adiastola* (pg. 456) and *P. pusilloides* (pg. 459) in southern Ontario, but both of these species have a white medulla instead of red-orange.

WOODLOT LICHENS

Physcia adscendens

Vernacular Names: Hooded rosette lichen (hood lichen)

Synonyms: *Parmelia anthelina*; *Parmelia stellaris* var. *adscendens*; *Physcia aipolia* f. *anthelina*

Form: Foliose

Photobiont: Green-algae, *Trebouxia*

Upper Cortex (thallus): Shades of grey-white to grey-green, darkening towards tips, greening when wet, irregular white maculae on older sections

Lower Cortex: White to grey-brown

Medulla: White

Lobes: Elongated, <1.0(–1.5) mm wide, inflated tips resembling hoods, ascending, marginal cilia often darkening towards tips

Soredia: Common, white to yellow-green, occurring on the interior of hoods at the lobe tips, granular, older specimens may produce laminal soralia

Isidia: None

Rhizines: Common, white, simple, sparse

Apothecia: Uncommon to common, disk red-brown to black-brown, occasionally pruinose, laminal, lecanorine, <2.0(–2.5) mm in diameter, sessile to stalked

Spores: Brown, 16.0–23.0 x 7.0–10.0 μm, 2-celled, 8 per ascus, thick walled

Pseudocyphellae: None

Chemistry: Upper cortex with atranorin

Substrates: Found on a variety of trees and wood throughout the Arboretum. Also known to occasionally occur on rocks

Comments: The hooded lobes and marginal cilia make this a distinctive species in southern Ontario. Tolerant to air pollution, nitrophilous and sensitive to fluoride.

WOODLOT LICHENS

Physcia millegrana

Vernacular Name: Mealy rosette lichen

Synonyms: None known

Form: Foliose

Photobiont: Green-algae, *Trebouxia*

Upper Cortex (thallus): Grey, irregular white maculae throughout

Lower Cortex: White

Medulla: White

Lobes: Ruffled, <2.0(-2.5) mm wide, mostly appressed, margins somewhat raised, finely divided tips

Soredia: Common, pale green to grey, marginal, granular

Isidia: None

Rhizines: Common, white, simple or forked

Apothecia: Common, disks dark brown, lecanorine, <1.0(-1.5) mm in diameter, with or without pruina

Spores: Brown, 2-celled, 8 per ascus, thick walled

Pseudocyphellae: None

Chemistry: Upper cortex with atranorin

Substrates: Found on exposed deciduous trees throughout the Arboretum. Also known to occasionally occur on rocks

Comments: In southern Ontario, it can appear similar to *P. subtilis*, which is distinguished by narrower lobes, soredia limited to lobe tips, and it only grows on rocks. Tolerant to air pollution.

WOODLOT LICHENS

Physcia stellaris

Vernacular Names: Star rosette lichen (black-eyed rosette, grey star lichen)

Synonym: *Lichen stellaris*

Form: Foliose

Photobiont: Green-algae, *Trebouxia*

Upper Cortex (Thallus): Shades of dark grey in the center to light grey towards the margins, can have blue-green to pale brown hue, inconspicuous maculae, smooth near the margins

Lower Cortex: White to brown

Medulla: White

Lobes: Radiating, <3.0(–3.5) mm wide, appressed, convex

Soredia: None

Isidia: None

Rhizines: Common, white to grey or brown, simple to branched

Apothecia: Common, disks dark brown, occasionally pruinose, laminal, lecanorine, <4.0(–4.5) mm in diameter, thick margins, concave to convex

Spores: Brown, 15.0–22.0 x 7.0–11.0 μm, 2-celled, 8 per ascus, thick walled

Pseudocyphellae: None

Chemistry: Upper cortex with atranorin

Substrates: Common on exposed trees and wood throughout the Arboretum. It typically grows on deciduous trees and is also known to occur on rocks

Comments: In southern Ontario, *P. aipolia* is very similar in appearance. *Physcia aipolia* can be distinguished by its flat to concave lobes, more conspicuous maculae, and a chemical difference that cannot be detected without chemical tests. *Physcia dubia* can also appear similar, but it has lip-shaped soredia on the lobe tips.

WOODLOT LICHENS

Physconia detersa

Vernacular Name: Bottlebrush frost lichen

Synonyms: *Parmelia pulverulenta* var. *detersa*; *Physcia detersa*

Form: Foliose

Photobiont: Green-algae, *Trebouxia*

Upper Cortex (thallus): Shades of brown or green-grey, smooth to wrinkled, shiny, white pruina on lobe tips

Exposed specimen

Lower Cortex: Dark to light brown, blackening toward the center, dull or shiny

Medulla: White, loose

Lobes: Rounded, <3.0(–3.5) mm wide, appressed tips occasionally slightly upturned

Soredia: Common, shades of brown or grey-green, marginal, granular becoming isidioid

Isidia: None

Rhizines: Common, black, simple to squarrose

Apothecia: Uncommon, disks dark brown and often with whitish pruina, laminal, lecanorine, <3.0(–4.0) mm in diameter

Spores: Brown, 2-celled, broad septum, 8 per ascus

Pseudocyphellae: None

Shaded specimen with pruina on the lobe tips

Chemistry: Usually without lichen substances; medulla and soredia may contain variolaric acid

Substrates: Found on trees of all types, wooden benches and rocks throughout the Arboretum

Comments: Can appear similar to three species of *Physconia* in southern Ontario: ***Physconia enteroxantha*** (next page) is distinguished by its yellow-tinted medulla and soredia; *P. grisea* by its pale and simple rhizines; and *P. leucoleiptes* by its lip-shaped soralia on the lobe tips and a chemical difference that cannot be detected without chemical tests.

WOODLOT LICHENS

Physconia enteroxantha

Vernacular Names: Yellow-edged frost lichen (bordered frost, yellow-cored frost lichen)

Synonyms: *Physcia enteroxantha*; *Physcia subdetersa*

Form: Foliose

Photobiont: Green-algae, *Trebouxia*

Upper Cortex (Thallus): Grey-brown to green-brown, pruinose, dull

Lower Cortex: Tan, darkening to black or brown towards the center, dull or slightly shiny

Medulla: Pale yellow to yellow-white

Lobes: Linear, <3.0 mm wide, partly overlapping, occasionally concave, adnate

Soredia: Very common, yellow-green, marginal and occasionally laminal, granular

Isidia: Uncommon

Rhizines: Common, black, squarrose

Apothecia: Uncommon, disks dark brown and often with whitish pruina, laminal, lecanorine, <2.0(−2.5) mm in diameter

Spores: Brown, 26.0–38.0 x 15.0–21.0 µm, 2-celled, broad septum, 8 per ascus

Pseudocyphellae: None

Chemistry: Medulla and soralia with secalonic acid

Substrates: Found on a mature *Populus × canescens* (grey poplar) in the Arboretum. Also known to occur on a variety of tree types and rocks

Comments: *Physconia detersa* (previous page) is very similar in appearance, but it has a white medulla. Two other species that can appear similar in southern Ontario are *P. grisea* and *P. leucoleiptes*. *Physconia grisea* is distinguished by its pale and simple rhizines, and *P. leucoleiptes* by its lip-shaped soralia at the lobe tips and by its white medulla.

WOODLOT LICHENS

Punctelia rudecta

Vernacular Name: Rough speckled shield lichen
Synonym: *Parmelia rudecta*
Form: Foliose
Photobiont: Green-algae, *Trebouxia*
Upper Cortex (thallus): Shades of grey-green to blue-grey, smooth near the margins, becoming wrinkled near the center
Lower Cortex: Pale brown or tan to cream coloured
Medulla: White
Lobes: Appressed, often overlapping, <6.0(-8.0) mm wide
Soredia: None
Isidia: Very common, cylindrical, laminal and marginal, dense towards the centre, sparse near the outer edges, occasionally branched
Rhizines: Common, white to tan, simple
Apothecia: Uncommon, disks brown, lecanorine, laminal
Spores: Colourless, 1-celled, ellipsoid, 8 per ascus
Pseudocyphellae: Common, white, occasionally coalescing, conspicuous, best seen near the lobe tips
Chemistry: Cortex with atranorin; medulla with lecanoric acid
Substrates: Found on deciduous trees in the Nature Reserve of the Arboretum. Known to occur on both coniferous and deciduous bark, occasionally on rocks
Comments: A distinctive and common species in southern Ontario. Its abundant pseudocyphellae and isidia along with its tan lower surface make it easy to identify in the field throughout this region.

WOODLOT LICHENS

Sarcogyne regularis

Vernacular Name: Frosted grain-spored lichen

Synonyms: *Sarcogyne pruinosa*; *Biatorella pruinosa*; *Lecanora pruinosa*

Form: Crustose

Photobiont: Green-algae, *Myrmecia*

Upper Cortex (thallus): Pale grey to pale beige, thin, often within the substrate and indistinguishable

Lower Cortex: Not applicable

Medulla: White

Lobes: Not applicable

Soredia: None

Isidia: None

Rhizines: Not applicable

Apothecia: Common, disks dark brown, margins black, lecideine, <1.5(–2.0) mm in diameter, flat to convex, disk usually with blue-grey pruina, pruina may occasionally occur on the margins

Spores: Colourless, ellipsoid, elongated, 2.5–5.5 x 1.0–2.0 µm, 1-celled, 100–200 per ascus

Pseudocyphellae: None

Chemistry: No known lichen substances

Substrates: Found in the Arboretum on exposed limestone rocks. Also known to occur on calcareous cement, mortar and seashells

Comments: *Porpidia albocaerulescens* and *P. cinereoatra* are similar in appearance in southern Ontario. Their apothecia are also heavily pruinose with black margins, but their thallus is often more conspicuous and thicker. Additionally, applying a drop of water to the apothecial disks of **S. regularis** will turn them chestnut brown, but this will not happen with *P. albocaerulescens* or *P. cinereoatra*.

A drop of water reveals the chesnut brown disks

WOODLOT LICHENS

Xanthomendoza fallax

Vernacular Names: Hooded sunburst lichen (powdered orange lichen)

Synonyms: *Xanthoria fallax*; *Physcia fallax*

Form: Foliose

Photobiont: Green-algae, trebouxioid

Upper Cortex (thallus): Orange to red-orange, can become pale yellow-green in shade, smooth, shiny

Lower Cortex: White, wrinkled

Medulla: White, loose, reticulate

Lobes: Rosette-forming, <2.0(–2.5) mm, raised margins, rounded

Soredia: Common, yellow to green-yellow, between upper and lower cortices in crescent shaped bird's nests along the margins, powdery, <50.0(–51.0) µm in diameter

Isidia: None

Rhizines: Common, white to yellow, simple

Apothecia: Uncommon, disk orange, margin same colour as thallus, laminal, lecanorine, <2.0(–2.5) mm in diameter, flat, dull

Spores: Colourless, ellipsoid, thick septum dividing two cell cavities connected by a narrow channel (polarilocular), 10.0–14.0 x 5.0–7.0 µm, 8 per ascus

Pseudocyphellae: None

Chemistry: Upper cortex with parietin, fallacinal, teloschistin, other anthraquinones

Substrates: Found in large colonies throughout the Arboretum, usually on mature deciduous trees. Also known to occur on wood and rocks

Comments: In the Arboretum, it is typically associated with **Candelaria concolor** (pg. 441), which is distinguished by smaller lobes, soredia on the surface of the lobe margins, and its yellow to yellow-green colour. **Xanthoria parietina** (pg. 471) and **Xanthoria polycarpa** (pg. 472) are similar in colour, but they lack soredia and typically have apothecia. **Xanthomendoza ulophyllodes** (next page) is also similar in appearance, but it has soredia on the upper surface of the lobe margins instead of between the cortices. Tolerant to air pollution.

WOODLOT LICHENS

Xanthomendoza ulophyllodes

Vernacular Names: Hooded sunburst lichen (powdered orange lichen)

Synonym: *Xanthoria ulophyllodes*

Form: Foliose

Photobiont: Green-algae, trebouxioid

Upper Cortex (thallus): Shades of yellow-orange to red-orange, can become quite yellow in shaded environments, smooth

Lower Cortex: White

Medulla: White

Lobes: Rosette-forming, radiating, branched, <1.2(–1.4) mm, flat and close to the substrate, slightly raised where soredia occur

Soredia: Common, shades of yellow-orange to red-orange, similar in colour to the thallus, mostly marginal, increasing in density towards the center

Isidia: None

Rhizines: Common, white to yellow, simple

Apothecia: Uncommon, disk orange, lecanorine, <2.5(–3.0) mm in diameter, laminal

Spores: Colourless, thick septum dividing two cell cavities connected by a narrow channel (polarilocular), 10.5–12.5(–14.5) x (5.5–)7–8 μm, 8 per ascus

Pseudocyphellae: None

Chemistry: Upper cortex with parietin, fallacinal, teloschistin, and other anthraquinones

Substrates: Found once in the Arboretum on cement. Also known to occur on deciduous trees and rocks

Comments: Most similar in appearance to the more common *Xanthomendoza fallax* (previous page), which can be distinguished by the cresent shaped bird's nests along its margins where soredia are produced between the upper and lower cortices. The soredia formed by *X. fallax* are produced from the medulla, making them more yellow to green-yellow. The soredia formed by *X. ulophyllodes* are on the surface and produced from the medulla and the cortex, which makes them more orange. *Xanthoria parietina* (pg. 471) and *X. polycarpa* (pg. 472) can also appear similar, but they lack soredia and typically have apothecia.

WOODLOT LICHENS

Xanthoparmelia cumberlandia

Vernacular Name: Cumberland rock-shield
Synonym: *Parmelia cumberlandia*
Form: Foliose
Photobiont: Green-algae, *Trebouxia*
Upper Cortex (thallus): Yellow-green, smooth to wrinkled, often peppered with black dots (pycnidia)
Lower Cortex: Pale to light brown
Medulla: White
Lobes: <3.0(–3.5) mm wide, tips round or somewhat toothed with lobules

Soredia: None
Isidia: None
Rhizines: Common, pale to light-brown, simple
Apothecia: Common, disk brown, laminal, lecanorine, <8.0(–10.0) mm in diameter, margins appearing toothed when mature
Spores: Colourless, 1-celled, ellipsoid, 6.0–13.0 x 4.0–8.0 µm, 8 per ascus

Pseudocyphellae: None

Chemistry: Cortex with usnic acid; medulla with norstictic and stictic acids

Substrates: Siliceous rocks, particulary granitic ones. In the Arboretum, siliceous rocks are uncommon, which is undoubtedly the reason this species is uncommon there

Comments: There are several species of *Xanthoparmelia* in southern Ontario, but they are all easily distinguished by the presence of isidia or a black lower surface, except one, *Xanthoparmelia viriduloumbrina*. The less common *X. viriduloumbrina* is very similar in appearance and can only be reliably distinguished by chemical tests; it contains salazinic acid in the medulla and lacks norstictic and stictic acids.

WOODLOT LICHENS

Xanthoria parietina

Vernacular Names: Maritime sunburst lichen

Synonyms: *Teloschistes parietinus*; *Lichen parietinus*

Form: Foliose

Photobiont: Green-algae, *Trebouxia*

Upper Cortex (thallus): Yellow-orange in the sun to grey in shade, smooth near the lobe margins, wrinkled towards the center, center often dying with age, oval to circular patches <15.0(–16.0) cm in diameter

Lower Cortex: White, wrinkled

Medulla: White, bunched, elongated hyphae

Lobes: Appressed, <3.5(–4.0) mm wide, margins scalloped, radiating from center, concave to flat, forming ruffled rosettes

Soredia: None

Isidia: None

Rhizines: True rhizines lacking, hapters (wide attachments that get thicker where they are connected) are common and white

Apothecia: Common, disk orange, laminal, lecanorine, <10.0(–11.0) mm in diameter, concave then becoming flat, smooth, occasionally with short stalks

Spores: Colourless, ellipsoid, thick septum dividing two cell cavities connected by a narrow channel (polarilocular), 11.0–17.0 x 6.0–10.0 μm, 2-celled, 8 per ascus

Pseudocyphellae: None

Chemistry: Upper cortex with parietin and other anthraquinones

Substrates: Found on a variety of deciduous trees throughout the Arboretum. Also known to occur on rocks and cement

Comments: In southern Ontario, there is a disjunct population. This species typically occurs along the east and west coasts of North America, except for this small inland population. The large wrinkled lobes, lack of soredia and abundance of apothecia distinguish this species in southern Ontario. Thought to have been used as a medieval treatment for jaundice.

WOODLOT LICHENS

Xanthoria polycarpa

Vernacular Names: Pincushion orange lichen (lumpy shore lichen, pincushion sunburst lichen, cushion lichen)

Synonyms: *Xanthoria ramulosa*; *Xanthoria alaskana*; *Xanthoria polycarpa var. maritima*; *Teloschistes polycarpus*

Form: Foliose

Photobiont: Green-algae, *Trebouxia*

Upper Cortex (thallus): Yellow-orange, occasionally green-yellow, smooth, shiny

Lower Cortex: White, wrinkled

Medulla: White

Lobes: Cushion forming, <0.7(–1.2) mm wide at the base, <0.3(–0.8) mm wide at the tips, finely divided, ascending, overlapping, flat to convex

Soredia: None

Isidia: None

Rhizines: True rhizines lacking, hapters (wide attachments that get thicker where they are connected) are common and white

Apothecia: Very common, crowded, disk orange, margin scalloped or smooth, laminal, lecanorine, <4.5(–5.0) mm in diameter, stalked, flat, round or irregularly shaped

Spores: Colourless, ellipsoid, thick septum dividing two cell cavities connected by a narrow channel (polarilocular), 11.0–15.0 x 5.5–8.0 μm, 8 per ascus

Pseudocyphellae: None

Chemistry: Upper cortex with parietin, fallacinal, emodin, teloschistin, and parietinic acid

Substrates: Found on a few exposed deciduous trees throughout the Arboretum. Also known to occur on rocks and cement

Comments: The pin-cushion growth form, lack of soredia, and crowded apothecia distinguish this species in southern Ontario. *Xanthomendoza hasseana* can appear similar, but it has abundant rhizines instead of hapters. This is a nitrophilous species with an intermediate sensitivity to air pollution.

WOODLOT LICHENS

Woodlot Birds

Chris Earley & Jose Maloles

Introduction
Southern Ontario is home to many species of birds that depend on woodlots to survive. This guide will help you learn how to identify the birds that come to forested areas. It also includes birds that prefer more open forest edge habitats, where the woods meet fields or swamps.

Identification
There are many things to watch for when trying to identify a bird. Look for some of the following clues:

1. Size (length from beak tip to tail tip)
2. Overall shape
3. Overall colours and markings
4. Song and/or call notes
5. Behaviour (bouncy flight, perches head down, usually on the ground, always in a flock)

When Is Each Species in Our Area?
Summer – during migration and in the summer only
Winter – during migration and in the winter only
Resident – all year round
Migration – spring and fall migration only (typically April–May and September–October)

WOODLOT BIRDS

Anatomy

Knowing the parts of a bird will help you find important field marks.

Head	– Is there a cap or a crest?
	– Are there stripes that go through the eye (eyeline) or over the eye (eyebrow)?
	– What is the beak shape and colour?
Body	– Does the back or the breast have streaks?
	– Is there a central spot on the breast? (the sparrow below has one)
	– What is the overall colour of the bird?
Wings	– Does the bird have wingbars?
Tail	– Is the tail long or short in relation to the bird's size?

Birdie Bits

 Adult   Juvenile   Male   Female

WOODLOT BIRDS

Predators

Predators are a healthy part of the forest ecosystem and can be exciting to watch. Sharp-shinned (pg. 478) and Cooper's (pg. 478) hawks are very manoeuvrable in forested areas, which allows them to catch small birds. These two species can be very hard to tell apart. If one of these hawks perches, try to look at its tail. The smaller sharp-shinned hawk has tail feathers that are all the same length when it is perched. In contrast, the outer tail feathers of Cooper's hawk are shorter than the central ones, and this can be seen on the underside of perched individuals. Another avian predator that may come to woodlots in winter is the northern shrike (pg. 498). Like most of the other small birds, the shrike is a songbird, not a hawk, but it is adapted to catching insects, mice, and other birds with its curved beak. Its other name is 'butcherbird' because it hangs its leftover meals from a thorn or crotch in a tree or shrub for later consumption. We also have owls in our wooded areas that hunt while the hawks and shrikes are asleep.

Red-tailed hawk
Buteo jamaicensis

Description: Stocky, broad-winged, and very large (length: 48 cm, wingspan: 125 cm). Upperparts brown, underparts white. Belly has dark streaks. Adult with red tail, juvenile with barred tail.

Woodlot Habits: Resident. Found in dry woodlands, forests, and farm woodlots. Nests 11–27 m above the ground in a tree. Nests constructed from sticks and twigs and lined with mosses, evergreen sprigs, and the inner bark of grapevines or cedars. Feeds mainly on rodents, hunting from a perch or by kiting.

WOODLOT BIRDS

Sharp-shinned hawk
Accipiter striatus

Description: Short-winged and relatively small, comparable to robins and pigeons (length: 28 cm, wingspan: 58 cm). Adults with greyish upperparts and rusty-barred underparts. Juveniles brown with streaked underparts.

Woodlot Habits: Resident. Found in woodlands. Nests 9–10.5 m from the ground in conifer trees and occasionally oak trees. The diet consists of a variety of small bird species.

Cooper's hawk
Accipiter cooperii

Description: Short-winged and medium in size, comparable to pigeons and crows (length: 42 cm, wingspan: 79 cm). Adults with greyish upperparts and rusty-barred underparts. Juveniles brown with streaked underparts.

Woodlot Habits: Resident. Found in woodlands, both coniferous and deciduous. Nests 6–18 m from the ground in deciduous or coniferous trees. Nests constructed from sticks and twigs and lined with chips or flakes of oak and pine. The diet consists mainly of a variety of bird species.

WOODLOT BIRDS

Great horned owl
Bubo virginianus

Description: Very large (length: 56 cm, wingspan: 112 cm). Ear tufts large, eyes yellow. Body brown, underparts barred. Juveniles more buff and fluffy.

Woodlot Habits: Resident. Found in forests and farm woodlots. Nests in cavities, caves, on cliff edges, or uses abandoned stick nests of hawks, crows, or herons. This owl eats almost anything it can catch. Large prey items include cottontails, skunks, opossums, house cats, ducks, and geese. Small prey items include meadow voles, mice, shrews, snakes, frogs, fish, crayfish, earthworms, and large insects.

Eastern screech-owl
Megascops asio

Description: Medium-sized (length: 22 cm, wingspan: 51 cm). Ear tufts small, eyes yellow, bill pale green. Underparts strongly barred. Most are mottled grey in our area, but some are rufous or brown.

Woodlot Habits: Resident. Found in forests and farm woodlots. Nests 1.5–9 m from the ground in crevices of buildings, wood duck boxes, or abandoned nesting holes of common flickers and pileated woodpeckers. The varied diet consists of small rodents, small birds, reptiles, amphibians, fish, crayfish, earthworms, and large insects.

WOODLOT BIRDS

Other Birds

Wood duck
Aix sponsa

Description: Large but much smaller than a mallard (length: 22 cm, wingspan: 76 cm). Breeding males are very colourful and have a crest. Head greenish with white lines, beak red, breast chestnut, back dark but iridescent, tail longish. Female mostly brown and mottled overall with a white eye ring and eyeline.

Woodlot Habits: Summer. Found in a variety of wooded habitats, sometimes fairly far from water. Often perches in trees, especially when looking for nest sites. Nests in cavities, often abandoned pileated woodpecker holes. Eats aquatic vegetation and aquatic insects like other ducks, but also eats fruits and nuts such as acorns and beech nuts.

Mallard
Anas platyrhynchos

Description: Large (length: 58 cm, wingspan: 89 cm). Breeding male with green head, white neck ring, yellow beak, and chestnut breast. Female is mottled brown overall with a mottled orange beak.

Woodlot Habits: Summer (though some stay over the winter if there is open water). Found mostly in open habitats, but will nest in wooded swamps and wet woods. Usually nests on the ground. Mallards eat mostly aquatic vegetation, grains, and aquatic invertebrates.

WOODLOT BIRDS

Canada goose
Branta canadensis

Description: Very large (length: up to 114 cm, wingspan: ip to 152 cm). Sexes similar. Neck long and black, cheek with a white patch, body brown.

Woodlot Habits: Summer (though some stay over the winter if there is open water). Found mostly in open habitats but will nest in wooded swamps and wet woods. Usually nests on the ground, often on islands. Canada geese are grazers eating lots of grasses but also eat aquatic vegetation and grains.

American woodcock
Scolopax minor

Description: Medium (length: 28 cm, wingspan: 46 cm). Sexes similar. This shorebird has a very plump body and very long beak. The body is mottled with a variety of browns, creams and greys. Eyes very large.

Woodlot Habits: Summer. Found in a variety of wooded or brushy areas or along the edges of swamps. The easiest way to find this species is to be out at dusk when the males can be heard making a "peent" call and doing a flight display. This usually occurs on the forest edge in a clearing. Woodcocks eats invertebrates, especially earthworms.

WOODLOT BIRDS

Ruffed grouse
Bonasa umbellus

Description: Large (length: 43 cm, wingspan: 56 cm). Sexes similar. Mottled brown overall with a black band at the end of the tail. Head small with a small crest.

Woodlot Habits: Resident. Usually in mixed woods. Males do a loud 'drumming' display. Nests on the ground. Eats a variety of leaves, buds, seeds, fruits, and insects.

Wild turkey
Meleagris gallopavo

Description: Very large (length: 117 cm, wingspan: 163 cm). Sexes similar but male much larger and has a more colourful head. Body dark brown and mottled. Tail with a dark band and a buffy band. Primary feathers dark brown and white striped. Head with wattles and warts, especially on male which can have a bright blue and red head.

Woodlot Habits: Resident. Recently reintroduced and numbers expanding in southern Ontario. Found in forest and forest edge habitats. Males make a loud gobbling sound. Nests on the ground. Turkeys eat acorns, nuts, seeds, grain, insects, and even salamanders and small snakes.

WOODLOT BIRDS

Mourning dove
Zenaida macroura

Description: Large (length: 32 cm, wingspan: 46 cm). Head and beak small, tail long and pointed. Adults and juveniles mostly brownish-grey with black spots on the wings and lower back.

Woodlot Habits: Resident. Found in open woods or roadside trees. Nests 3–7.5 m from the ground, typically in evergreen trees or in tangles of shrubs or vines. Nest constructed of twigs made into a platform. The diet consists of various grasses and grains (wheat and corn) and insects.

Black-billed cuckoo
Coccyzus erythropthalmus

Description: Similar in size to mourning dove (length: 30 cm, wingspan: 44 cm). Sexes similar. Upperparts brown and underparts white. Tail long with small white feather tips on underside. Bill black and eye ring red.

Woodlot Habits: Summer. Forest and scrubby areas. Nests 1–5 m up in a dense shrub or tangle. These cuckoos mostly eat insects and are well known for eating tent caterpillars.

WOODLOT BIRDS

Downy woodpecker
Picoides pubescens

Description: Relatively small (length: 17 cm, wingspan: 30 cm). Beak small. Plumage black and white. Males with red spot behind head.

Woodlot Habits: Resident. Found in open forests of mixed growth. Nests 1–15 m above ground in holes dug in live or dead trees, stumps, or stubs. Forages along small twigs or weed stalks. The diet consists of insects (wood-boring larvae of beetles and moths, adult beetles, and ants), spiders, snails, and wild fruits (poison ivy, serviceberry, and apple).

Hairy woodpecker
Picoides villosus

Description: Relatively large (length: 24 cm, wingspan: 38cm). Bill long and sturdy. Plumage black and white. Males with red spot behind head.

Woodlot Habit: Resident. Found in coniferous and deciduous forests or wooded swamps. Favours mature woods and larger branches; it rarely forages on weed stalks. Nests 1.5–9 m above the ground in holes dug in live or dead trees. The diet consists of insects (beetles and larvae, ants, and caterpillars), spiders, and fruits (cherry, serviceberry, blackgum, and apple).

WOODLOT BIRDS

Downy vs. Hairy Woodpecker

These two woodpeckers are always a challenge, even for experienced birders. The field marks to look for are in the tail feathers: downy woodpeckers usually have black markings on their outer tail feathers and hairy woodpeckers don't. Downy woodpeckers also have a relatively short beak and hairy woodpeckers have a long beak. Once you get used to them, you will find that the difference in overall size and shape will be your clues to identifying them.

Downy

Hairy

Pileated woodpecker
Dryocopus pileatus

Description: Large, comparable to crows (length: 42 cm, wingspan: 74 cm). Long-necked, broad-winged, and long-tailed. Body black with white wing patches and a red crest. Female with black forehead.

Woodlot Habits: Resident. Found in mature coniferous and deciduous forests and mixed woodlands. Nests 4.5–21 m above ground. Digs large rectangular holes in dead stubs or live trees. The diet consists of insects (mainly ants and beetle larvae) and fruit (grape and blackgum).

485

WOODLOT BIRDS

Red-bellied woodpecker
Melanerpes carolinus

Description: Medium-sized (length: 24 cm, wingspan: 41 cm). Beak long, wings short, body heavy. Males with red cap, females with red nape only. Upperparts barred black and white. Underparts buffy with reddish belly.

Woodlot Habits: Resident. Found in mixed forests. Nests 1.5–23 m above ground in holes dug in live trees, dead trees, or stumps. The diet consists of various insects (beetles, ants, grasshoppers, crickets, and caterpillars), grasses (corn) and woody plant fruit and seeds (oak, mulberry, cherry, and pine).

Northern flicker
Colaptes auratus

Description: Large (length: 32 cm, wingspan: 51 cm). Beak long. Plumage brown with black bars and spots. Breast with a black patch. Tail and wing feathers yellow. Rump white.

Woodlot Habits: Summer. Found in farm woodlots and open deciduous, coniferous, or mixed woods. Nests 0.5–18 m above ground in holes dug in live trees, dead trees, or stubs. The diet consists of various insects (beetles, ants, grasshoppers, crickets, and caterpillars) and wild fruits (poison-ivy, blackgum, blackberry, blueberry, and serviceberry).

WOODLOT BIRDS

Eastern phoebe
Sayornis phoebe

Description: Small (length: 18 cm, wingspan: 27 cm), but relatively long-tailed. Upperparts brown, underparts white or cream-coloured. Often pumps its tail.

Woodlot Habits: Summer. Often found in suburban areas. Nests on shelf-like projections (windows, rafters, girders, and trestles). Nest constructed of weeds, grasses, fibres, and mud, covered with mosses, and lined with grasses and hair. The diet consists of insects (bees, wasps, and ants) and fruits (elderberry, hackberry, blueberry, and blackberry).

Eastern wood-pewee
Contopus virens

Description: Small (length: 16 cm, wingspan: 25 cm). Upperparts brown, underparts gray. Crest slight and wingbars whitish.

Woodlot Habits: Summer. Found in mature forests or farm woodlots. Nests 4.5–20 m above ground on horizontal tree limbs far out from trunk. Nests in trees such as oaks, pines, birches, and maples. Nest constructed from grasses, weed stems, plant fibres, spider webs, and hair and covered on the outside with lichens. The diet consists of insects and other anthropods.

WOODLOT BIRDS

Great crested flycatcher
Myiarchus crinitus

Description: Medium-sized (length: 22 cm, wingspan: 33 cm). Wings and tail rufous. Throat white, belly and undertail coverts yellow. Face and breast dark gray, wingbars white.

Woodlot Habits: Summer. Found in woodlands, parks, and edges of clearings. Nests 1–23 m above ground in natural cavities or abandoned woodpecker holes in live or dead trees. Will use nest boxes. Nest constructed from twigs, leaves, hair, feathers, bark fibres, rope and other trash. Nests often includes cast-off snakeskin or similar looking piece of cellophane, which is possibly included to deter nest predators. The diet consists of insects (mainly moths, caterpillars, beetles, grasshoppers, and crickets) and fleshy fruits (mainly sassafras).

Eastern kingbird
Tyrannus tyrannus

Description: Medium-sized (length: 22 cm, wingspan: 38 cm). Upperparts dark grey, underparts white. Tail dark with a white tip. Dark crown feathers conceal an orange or red median crown-stripe that is only exposed during displays.

Woodlot Habits: Summer. Found in open woods and forest edges. Nests 3–6 m above the ground on tree limbs away from the trunk and often over water. The nest is a bulky cup constructed from weed stalks, grasses, and mosses and lined with fine grasses. The diet consists of insects (honeybees, ants, grasshoppers, and various beetles) and fruits (sassafras, dogwood, and wild cherry).

WOODLOT BIRDS

Ruby-throated hummingbird
Archilochus colubris

Description: Very small (length: 10 cm, wingspan: 11 cm). Bill thin. Upperparts shiny green, underparts white. Males with red throat. Can hover in flight (wings blurred).

Woodlot Habits: Summer. Found in mixed woodlands. Nests 3–6 m above ground in a variety of trees. Nests attached to twigs or small branches that slant downward from the tree, sheltered above with leafy branches. Nests constructed from plant down, fibres, and bug scales, lined with soft plant down, and covered outside with greenish-grey lichens. The diet consists of nectar and insects (small flies, ants, bees, and beetles).

Tree swallow
Tachycineta bicolor

Description: Small (length: 15 cm, wingspan: 37 cm). Tail short and forked. Upperparts bright greenish-blue, underparts bright white. Juveniles with grey-brown upperparts and pale greyish breastband.

Woodlot Habits: Summer. Found in wooded swamps and open woods. Nests usually in tree cavities and old woodpecker holes. Will use nest boxes. The nest is constructed from the accumulation of dry grass and lined with feathers. The diet consists of insects (mainly flies, beetles, ants, bees, and wasps) and fruits (bayberry and wax-myrtle).

WOODLOT BIRDS

Red-eyed vireo
Vireo olivaceus

Description: Small (length: 15 cm, wingspan: 25 cm), relatively long-billed. Eyes red with black eyeline. Eyebrow white with a black band above eyebrow. Upperparts green, underparts white.

Woodlot Habits: Summer. Found in open deciduous woods with thick undergrowth, sometimes in mixed woods. Nests 1.5–3 m above ground on horizontal forks of slender trees. Nest is a deep-cupped hanging structure constructed from grasses, paper, bark strips, rootlets, and vine tendrils, decorated with lichens. The diet consists of insects (caterpillars and moths, beetles, ants, and bees) and fleshy fruits (dogwood and virginia-creeper).

Warbling vireo
Vireo gilvus

Description: Small (length: 14 cm, wingspan: 22 cm). Eyes dark with light eyebrows and faint eyeline. Upperparts grayish, underparts white to yellow. Overall, plain in colouration.

Woodlot Habits: Summer. Found in open, mixed, or deciduous woods. Nests 6–27.5 m above ground on horizontal forks of slender branches away from trees, often on broadleaved trees such as poplar. Nest is a neat cup closely built to the branch, constructed from bark strips, leaves, grasses, feathers, and plant down, woven with spider webs, and lined with fine plant stems and horsehair. The diet consists of insects (mostly caterpillars, followed by beetles) and fruits (bunchberry and cherry).

WOODLOT BIRDS

Blue jay
Cyanocitta cristata

Description: Broad-winged and relatively large (length: 28 cm, wingspan: 40 cm). Head with crest on top and dark band around neck. Upperparts blue with black and white markings. Underparts white.

Woodlot Habits: Resident. Nests 3–7.5 m above ground, well hidden within trees or the outer branch of coniferous or deciduous trees. Nest constructed from thorny twigs, bark, mosses, string, and leaves and lined with rootlets. The diet consists of large insects (caterpillars, grasshoppers, and beetles) and acorns, or occasionally larger animals (frogs and mice).

Black-capped chickadee
Poecile atricapilla

Description: Very small (length: 13 cm, wingspan: 20 cm). Head large, throat and cap black, cheeks white. Upperparts gray, underparts white to cream-coloured.

Woodlot Habits: Resident. Found in coniferous and deciduous forests and rural woodlands. Nests 1–3 m above ground in natural cavities or holes in the soft, rotting wood of dead stubs. Nest lined with wool, hair, fur (rabbit), mosses, feathers, insect cocoons, and cottony fibres. Winter diet consists of insects (moth eggs, aphids, and katydids) and spiders. Summer diet consists of insects (mainly moths, caterpillars, beetles, and wasps), spiders, seeds (sunflower), and berries (blueberry, serviceberry, and bayberry).

WOODLOT BIRDS

Red-breasted nuthatch
Sitta canadensis

Description: Very small (length: 11 cm, wingspan: 22 cm). Males with black cap and eyeline, females with grayish cap and eyeline. Upperparts bluish, underparts pale to light orange.

Woodlot Habits: Resident. Found in coniferous and mixed forests. Nests 4–5 m above ground in excavated cavities in rotten stubs or branches of dead trees. Dabs pitch around hole to prevent insects and small animals from entering. Nest lined with bark shreds, grasses, mosses, and feathers. The diet consists of insects (common bark inhabitants) and seeds (pine).

White-breasted nuthatch
Sitta carolinensis

Description: Small (length: 15 cm, wingspan: 28 cm). Bill upturned, tail short and broad. Males with black cap, females with grey cap. Face white. Upperparts blue, underparts white.

Woodlot Habits: Resident. Found in open and mixed woodlands with mature trees (most often oaks and pines). Nests 4.5–15 m above ground in natural cavities of trees. Nest lined with bark shreds, twigs, grasses, rootlets, fur, and hair. The diet consists of insects (weevils, ants, grasshoppers, and moths), spiders, and nuts (particularly acorns).

WOODLOT BIRDS

American crow
Corvus brachyrhynchos

Description: Large (length: 44 cm, wingspan: 99 cm). Tail short and rounded, wings broad. Plumage black and eyes dark.

Woodlot Habits: Resident. Found in forests and farm woodlots. Nests 3–21 m above ground, close to the trunks of deciduous or coniferous trees. Forages mainly on the ground. The nest is a basket of twigs, bark, and vines, lined with shredded bark fibres, mosses, grasses, feathers, fur, hair, roots, and leaves. The diet consists of insects (grasshoppers, beetles, carrion, and caterpillars), crustaceans, amphibians, reptiles, agricultural crops (corn and wheat), and fruit (mulberry and cherry).

Brown creeper
Certhia americana

Description: Very small (length: 13 cm, wingspan: 20 cm). Tail long and thin, bill thin and curved. Upperparts with mottled brown plumage, underparts white.

Woodlot Habits: Resident. Found in mature coniferous, deciduous, or mixed woods. Nests 1.5–4.5 m above ground behind loose slabs of bark still attached to living or dead trees. Nest constructed from twigs, leaves, and bark shreddings and lined with finer bark shreds, grasses, feathers (seldom), and mosses. The structure of nest depends on where it is placed, although the centre is neatly cupped with a half-moon shape. Brown creepers forage by clinging to tree trunks, spiralling upward searching for insects in crevices of the bark. The diet consists of spiders, insects (beetles, caterpillars, ants, and small grasshoppers), and seeds (pine and corn).

WOODLOT BIRDS

House wren
Troglodytes aedon

Description: Very small (length: 12 cm, wingspan: 15 cm). Beak long and thin. Tail long, sometimes held vertically. Plumage generally brown with faint barring. Face with weakly marked pattern.

Woodlot Habits: Summer. Found in open forests. Nests in old woodpecker holes, birdhouses, and natural cavities of trees, stubs, and fenceposts. The male establishes the territory first and builds a dummy nest of twigs. If the female accepts the chosen location, she builds a cup of grasses, plant fibres, rootlets, feathers, and hair on the male's pile of twigs. The diet is mainly composed of spiders and insects; fruits and seeds are consumed to a limited extent.

Winter wren
Troglodytes troglodytes

Description: Very small (length: 10 cm, wingspan: 14 cm) Beak thin. Tail short. Plumage generally dark brown with faint barring. Eyebrow light.

Woodlot Habits: Resident. Found in coniferous forests or heavily wooded swamps. Nests in well-hidden cavities in upturned roots of fallen trees, under stumps or live tree roots, and in old woodpecker holes, mossy hummocks, or rock crevices. For the nest, the cavity is filled with mosses, grasses, weed stems, fine twigs, and rootlets and lined with hair and feathers. The diet is mainly composed of spiders and insects; fruits and seeds are consumed to a limited extent.

WOODLOT BIRDS

Hermit thrush
Catharus guttatus

Description: Medium-sized (length: 17 cm, wingspan: 29 cm). Upperparts brown, underparts white with smudgy dark spots. Tail reddish.

Woodlot Habits: Migration. Found in open, brushier habitats of coniferous or mixed forests. Nest typically on the ground, well hidden under trees, bushes, or ferns. Nest is a bulky structure of twigs, bark fibres, grasses, ferns, and mosses, lined with pine needles, plant fibres, and rootlets. The diet consists of insects (beetles, ants, caterpillars, and flies) and an assortment of fruits (holly, grape, sumac, serviceberry, and greenbrier).

Wood thrush
Hylocichla mustelina

Description: Medium-sized (length: 20 cm, wingspan: 33 cm). Upperparts rich reddish-brown, underparts white with large dark spots.

Woodlot Habits: Summer. Found in shady woods (mainly deciduous) with leafy understory. Prefers cool, humid forests. Nests 2–15 m above the ground in forks or horizontal limbs of trees. Nest structure is a firm, compact cup constructed from grasses, bark, mosses, paper, and mud, lined with rootlets. Forages mainly on the ground. The diet consists of insects (beetles, ants, grasshoppers, crickets, and caterpillars), spiders, and fruits (spicebush, dogwood, cherry, grape, blackberry, blackgum, mulberry, and virginia-creeper).

WOODLOT BIRDS

American robin
Turdus migratorius

Description: Large (length: 25 cm, wingspan: 43 cm). Males with orange breast and dark gray upperparts. Females with light orange breast and brownish gray upperparts. Beak yellow, eyes with white border.

Woodlot Habits: Resident. Found in a variety of habitats. Nests in shrubs, tree forks and horizontal branches, or on ledges. The nest is a deep cup structure constructed from grasses, weed stalks, strips of cloth, and string, held together by wet or soft mud, and lined with fine grasses. The diet consists of insects (caterpillars and beetles, particularly ground beetles, weevils, and dung beetles), earthworms, and fruits (cherry, dogwood, sumac, blackgum, grape, virginia-creeper, and blackberry).

Eastern bluebird
Sialia sialis

Description: Medium-sized (length: 18 cm, wingspan: 33 cm). Male with bright blue upperparts and white underparts with a bright orange breast. Female similar but paler in colour.

Woodlot Habits: Summer. Found in small groups in fields or open woods, often perched on wires or fences. Nests in natural cavities of trees or old woodpecker holes. Will nest in nest boxes. Nest is carelessly arranged and forms a loosely built cup. Nest is constructed from fine grasses and weed stalks. The diet consists of insects (beetles, grasshoppers, crickets, and caterpillars) and fleshy fruits (dogwood, red cedar, sumac, bayberry, and virginia-creeper).

WOODLOT BIRDS

Cedar waxwing
Bombycilla cedrorum

Description: Medium-sized (length: 18 cm, wingspan: 30 cm). Tail short, wings pointed. Bill stubby. Face with a black mask and a brown crest. Both belly and tip of tail yellow. Adults with red 'drops' on wings.

Woodlot Habits: Resident. Found in open woods (avoids dense woods). Nests 2–6 m above ground on horizontal limbs of trees. Nests constructed from loosely woven grasses, twigs, weed stems, cottony fibres, string, and yarn and lined with rootlets, fine grasses, and plant down. The small amount of space required for nesting territory allows for colonial nesting.

Bohemian waxwing
Bombycilla garrulus

Description: Medium-sized (length: 21 cm, wingspan: 37 cm). Small-headed and round-bodied. Similar in appearance to cedar waxwing but wings with yellow stripe and white patches. Also, undertail chestnut brown.

Woodlot Habits: Winter. The diet consists of berries supplemented by insects.

WOODLOT BIRDS

European starling
Sturnus vulgaris

Description: Medium-sized (length 22 cm, wingspan 41 cm). Tail short, wings pointed and triangular, bill long, straight, and pointed. Plumage black with light speckles, summer plumage with fewer speckles.

Woodlot Habits: Resident. Found in wooded farmlands. Nests 3–7.5 m above ground in cavities with preference to natural cavities in trees or new woodpecker holes. Nests solitary or in colonies. Nest composed of grasses, weed stems, twigs, corn husks, dried leaves, cloth, and feathers and lined with fine grasses and feathers. The diet consists of insects (beetles, grasshoppers, millipedes, and caterpillars) and fruits (cherry, sumac, bayberry, mulberry, and elderberry).

Northern shrike
Lanius excubitor

Description: Large (length: 25 cm, wingspan: 37 cm). Bill hooked, wings and tail relatively long. Face with dark mask. Wings and tail black. Adult with gray upperparts, juveniles with brown upperparts.

Woodlot Habits: Winter. Found in grasslands or open forests. Nests 2–16 m above ground in deciduous or coniferous trees. The surrounding branches can be decorated with mistletoe or common vines. The diet is almost exclusively animal life. Large insects such as grasshoppers, beetles, caterpillars, and wasps are prominent in the diet. Small rodents, birds, and large invertebrates are also consumed. Prey can be found impaled and left hanging on barbed wires or thorns or wedged between forked branches.

WOODLOT BIRDS

Grey catbird
Dumetella carolinensis

Description: Medium-sized (length: 22 cm, wingspan: 28 cm). Tail long. Uniform dark gray, distinctive. Head with black cap. Undertail coverts reddish-brown.

Woodlot Habits: Summer. Found in woodland undergrowth. Nest built 1–3 m above ground in dense thickets, briars, vine tangles, shrubs, or low trees. Nest constructed from twigs, grapevines, leaves, grasses, paper, and weed stems and neatly lined with rootlets. The diet consists of insects (ants, beetles, caterpillars, and grasshoppers) and fleshy fruits (blackberry, cherry, elderberry, greenbrier, and grape).

Brown thrasher
Toxostoma rufum

Description: Large (length: 29 cm, wingspan: 33 cm). Bill long, tail very long. Upperparts reddish brown, underparts white with dark streaks.

Woodlot Habits: Summer. Found in dense, tangled thickets or woodland edges. Nest built on ground or 0.5–2 m above ground in trees, shrubs, or vines. Nest constructed from dry leaves, small twigs, grass stems, grapevines, or inner bark and lined with rootlets. Forages on the ground, turning over leaves and debris. The diet consists of insects (mainly beetles, followed by grasshoppers, crickets, and caterpillars), spiders, and fruits (blackberry, cherry, and elderberry).

WOODLOT BIRDS

Yellow warbler
Dendroica petechia

Description: Very small (length: 13 cm, wingspan: 20 cm). Tail short. Eyes dark. Generally has yellow body, with males having reddish streaks down the breast.

Woodlot Habits: Summer. Found in low trees and woodland edges, especially willows in wet areas. Nests built in trees 1–2.5 m above ground, often in colonies in ideal habitat. Nest is a strong compact cup composed of milkweed fibres, hemp, grasses, and plant down and lined with plant down, hair, and fine grasses. The diet consists mainly of insects (beetles, and caterpillars for nestlings) and some berries.

Pine warbler
Dendroica pinus

Description: Small (length: 14 cm, wingspan: 22 cm). Head round, bill stout. Cheeks dark, neck patch distinctive and pale. Upperparts olive-green with white wingbars, underparts yellow. Male streaked on sides of breast.

Woodlot Habits: Summer. Found in open pine and mixed woods. Nests built 9–15 m above ground on horizontal limbs of pine trees away from the trunk. Nest inconspicuous from below, compactly made from weed stems, bark strips, pine needles, pine twigs, and spider webbing and lined with fern down, hair, pine needles, and feathers. The diet consists of insects (ants, beetles, caterpillars, and grasshoppers) and seeds (pine).

WOODLOT BIRDS

Common yellowthroat
Geothlypis trichas

Description: Very small (length: 13 cm, wingspan: 17 cm). Bill small, tail rounded. Underparts yellow and grey. Male face has black mask bordered with white band above. Female duller in colour and with no mask.

Woodlot Habits: Summer. Found in areas with dense low cover, brushy thickets, wet bottomlands, or swamps. Nest securely built on or near ground in surrounding vegetation. Nest constructed from coarse grasses, reed shreds, leaves, and mosses and lined with grasses, bark fibres, and hair.

Indigo bunting
Passerina cyanea

Description: Very small (length: 13 cm, wingspan: 20 cm). Tail short, bill small. Male bright blue, darker blue on head. Female light brown with faint streaks on the breast and blue highlights on the tail.

Woodlot Habits: Summer. Found in open brushy fields, roadside thickets, and forest edges; avoids mature forests. Nest built 0.5–3.5 m above ground in dense cover (bushes, shrubs, low trees, or tangle of blackberries). Nest is constructed from dried grasses, bark strips, and twigs and lined with fine grasses, cotton, rootlets, and occasionally hair and feathers. The diet consists largely of insects (beetles, caterpillars, and grasshoppers), followed by seeds (ragweed, bristlegrass, and farm grains) and berries (blackberry and elderberry).

WOODLOT BIRDS

Northern cardinal
Cardinalis cardinalis

Description: Medium-sized (length: 22 cm, wingspan: 30 cm). Bill large and triangular, red or orange. Crest on head and face with black mask. Male red, females brown with red highlights.

Woodlot Habits: Resident. Found in thickets and forest edges; avoids deep forests. Nest generally built less than 3 m above ground in dense shrubbery, small deciduous or coniferous trees, thickets, or vines. Nest is loosely built of twigs, vines, some leaves, bark strips, grasses, weed stalks, and rootlets and lined with fine grasses and hair. The diet consists of insects (caterpillars, grasshoppers, and beetles), wild fruits (grape and mulberry), weed seeds (smartweeds and sedges), and cultivated grains (corn and oats).

Rose-breasted grosbeak
Pheucticus ludovicianus

Description: Medium-sized (length: 21 cm, wingspan: 32 cm). Head and bill large. Bill pink and breast coarsely streaked. Underwing coverts buffy yellow or pink. Male plumage black and white with red breast. Female similar in appearance to a large sparrow.

Woodlot Habits: Summer. Found in moist deciduous second-growth woods or thickets. Nests built 3–4.5 m above ground in forks of deciduous trees or shrubs (sometimes conifers). The nest is flimsy and built using twigs. The diet consists of insects (mainly beetles, followed by ants, bees, wasps, and caterpillars) and fruits (elderberry and cherry).

WOODLOT BIRDS

Fox sparrow
Passerella iliaca

Description: Medium-sized (length: 18 cm, wingspan: 27 cm). Upperparts red, underparts white with large reddish spots. Breast with central spot.

Woodlot Habits: Migration. Found in wooded areas. Nests built close to the ground or in low shrubs and trees. The diet consists of insects (mainly ground beetles and millipedes), fruits (blackberry), and extensively of weed seeds (smartweed and ragweed).

Song sparrow
Melospiza melodia

Description: Small (length: 16 cm, wingspan: 21 cm). Tail fairly long, head rounded. Upperparts brown, underparts white with brown spots and coarsely streaked. Breast with central spot.

Woodlot Habits: Summer. Found in brushy fields, thickets, and woodland edges. Nests built hidden on the ground under tufts of grasses or in a low bushes or trees 0.5–1 m above ground. Nests constructed from grasses, weed stems, leaves, and bark fibres and lined with fine grasses, rootlets, and hair. The diet consists of insects (beetles, grasshoppers, crickets, caterpillars, and ants) and weed seeds (smartgrass, bristlegrass, ragweed, and panicgrass).

WOODLOT BIRDS

White-throated sparrow
Zonotrichia albicollis

Description: Medium-sized (length: 17 cm, wingspan: 23 cm). Wings rufous. Head with either black and white stripes or brown and tan stripes. Eye with yellow spot in front, throat white.

Woodlot Habits: Migration. Found in undergrowth or edges of coniferous and northern deciduous forests. Nests built close to the ground in grass hummocks, in flat tree branches, or in mats of dead bracken fern or dead grasses. Nest constructed of coarse grasses, rootlets, pine needles, twigs, bark fibres, and mosses and lined with fine grasses, rootlets, and hair. The diet consists of insects (ants, beetles, flies, caterpillars) and weed seeds (mainly ragweed and smartweed).

White-crowned sparrow
Zonotrichia leucophrys

Description: Medium-sized (length: 18 cm, wingspan: 24 cm). Adult with black and white stripes on head. Juveniles with rusty and cream stripes. Bill pale pink or pale yellow.

Woodlot Habits: Migration. Found in brushy areas to breed. Nests built low in bushes or close to the ground under shrubs. The diet consists of insects (parasitic *Hymenoptera*, ants, caterpillars, beetles, and grasshoppers), grains (oats), and weed seeds (bristlegrass, panicgrass, and smartweed).

WOODLOT BIRDS

Chipping sparrow
Spizella passerina

Description: Very small (length: 14 cm, wingspan: 22 cm). Tail long. Head with reddish cap. Eyeline black, brownish in fall. Eyebrow white. Breast and rump grayish in colour. Eyeline and cap brownish in winter.

Woodlot Habits: Summer. Found in open woodlands and conifer plantings. Nests 1–3 m above ground in trees (often conifers), shrubs, or vines. Nests constructed of fine grasses, weed stalks, and rootlets, lined with hair and fine grasses. The diet consists of a variety of insects (grasshoppers, caterpillars, beetles, ants, and wasps), and seeds (crabgrass, bristlegrass, and panicgrass).

American tree sparrow
Spizella arborea

Description: Small (length: 16 cm, wingspan: 24 cm). Head with reddish cap and eyeline. Bill bicoloured. Wingbars white, rump brown. Breast with central dark spot.

Woodlot Habits: Winter. Breeds in the tundra and subarctic. Weed seeds (bristlegrass, crabgrass, panicgrass, and sedges) constitute more than nine-tenths of the diet. The small animal diet consists of small insects (beetles and ants).

WOODLOT BIRDS

Dark-eyed junco
Junco hyemalis

Description: Small (length: 16 cm, wingspan: 24 cm). Bill pink. Underparts and outer tail feathers white. Males with gray upperparts, females with brownish-gray upperparts.

Woodlot Habits: Winter. Found in coniferous or mixed forests and forest edges. Nests commonly on the ground, hidden under weeds and grasses, in tree roots, or under fallen trees and logs. Nest is built as a compact structure of grasses, rootlets, bark shreds, mosses, and twigs and lined with finer grasses, rootlets, and hair. The diet consists of insects (caterpillars, beetles, and ants), and weed seeds (ragweed, bristlegrass, crabgrass, and dropseedgrass).

Pine siskin
Carduelis pinus

Description: Very small (length: 13 cm, wingspan: 23 cm). Upperparts streaked brown, underparts white with dark streaks. Males with yellow wingbars.

Woodlot Habits: Winter. Found in coniferous and mixed forests. Nests built 2–10 m above ground, concealed in horizontal conifer branches away from the trunks of pine trees. Nests are solitary or in loose colonies. Nest is a shallow cup constructed from twigs, grasses, mosses, lichens, bark strips, and rootlets, lined with mosses, rootlets, hair, fur, and feathers. The diet consists of insects (caterpillars and aphids), spiders, and seeds (pine and alder).

WOODLOT BIRDS

American goldfinch
Carduelis tristis

Description: Very small (length: 13 cm, wingspan: 23 cm). Upperparts brown, underparts light gray. Wings dark with wingbars. Breeding male yellow with black cap.

Woodlot Habits: Resident. Found in overgrown fields with scattered trees. Nest built 0.5–10 m above the ground in the fork of a tree. Nest is stout, woven with fine vegetable fibres, and lined with thistle and cattail down. Insects (aphids and caterpillars) make a small portion of the diet. The primary diet consists of seeds (ragweed, thistle, shepherd's purse, and sweetgum).

Common redpoll
Carduelis flammea

Description: Very small (length: 13 cm, wingspan: 23 cm). Head with small red cap. Face dark, bill yellow. Upperparts brown and streaked. Underparts streaked. Male with pink breast, females with white breast.

Woodlot Habits: Winter. Found in woodlands, usually with birch trees. The diet is almost entirely vegetarian during the winter, specifically weed seeds (ragweed, smartweed, goosefoot, and pigweed) and seeds of alder and birch. In the summer, the diet includes insects (ants and flies) and spiders.

WOODLOT BIRDS

Purple finch
Carpodacus purpureus

Description: Small (length: 15 cm, wingspan: 25 cm). Male with extensive red colour on head and back, appearing washed with raspberry. Female with boldly patterned face. Upperparts of female brown, underparts white with blurry streaks, eyebrows white.

Woodlot Habits: Resident. Found in coniferous and mixed forests or roadside conifers. Nests built on horizontal branches of conifer trees, 1.5–18 m above ground, well hidden. Nest is a shallow cup comprised of grasses, twigs, weed stems, bark strips, and rootlets, lined with fine grasses and hair. Insects contribute minimally to the diet (aphids or caterpillars). The diet mainly consists of buds, fruits, and seeds (elm, tuliptree, apple, cherry, and peach).

House finch
Carpodacus mexicanus

Description: Small (length: 15 cm, wingspan: 24 cm). Forehead, throat, and rump of male mostly red. Female gray-brown with blurry streaks and weak face pattern.

Woodlot Habits: Resident. Found in open woods. Nests built in tree cavities, constructed with twigs, grasses, and debris. Insects contribute minimally to the diet (aphids or caterpillars). Although they consume cultivated foods, the main diet consists of weed seeds (filaree, mustard, knotweed, and pigweed).

WOODLOT BIRDS

Purple vs. House Finch

These finches tend to give beginners some trouble. Try to think of the male purple finch as a bird dipped in raspberry juice, whereas the male house finch's colour is redder and is mostly confined to the forehead, throat and rump. The male house finch also has more streaking on its underparts. For the females, the purple finch is bright white below with dark streaks. It also has a distinctive white eyebrow. The female house finch has creamy underparts with streaks and no white eyebrow; it is not as crisp-looking as the female purple finch.

Purple   House

Evening grosbeak
Coccothraustes vespertinus

Description: Medium-sized (length: 20 cm, wingspan: 36 cm), beak large. Male with yellow eyebrow and belly and black and white wings. Females duller in colour with hints of yellow.

Woodlot Habits: Winter. Found in coniferous forests. Nest built in conifers 6–18 m above the ground, occasionally in deciduous trees. Nest is a frail, loosely constructed cup made from twigs interwoven with mosses and lichens. The diet is almost exclusively fruits (cherry) and tree seeds (maple and dogwood). The diet also consists of a small portion of insects (mainly beetles and caterpillars).

509

WOODLOT BIRDS

Pine grosbeak
Pinicola enucleator

Description: Medium-sized (length: 23 cm, wingspan: 37 cm). Male plumage bright pink and grey. Female and young male plumage grey with orange-yellow head. Wingbars white.

Woodlot Habits: Winter. Found in spruce forests. Nests built low in conifer trees or underbrush of coniferous forests. Nest is constructed with mosses, twigs, grasses, and lichens and lined with hair. The diet consists in part of insects (beetles, caterpillars, and *Hymenoptera*) and invertebrates, but they feed largely on fruits (mountain ash, pine, maple, and blackberry) and occasionally weed seeds (ragweed).

Baltimore oriole
Icterus glabula

Description: Medium-sized (length: 22 cm, wingspan: 29 cm). Wings black with white wingbars. Male plumage black and orange. Female and young male plumage orange-yellow without black head.

Woodlot Habits: Summer. Found in the edges of mixed and deciduous forests to breed. Nests built 7.5–9 m above the ground in elms, maples, willows, and apple trees. Nest is an intricately woven deep pouch of milkweed fibres, hair, yarn, string, and grapevine bark, lined with hair, wool, fine grasses, and cottony materials.

WOODLOT BIRDS

Rusty blackbird
Euphagus carolinus

Description: Medium-sized (length: 23 cm, wingspan: 36 cm). Eyes light with dark patch around eyes. Plumage black or dark grey. Plumage rusty-coloured with brownish eyebrow during fall and winter.

Woodlot Habits: Migration. Found in marshes bordered with trees and swampy woodlands. Nests are solitary, built 0.5–6 m above water or the ground, in thick growth of evergreen trees (balsam and spruce). The nest is bulky in structure, made with twigs, lichens, leaves, grasses, and rotting plant material and lined with grasses and fine twigs. The animal diet consists of aquatic beetles (and larvae), grasshoppers, and caterpillars. The plant diet consists of grains (corn, wild rice, oats, and wheat) and weed seeds (bristlegrass).

Red-winged blackbird
Agelaius phoeniceus

Description: Medium-sized (length: 22 cm, wingspan: 33 cm). Male plumage black with red with yellow coverts. Female brown and streaked.

Woodlot Habits: Summer. Found in swamps, streamside bushes, and dry fields. Nests built in loose colonies, 7 cm–4 m above the ground. Nests placed near or over water in cattails, rushes, sedges, reeds, or bushes (alders and willows). Nest constructed with sedge leaves, grasses, rushes, rootlets, and mosses, bound with milkweed fibres, and lined with fine grasses. The diet consists of insects (weevils, beetles, caterpillars, grubs, cankerworms, grasshoppers, and ants), weed seeds (ragweed, bristlegrass, and smartweed), and farm crops (corn and oats).

511

WOODLOT BIRDS

Common grackle
Quiscalus quiscula

Description: Large (length: 32 cm, wingspan: 43 cm). Tail keeled and relatively long. Bill long, heavy, and sharp. Plumage dark with purple, blue, and bronze iridescence. Eyes yellow.

Woodlot Habits: Summer. Found in marshes, swampy thickets, and conifer groves. Nests built in small colonies in coniferous trees up to 18 m above ground. Also nests in natural cavities and cattail marshes. Nest is a loose bulky structure of weed stalks, grasses, and debris, strengthened with mud, and lined with grasses, feathers, and fine debris. The diet consists of insects (grasshoppers, bees, and crickets), and farm crops (corn, oats, and wheat).

Brown-headed cowbird
Molothrus ater

Description: Medium-sized (length: 19 cm, wingspan: 30 cm). Male plumage black with green iridescence, head brown. Female plumage light brown with whitish throat.

Woodlot Habits: Summer. Found in open deciduous and coniferous woods and forest edges. A parasitic species: lays eggs in nests of other birds (predominantly warblers and sparrows). Robins and catbirds destroy and remove the eggs. Other birds (like some vireos and warblers) may cover the egg with additional nest floor. The diet mainly consists of insects (mostly grasshoppers) and weed seeds (bristlegrass, ragweed, and oats).

WOODLOT BIRDS

WOODLOT MAMMALS

Woodlot Mammals

Chris Earley, Brian Lacey & Steven Newmaster

Introduction
Our fellow mammals are fascinating creatures. Mammals all have hair and feed their young milk, but they come in many different shapes and sizes, each specialized to fill their niche in the forest community. Many woodlot mammals are described here, but you may see signs of others using woodlots as habitat. Unfortunately for us, many woodlot mammals are nocturnal, meaning they are only active at night. But if you can learn their tracks, you can still discover what each mammal has been up to and thus learn about its behaviour and lifestyle.

Through the vast majority of human history our survival was heavily dependent on the ability to effectively track animals. Out of necessity, 'primitive' hunters raised tracking to an art form, gathering valuable information from extremely subtle features of tracks. Based on an animal's tracks alone, hunters could determine its weight, sex and age, how long ago it passed, and where its attention was focused as it moved. Even an animal's thoughts could be read in the hesitations, stutters, and sudden turns which reveal themselves in the tiniest ripples of its tracks. This seems hard for us to imagine because we have lost the ability to read the elaborate language written by animals as they move through their world. Those who know how to read it mystify us, just as your ability to draw meaning from this text would mystify someone with no understanding of the written word. But rest assured that the language is there for you to learn, speaking volumes about the lives of all the woodlot creatures.

If you visit a woodlot after a snowfall you will encounter a multitude of mammal tracks providing evidence of hunting, feeding, rearing young, or even territorial disputes. So, get your Sherlock Holmes hat on and try to sleuth out if any woodlot mammals visited your nearest woodlot last night.

Tracks
The consistency and depth of the snow (or mud) can make tracks quite variable. For example, the photo on the right (below) shows the details of a cottontail's toes, whereas the photo on the left doesn't. Arrows show the direction the mammal was travelling.

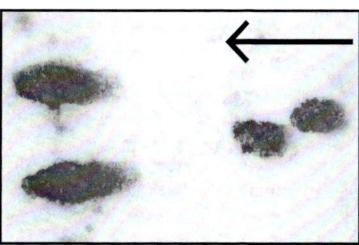 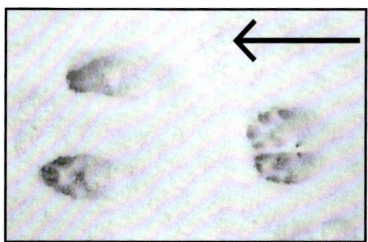

WOODLOT MAMMALS

On individual tracks, note the shape, size (W – width, L – length), number of toes (F5, H4 – front foot has five toes, hind foot has four), and presence of claw marks. On trails, note the width and pattern.

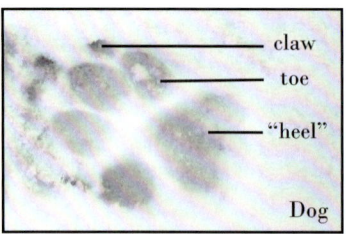
Dog

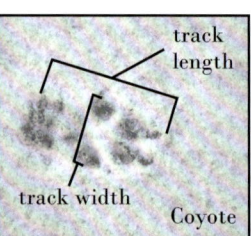
Coyote

Grey squirrel

We consider a 'trail' to be the pattern that a mammal leaves as it moves along. There are 5 main categories of trails. The focus here is on the 'regular' gait of the mammal, not when it is running at full speed or only snuffling around slowly.

I. JUMPERS: RABBITS, SQUIRRELS, AND DEER MICE

These are some of the most common tracks that you may come across. These mammals often leave tracks where the hind foot tracks are in front of the front foot tracks. The following cartoon shows how this works.

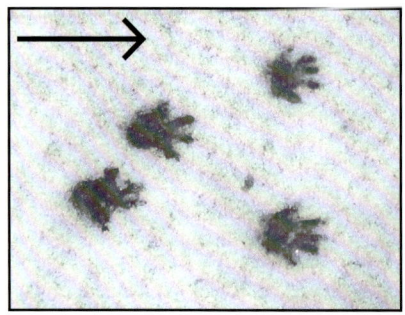

1. A rabbit has just jumped and is landing with its front feet.
2. It then brings its hind feet in front of its front feet by crossing them to the outside of its front legs.
3. Now it springs off of its hind feet for its next jump, leaving its track pattern behind.

WOODLOT MAMMALS

II. WALKERS: COYOTES, FOXES, DOGS, CATS, GROUNDHOGS, DEER, PORCUPINES, OPOSSUMS AND US

These are animals that leave an alternating pattern as they walk. Two-legged humans leave this pattern, but our four-legged friends can also by putting their hind foot in the same spot that their front foot was. Foxes, coyotes and cats tend to register their hind track almost directly over their front track, which makes it appear that a two-legged animal has walked by. Domestic dogs are a bit sloppier, often leaving their hind track beside the front track and, thus, losing a lot of the bipedal effect that their wild cousins have. When any of these animals break into a trot or a run, the alternating pattern is lost and becomes quite variable depending on the species of animal and the speed at which it is going.

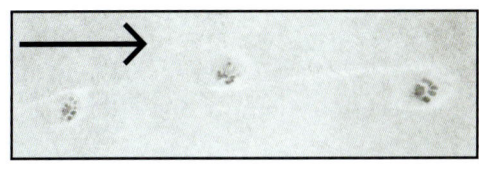

III. LEAPERS: WEASELS AND MINK

A weasel's most common trail pattern is two tracks beside one another. They do this by bringing their hind feet into the same spot that their front feet were.

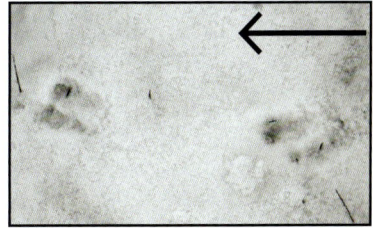

IV. AMBLERS: RACCOONS AND SKUNKS

These animals are short-legged and so they often appear to just 'amble' along. The trail patterns they leave can be quite variable (including a walk), depending on the species and how fast they are going. Raccoons and skunks do a lot of sleeping in the winter, but will often come out on warmer nights in early winter or late winter/early spring.

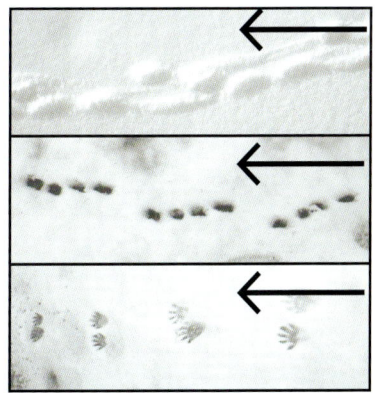

V. TUNNELERS: VOLES AND SHREWS

You might find small tunnels on the snow surface on your next hike. These are usually made by meadow voles (sometimes called field mice) or shrews. Distinguishing between the two can be very difficult. When out of their tunnels, these mammals often move in a walking or leaping pattern but may also leave a furrow in the snow. Red squirrels and mink may leave large tunnels in the snow, too.

WOODLOT MAMMALS

Virginia opossum
Didelphis virginiana

Description: Fur greyish-white. Moderate body size (40–50 cm) with a long tail (25–50 cm). The ears, toes, and tail are bare and pinkish. Appearance is similar to a very large rat.

Habits: Diet of plants (acorns, corn, and fleshy fruits) insects, aquatic invertebrates, small mammals and birds, carrion, and eggs. The saying 'play possum' refers to the possum's habit of feigning death when threatened.

Tracks F5, H5: Front track is star-shaped (4.4–6 cm W), hind track shows a distinctive 'thumb' (4.4–7.2 cm W). The tail may drag or intermittently slap.

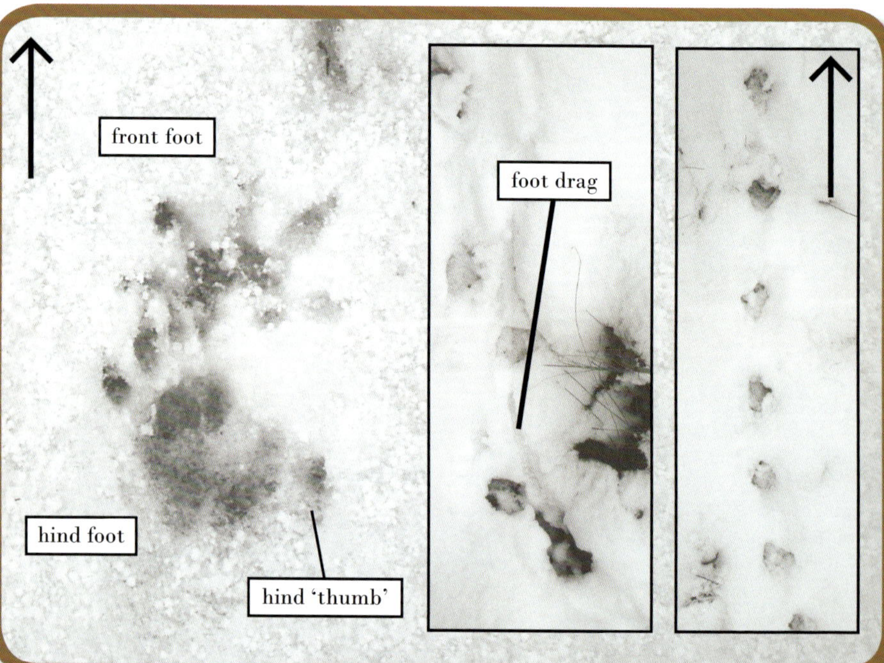

WOODLOT MAMMALS

Northern short-tailed shrew
Blarina brevicauda

teeth

Description: Fur grey. Small body size (7.5–10 cm) with a short tail (2–3 cm). The small eyes and ears are hidden within its fur. Teeth pointed and reddish in colour. The saliva is poisonous.

Habits: Diet consists mainly of insects, but can also include earthworms, larvae, small mice, seeds, and plants.

Tracks F5, H5: Variable trail patterns similar to meadow vole. Sometimes leaves a tail drag and often tunnels under the snow (see also meadow vole, pg. 524). Trail 2.5–3.5 cm W.

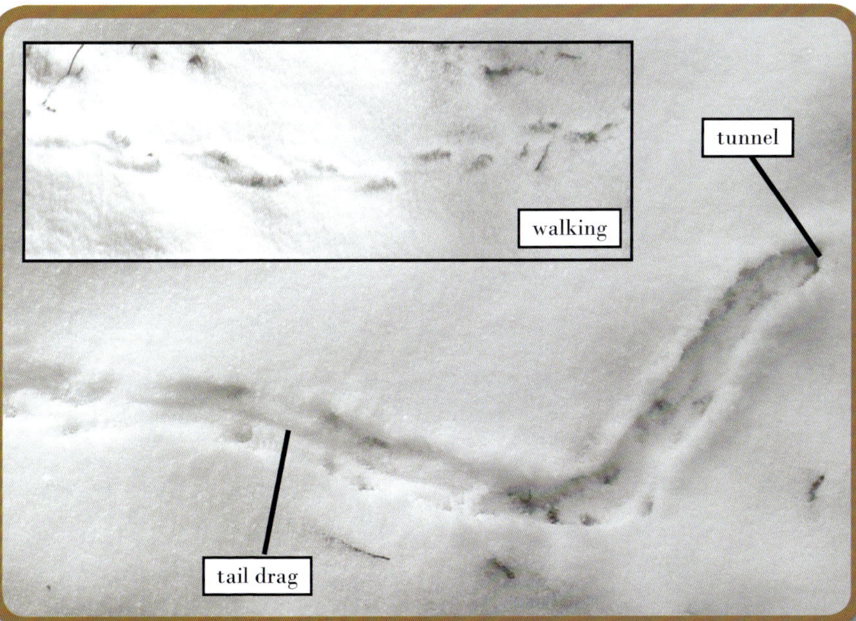

walking

tunnel

tail drag

WOODLOT MAMMALS

Masked shrew
Sorex cinereus

Description: Fur brown above, whitish-grey below. Small body size (5–6.5 cm) with a long tail (3–5 cm).

Habits: Diet consists mainly of insects, but can also include earthworms, larvae, small mice, and plants. Can eat more than its body weight every day.

Tracks F5, H5: In snow, the leaping pattern is most commonly seen. Note the tail drag (not always seen). Trail 1.9–2.5 cm W.

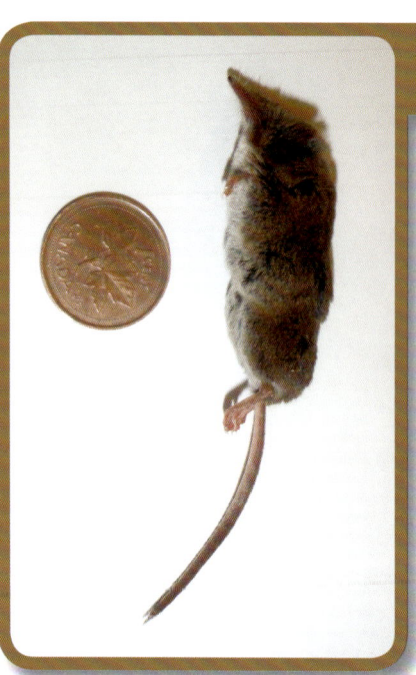

tail drag

tunnel

penny

WOODLOT MAMMALS

Eastern red bat
Lasiurus borealis

Description: Fur reddish, especially on males. Body and tail 9.5–12.5 cm long.

Habits: Summer. This is a migratory bat species that flies south for the winter months. It is rare in our area but when found, it is usually hanging in a bunch of leaves. It eats insects.

Little brown bat
Myotis lucifuga

Description: Fur brown, wings and face dark. The hair on the back is longer with glossy tips. Body and tail length moderate for a bat (8.5 cm). 38 sharp teeth.

Habits: Flies at dusk and at night. Some semi-hibernate, but most migrate south in winter. Echolocates to find prey, catching small insects in its mouth and larger prey with its wingtips, which it then transfers to a cup formed by the tail to be eaten at a later time. Diet of insects along the forest edge including moths, June bugs, flies, mosquitoes, and caddisflies. An adult can sometimes fill its stomach in less than 20 minutes.

WOODLOT MAMMALS

Star-nosed mole
Condylura cristata

Description: Fur black. Body small (11.5-12.5 cm) with a long tail (7.5-9 cm). Large front feet are paddle-shaped for digging. There are 22 finger-like projections on the nose that are extremely sensitive tentacles with touch receptors. These touch receptors quickly identify edible items in the subterranean and nocturnal environments where moles feed.

Habits: Quite gregarious and can be found wandering around at night above the ground. Diet of earthworms and grubs (larvae) but may also include insects, spiders, and centipedes.

Tracks F5, H5: Tunnels deep enough that it does not form large mounds. Although it is difficult to find mole tracks above ground, look for 5-clawed prints with small elongated hind prints and large broad fore prints.

Woodland jumping mouse
Napaeozapus insignis

Description: Dark back, light sides and white underparts. Small body (7.5-10 cm) and very long tail (10-16 cm). Hind feet large.

Habits: Summer only (hibernates for about 6 months). Nocturnal. Can be quite noisy as it hops across the leaves on the forest floor. Eats fungi, seeds, fruit, and insects.

Tracks: F5 (but one usually doesn't register), H5. Front foot has long thin toes that are very spread apart. Trail 3.5-5.5 cm wide. Some jumps can be over 3.5 m long! Not found in snow because this species hibernates.

WOODLOT MAMMALS

North American deer mouse
Peromyscus maniculatus

Description: Fur brown above, white below. Small body size (7–10 cm) with a long tail (5–12 cm). Eyes large. Very similar to white-footed mouse (*Peromyscus leucopus*).

Habits: Stores food, particularly in the fall. The diet is mainly seeds (grass, herbs, maple), nuts (oak), and bark, but also includes fruits (strawberries, blueberries), leaves, and insects.

Tracks F4, H5: A jumper, like a squirrel, but leaves a smaller trail, 2.5–4.5 cm W. Often leaves a tail mark, and is less likely to tunnel than voles and shrews.

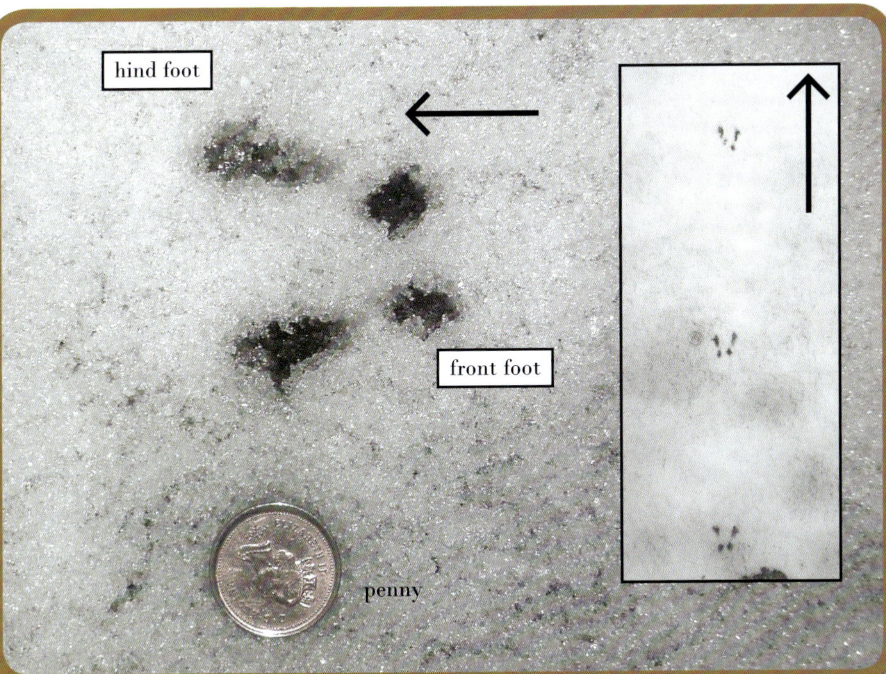

WOODLOT MAMMALS

Meadow vole
Microtus pennsylvanicus

Description: Fur greyish-brown above, lighter below. Small body size (9–12 cm) with short tail (3.5–6.5 cm).

Habits: Can be seen at any time of day and is a good swimmer and fighter. Diet is quite variable including grasses, herbs, tree bark, and insects.

Tracks F4, H5: Tunnels under snow. Less likely to leave a 4 print jump pattern than a deer mouse (previous page). Trail 2.5–5 cm W. See also northern short-tailed shrew (pg. 519).

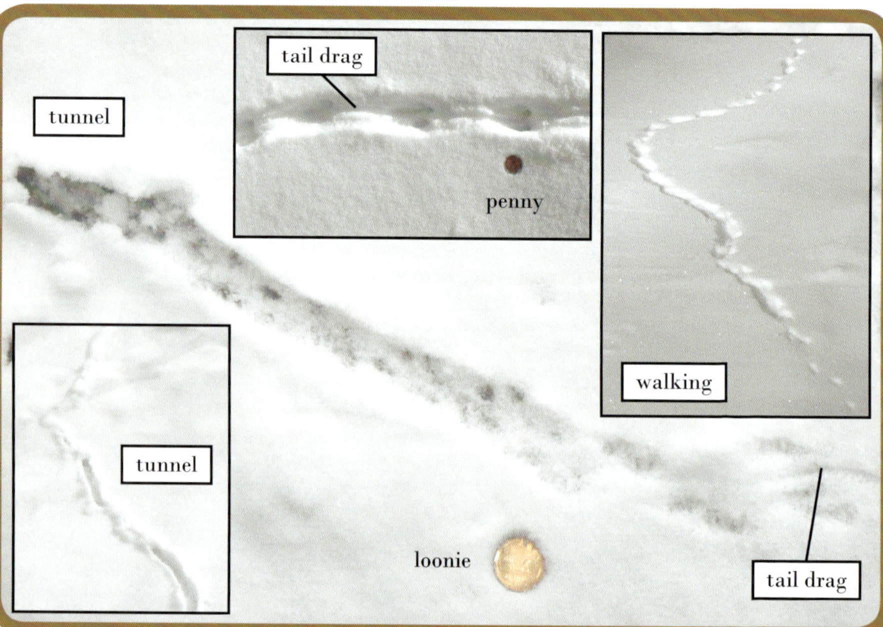

524

WOODLOT MAMMALS

Norway rat
Rattus norvegicus

Description: Fur brown. Small body size (18–25 cm) and long naked tail (13–20 cm). Vision is poor, but long whiskers provide a navigational aid, conveying the locations of nearby objects.

Habits: Introduced. More commonly found in urban woodlots. Diet includes just about everything from insects to other rodents, garbage, and plants.

Tracks F4, H5: Often leave a walking or bounding pattern with a tail drag, particularly in deep snow. Trail 4–8 cm W.

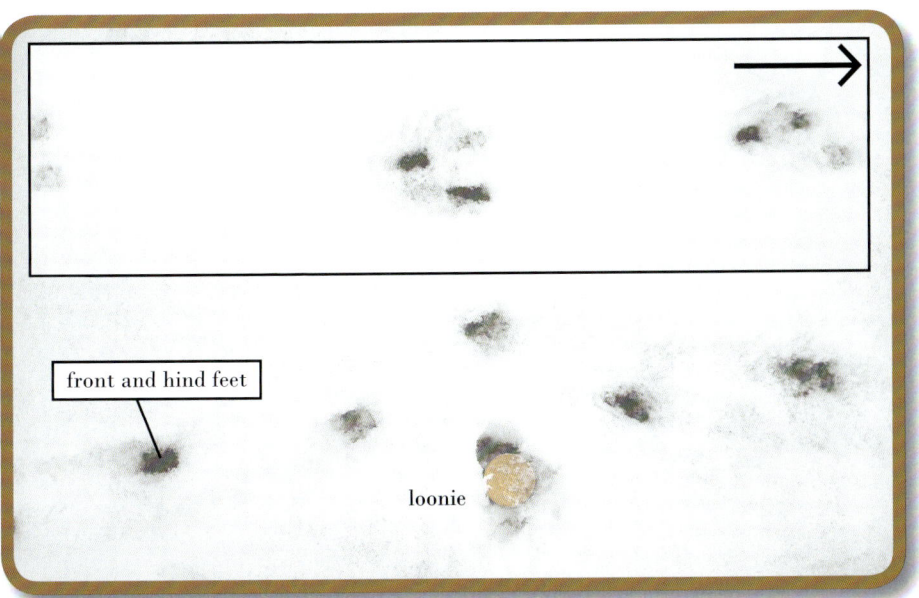

front and hind feet

loonie

WOODLOT MAMMALS

Eastern grey squirrel
Sciurus carolinensis

Description: Fur with two colour phases: 1) grey phase – grey above and white underneath, and 2) black phase – all black. Some individuals have brownish colouration, especially on the belly or tail; some have a white tip on the tail. Small body size (20–25 cm) with fluffy tail (20–25 cm).

Habits: Stores lots of food for winter. The diet is mainly seeds and nuts from trees (e.g., oak, beech, pine, spruce) but also includes insects and fungi.

Tracks: F5 (often only 4 register), H5: Trail 9–14 cm W.

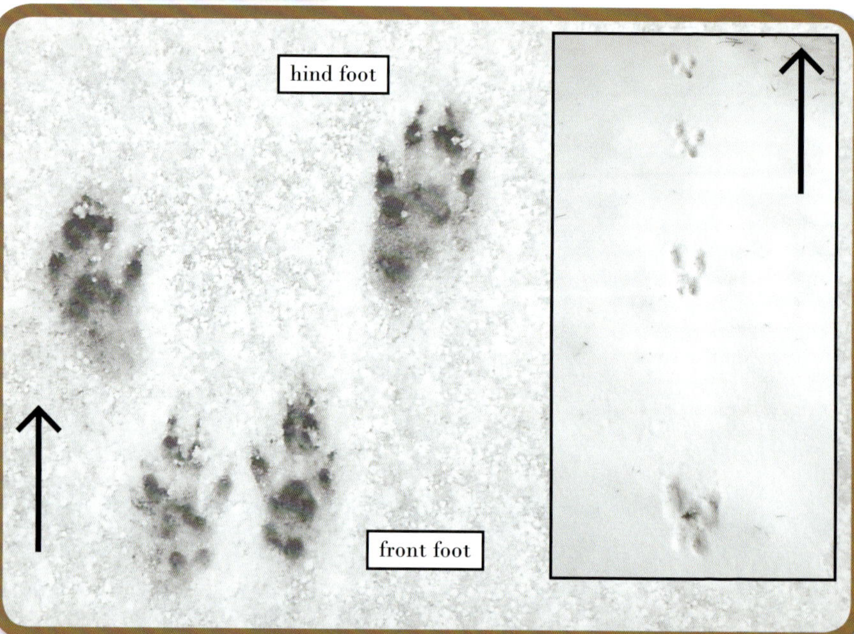

WOODLOT MAMMALS

Red squirrel
Tamiasciurus hudsonicus

Description: Fur reddish on upperparts and white below. Small body size (18–20 cm) with a fluffy tail (10–15 cm). Often the noisiest creature in the forest.

Habits: Diet includes young birds, eggs, insects, seeds (beech, oak, spruce, fir, hemlock), fruits (apple, cherry, serviceberry), and fungi. Some have been seen eating hallucinogenic mushrooms and subsequently displaying strange behaviour.

Tracks F5 (often only 4 register), H5: Trail 7–11 cm W. Sometimes tunnels in the snow. Note: tracks shown are in slushy snow.

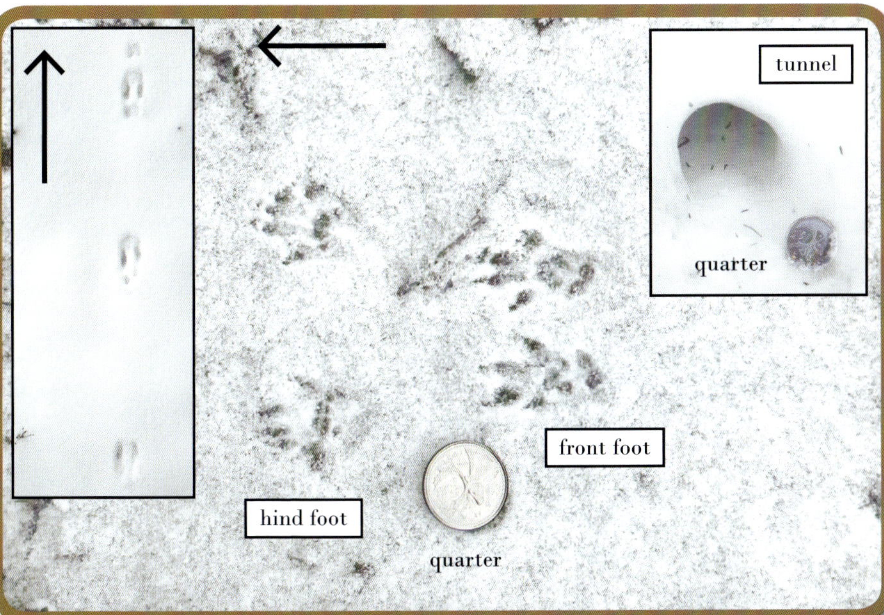

WOODLOT MAMMALS

Eastern chipmunk
Tamias striatus

Description: Fur rusty brown with black and white back stripes. Small body size (12–15 cm) and long fluffy tail (7.5–10 cm).

Habits: Hibernates (usually no tracks in the winter), storing lots of food underground. Diet consists mainly of seeds or nuts (maple, oak, basswood, dogwood, cherry), but can include insects, mice, small birds, and snakes.

Tracks F5 (often only 4 register), H5: Trail 4.5–7 cm W. Tracks most commonly seen in snow after spring snowfall.

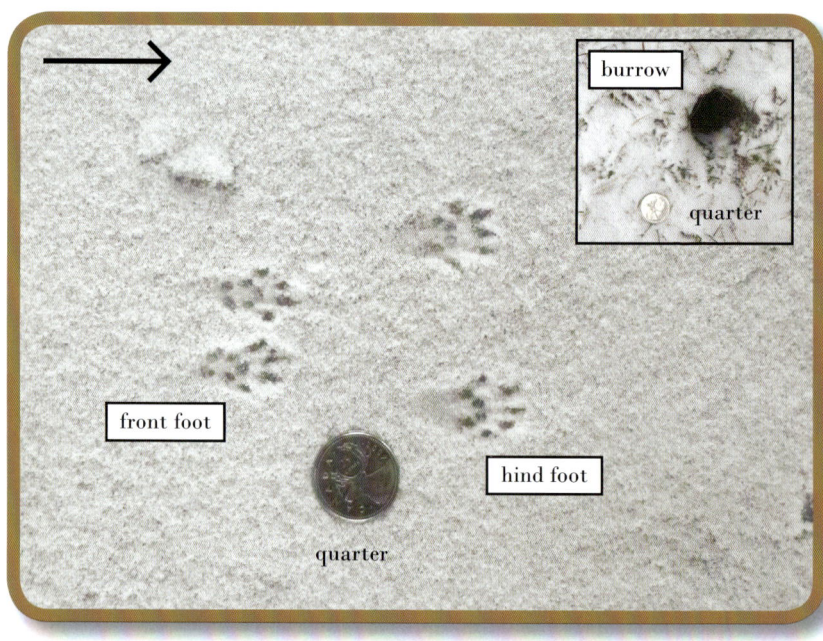

WOODLOT MAMMALS

Groundhog (woodchuck)
Marmota monax

Description: Fur usually grizzled brown. Moderate body size (40–50 cm) with short tail (10–20 cm). Lives in burrows, but can also climb trees.

Habits: Hibernates (sometimes comes out to check for shadow). A vegetarian that eats leaves and seeds (grass, clover, honeysuckle).

Tracks F5 (often only 4 register), H5: Typically a walker, but jumps to move quickly. Front track 2.5–5 cm W, hind track 3.5–5 cm W.

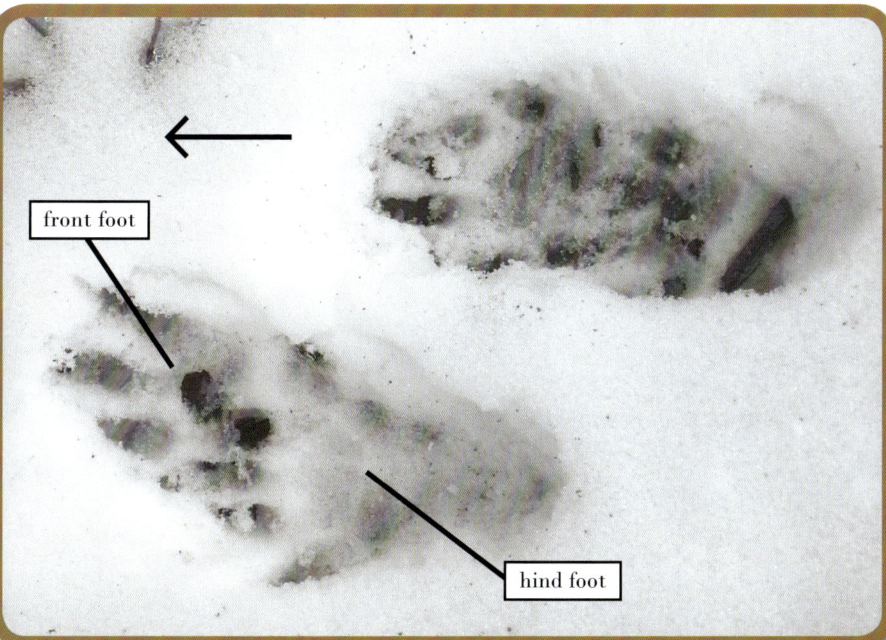

front foot

hind foot

WOODLOT MAMMALS

North American porcupine
Erethizon dorsatum

Description: Fur black with some brown highlights or hair with white tips. As many as 30,000 long, barbed, sharp spines, or 'quills', provide defense against predators. Quills are modified hairs that have hollow shafts with solid tips and bases. They have numerous backwards-pointing barbs that swell when wet and are coated in a waxy antibiotic substance. Quills are released by direct contact or may drop out when the porcupine shakes its body. Contrary to popular belief, quills cannot be projected at attackers. New quills grow to replace lost ones. The body size is moderate (46–56 cm) with short legs, a chunky torso, and a long tail (18–23 cm). Adults typically weigh 4–7 kg.

Habits: May wander at night making grunting noises or even high-pitched cries, especially during breeding season. The diet consists mainly of tree bark but also includes woody buds and twigs (hemlock, basswood, poplar, pine, maple, birch, beech, ash, fir, spruce, and oak).

Tracks F4, H5: Very long claw marks. Trail 13–22 cm W. In deep snow they may leave a trough with quill marks made by the tail.

WOODLOT MAMMALS

American beaver
Castor canadensis

Description: Fur rich brown with broad scaly tail. Body size moderate (60–75 cm) with long tail (20–25 cm). Adults typically weigh 15–30 kg.

Habits: Builds dams of mud, branches, and rocks to raise water levels to more easily forage on plants growing next to water. Diet of water lilies and bark/wood from poplar, willow, birch, alder, and maple. Remains active throughout winter by feeding on water-lily roots and piling up branches in front of lodge entrance.

Tracks F5, H5: Front 7–9 cm W, hind 8–13 cm W. Often only 3 toes will register and the hind track (webbed) may cover front track. Frequently make 'push-ups' of vegetation on the banks which are used to mark territory with a secretion from the castor glands (you may find that these vegetation piles smell like lilies).

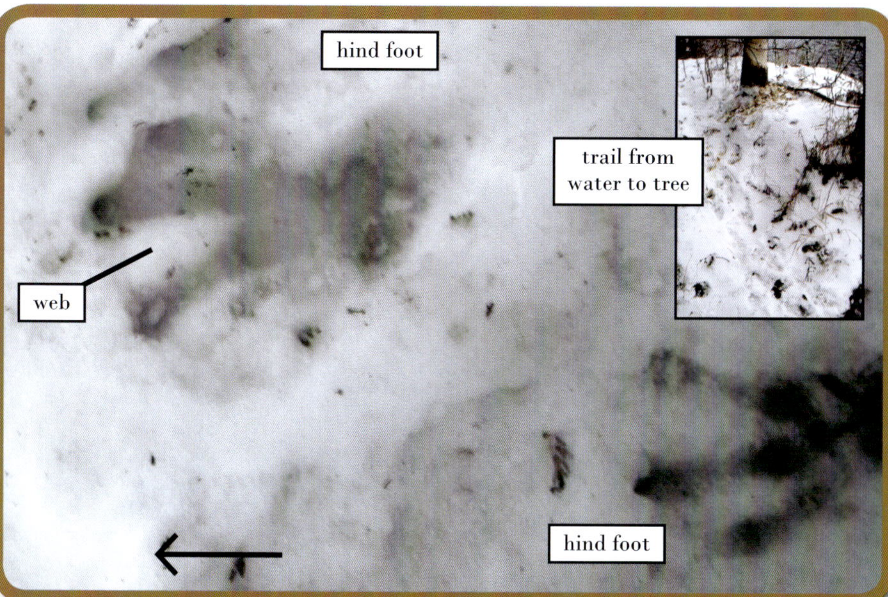

531

WOODLOT MAMMALS

White-tailed deer
Odocoileus virginianus

Description: Fur reddish-brown in summer, greyer in winter; young (fawns) are spotted. Tall, with a height at the shoulder of 100–130 cm. Adult females typically weigh 50–100 kg, and males 75–150 kg. The white, flag-like underside of the tail and simply branching antlers of the males differentiate them from all other North American members of the deer family.

Habits: Diet consists mainly of twigs and buds from trees and shrubs (maple, poplar, birch, oak, pine, hemlock, yew, and dogwood), fruits (apple, pear, blueberry, blackberry, and raspberry) and various grasses and grains (corn, wheat). Have also been documented eating eggs and nestlings of various birds.

Tracks F4, H4: Typically only two hooves are evident in tracks but if running or on slippery ground there may be marks from its two 'dew claws' that are located behind each hoof. Tracks 3.4–7.2 cm W with distinctive, pointed toes of a cloven hoof.

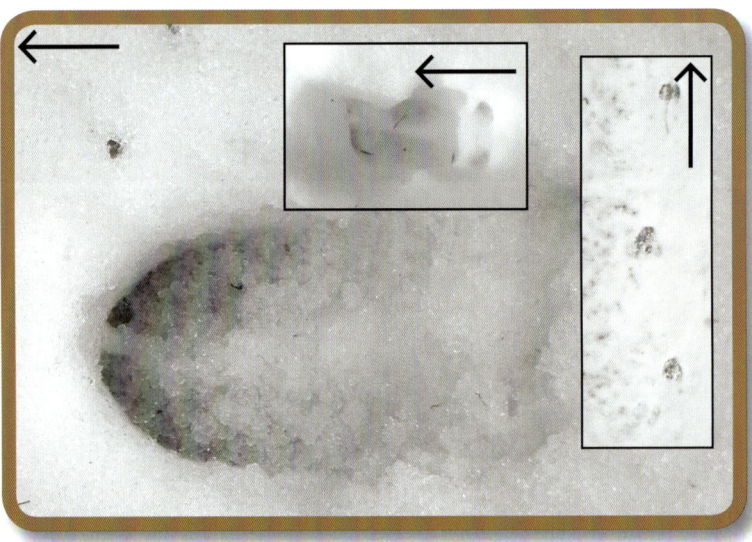

WOODLOT MAMMALS

Eastern cottontail
Sylvilagus floridanus

Description: Fur brown above and white below with a small fluffy white tail and long ears. Moderate body size (35–45 cm).

Habits: Active in the early morning and late evening. The diet consists of leaves in the summer (grass, clover, plantain, goldenrod, and alfalfa) and bark in the winter (maple, birch, cherry, and dogwood).

Tracks F5 (often only 4 register), H4: Front track 2–3.5 cm W, hind track 2.5–4.5 cm W. Front tracks are often, but not always, staggered.

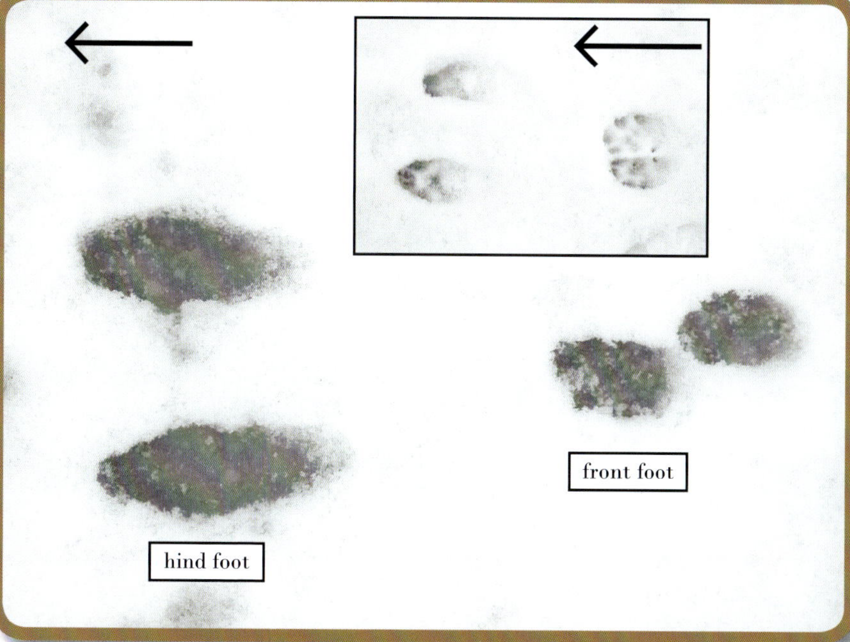

front foot

hind foot

533

WOODLOT MAMMALS

Northern raccoon
Procyon lotor

Description: Fur grizzled grey and brown with a black mask on the face. The body size is moderate (45–70 cm) with long ringed tail (20–30 cm). Adults weigh 5–15 kg.

Habits: Sleeps a lot in the winter, but does not hibernate. Diet consist of crayfish, bird eggs, insects, fruits (e.g. grape, cherry), and nuts (e.g. oak, beech). Often washes its food before consuming.

Tracks F5, H5: Front track 4–8 cm W, hind track 4–7 cm W.

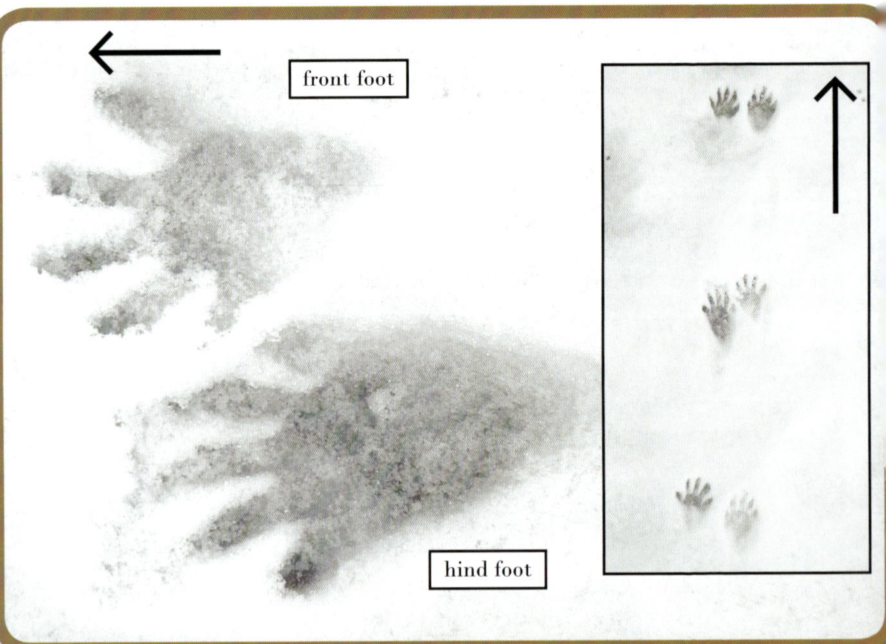

WOODLOT MAMMALS

Red fox
Vulpes vulpes

Description: Fur reddish. Legs dark. Belly, throat, and tail tip white. Some with more (or all) black. Moderate body size (56–63 cm) with long tail (35–41 cm).

Habits: Diet of rodents and rabbits, but also eats fruit (apple, cherry, raspberry, blueberry) grass, and seeds (cedar, beech, oak, hawthorn).

Tracks F4, H5: Often walks with direct register of hind foot over front foot. Distinguished from coyote (next page) by long, thin individual toes, and shorter track length: front 5–7 cm L, hind 4–6.5 cm L.

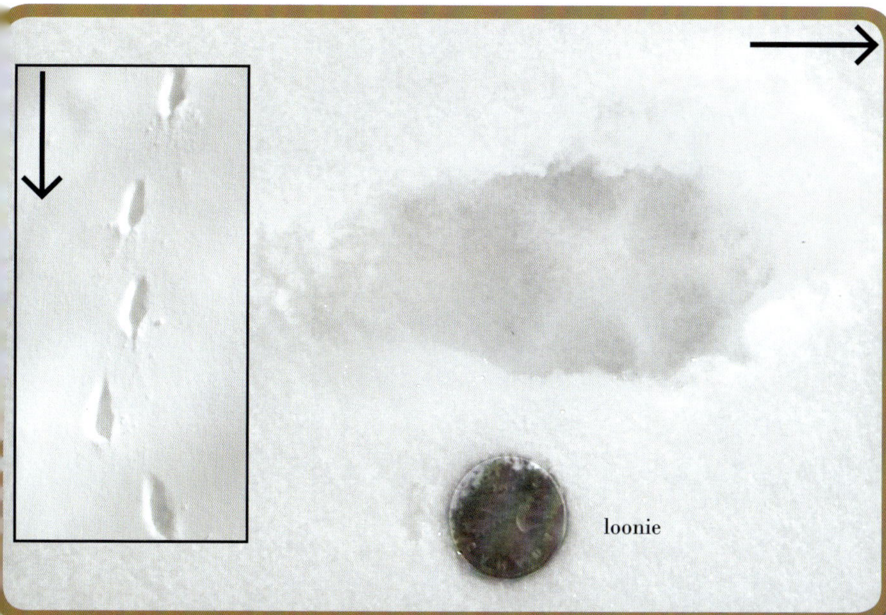

loonie

535

WOODLOT MAMMALS

Coyote
Canis latrans

Description: Fur reddish, brown, or grey. Moderate body size (81–94 cm) and weight (9–22 kg) with a long tail (28–40 cm).

Habits: A predator and scavenger that eats most things including numerous small mammals, snakes, frogs, deer, fruits, nuts, and grass. Often competes with the red fox for resources, forcing the fox to relocate to a new territory.

Tracks F4, H5: The hind print is smaller than the front one and the inner two toes are smaller than the outer two. Often walks with direct register of hind foot over front foot. Distinguished from the fox (previous page) by track length: front 6.5–9 cm L, hind 6–8 cm L.

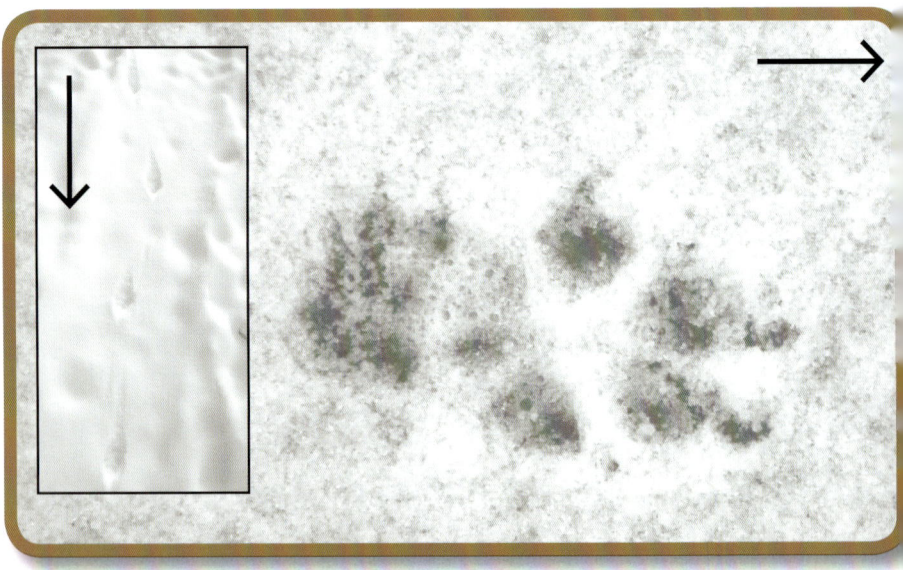

WOODLOT MAMMALS

Domestic dog
Canis familiaris

Tracks F5 (often only 4 register), H4: Most dogs do not directly register the hind track over the front and tracks are less oval than those of the coyote, but at times the two may be indistinguishable. Have been known to interbreed with coyotes.

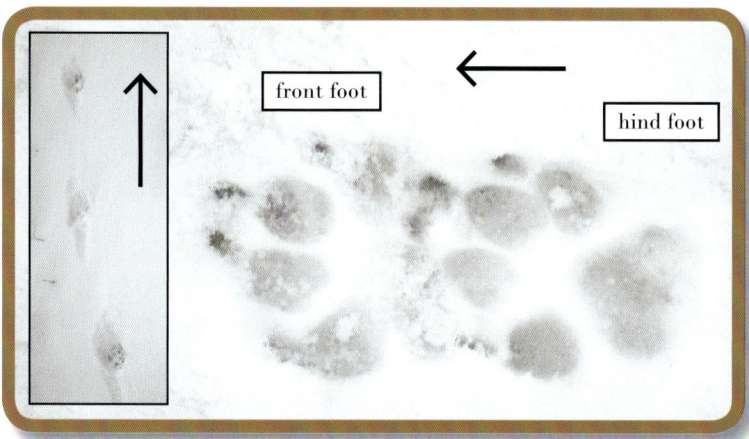

Domestic cat
Felis catus

Tracks F5 (often only 4 register), H4: Front and hind tracks are the same size (2.5–4.5 cm W). Retractable claws don't usually show in tracks.

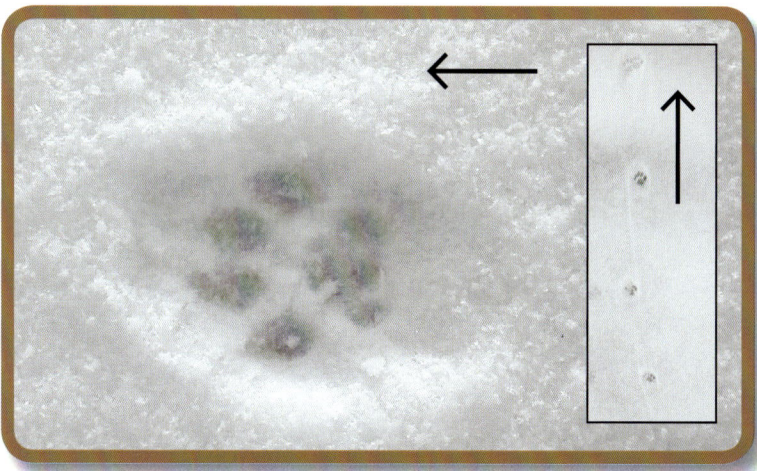

WOODLOT MAMMALS

Striped skunk
Mephitis mephitis

Description: Fur all black with variable amounts of white on head, back, and tail, and a characteristic stripe on the forehead and nape. Moderate body size (33–46 cm) and distinctive tail (18–25 cm).

Habits: Hunts at night. Often seen digging for grubs, but also eats mice, eggs, insects, and berries (strawberries, blueberries, cherries, nightshade, grapes).

Tracks F5, H5: Long claws usually show in front tracks. Ambling pattern may result in an odd line of 4 tracks. Front track 2.5–3 cm W, hind track 2.5–3 cm.

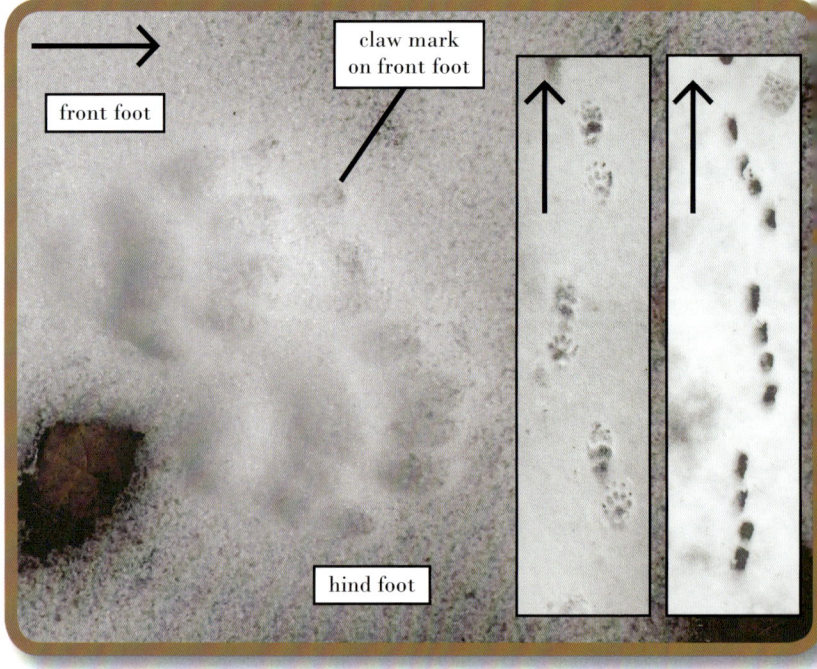

WOODLOT MAMMALS

American mink
Mustela vison

Description: Fur dark brown, sometimes with white on the chest. The body is small (30–43 cm) with a long tail (13–23 cm). Very good swimmer and climber. Weighs 500–900 g.

Habits: Diet mostly small mammals, birds, eggs, frogs, snakes, crayfish, and fish.

Tracks F5, H5: Trail 5–9 cm W. A leaper, but may leave tunnels in snow and sometimes slides down snowy slopes.

hind foot over front foot

trail in deep snow showing tunnels

loonie

539

WOODLOT MAMMALS

Long-tailed weasel
Mustela frenata

Description: Fur brown in summer, white in winter with a black tip on the tail. Small body (20–26 cm) with long tail (7.5–15 cm). The ermine (*Mustela erminea*) is very similar, but is smaller in size and is sometimes referred to as the short-tailed weasel.

Habits: Sometimes climbs trees but most often seen on the ground. A predator that hunts rodents (mostly mice), rabbits, birds, snakes, and frogs. Also eats eggs.

Tracks F5, H5: Trail 3.5–7 cm W. The size of long-tailed weasel and ermine tracks may overlap, but generally ermine tracks are smaller.

hind foot over front foot

loonie

WOODLOT MAMMALS

Ermine
Mustela erminea

Description: Fur brown in summer, white in winter with a black tip on the tail. Body 13–23 cm, tail 5–10 cm. Sometimes referred to as the short-tailed weasel. The long-tailed weasel (*Mustela frenata*) is larger in size.

Habits: Same as long-tailed weasel.

Tracks F5, H5: Trail 2.2–5 cm W. Leaper.

541

Glossary

Achene: A dry, single-seeded fruit.

Acrocarpous: A moss gametophyte that produces sporophytes at the apex of a stem or main branch. Acrocarpous mosses generally grow erect in tufts (rather than mats) and are sparsely branched (opposed to **Pleurocarpous**).

Actinomorphic: a radially symmetric flower with all of the petals similar in size and shape.

Acuminate: A leaf tip tapering to a point with somewhat concave sides.

Acute: A pointed leaf tip with straight sides.

Adventitious: In reference to roots, a root that grows from somewhere other than the primary root, for example, roots that arise from stems or leaves.

Aggregate fruit: A fruit composed of numerous smaller fruits collected on a single receptacle, e.g. a raspberry.

Allotetraploid: Possessing four chromosome sets that are derived from several species (see also **Autotetraploid**).

Alternate

Alternate: When a single leaf is attached at each node. See also **Opposite** and **Whorled**.

Alternation of generations: Life cycle in which haploid and diploid generations alternate with each other.

Androecium: collective term for all of the male reproductive organs (stamens) in a flower.

Anemophilous: Seed plants that are pollinated by wind.

Angiosperm: A group of plants that produce seeds enclosed within an ovary, which may mature into a fruit; flowering plants.

Annual: A plant which completes its life cycle in one year or less.

Annulus (pl. annuli): A zone of variously differentiated cells between the moss capsule urn and operculum, facilitating opening of the capsule.

Anther: The pollen producing tip of a stamen; part of a flower.

Antheridium: The organ on a gametophyte plant which produces the sperm cells.

Apocarpous: A gynoecium composed of multiple distinct, unfused carpels. See also **Monocarpous** and **Syncarpous**.

Apogamous: In ferns, when a sporophyte develops from the gametophyte without fertilization.

Apomixis: A method of asexual reproduction in angiosperms; seeds develop directly from ovule without requiring fertilization.

Apothecia: Reproductive structures where fungal spores are produced (see diagrams at the top of the next page)

   Biatorine – pale margins (not carbonized or black) without photobiont cells and usually similar in colour to the disk

   Lecanorine – margins the same colour as the thallus and usually a different colour than the disk, margins contain photobiont cells

   Lecideine – margins black (carbonized) and do not contain photobiont cells, margins and disks are usually the same colour

Lirellae – narrow and elongated, branched or not
Stalked – spores produced in cups at the tips of stalks

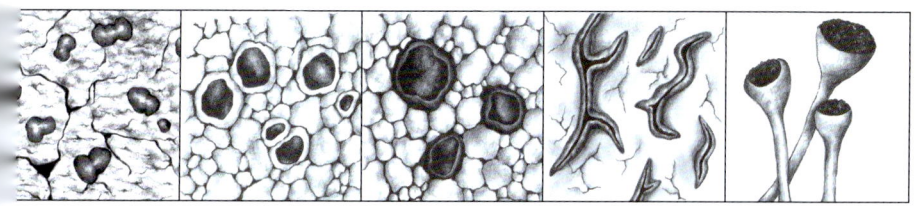

Biatorine   Lecanorine   Lecideine   Lirellae   Stalked

Appessed: Pressed close against the stem.
Aril: An outgrowth of the stalk connecting the ovule to the placenta, creating a partial or complete fleshy coating around the seed.
Archegonium: The organ on a gametophyte plant that produces the egg cell and nurtures the young sporophyte.
Armed: Bearing thorns, spines, and/or bristles.
Aromatic: Fragrant.
Ascending: At an upward angle.
Astringent: Bitter; also, something that causes tissues to contract.
Attenuate: Gradually tapering to a slender point.
Autotetraploid: Possessing four chromosome sets that are all derived from the same species (see also **Allotetraploid**).
Auricle: An ear-shaped lobe.
Axil: The angle formed between the base of an organ and the structure from which it originated; upper angle between the leaf base and the stem.
Axillary: Contained in or originating from an axil.
Axis: Pertaining to the central stem of an inflorescence.
Basal: Leaves that all arise at ground level.
Beard: Hairs on the inside throat of flowers, e.g. some violets.
Berry: A fleshy or pulpy fruit with two or more single ovule seeds.
Biatorine: See **Apothecia**.

Axillary

Biennial: A plant that lives for two growing seasons, usually flowering in the second year before dying.
Bilateral: In reference to flower symmetry, a flower with only one plane of symmetry.
Bipinnate: Two times pinnate; a pinnately compound leaf where each primary leaflet is further divided into secondary leaflets. See **Compound**.
Bipinnatifid: Two times pinnatifid; a pinnatifid leaf where each primary lobe is further divided into secondary pinnatifid lobes.
Bisporangiate: A flower or cone that produces both megaspores and microspores.
Biternate: Two times ternately compound.

Blade: The expanded part of a leaf or fern frond.
Bloom: A waxy or powdery whitish or bluish coating on plant parts.
Bract: Any reduced leaf-like structure associated with a cone or flower.
Bryophytes: Mosses, liverworts, and hortworts; non-vascular plants.
Bud: Very young developing tissue enclosed in scales or valves; can refer to a leaf, branch, or flower bud.
Bulblet: A small, bulb-like structure.
Caducous: Dropping or shedding at an early stage.
Calcareous: Composed of calcium carbonate or containing lime; often chalky.
Calyptra: In mosses, the protective cap or hood covering the sporophyte capsule.
Calyx: The collective group of sepals, the outermost section of a flower.
Cane: A supple woody stem.
Capsule: A dry dehiscent fruit composed of two or more carpels; in mosses, a structure that contains spores and is capped with an operculum.
Carpel: Collective term for an ovary, stigma, and style; may be solitary in the flower (see **Monocarpous/Unicarpellate**), several but distinct in a flower (see **Apocarpous**), or several fused into a single structure (see **Syncarpous**). See also **Pistil**.
Catkin: A dry, usually elongate, often drooping, scaly spike bearing imperfect flowers on woody plants.
Cauline: Leaves arising on the stem as opposed to being basal.
Cilia: Lichen hair-like fungal growth on lobe margins.
Ciliate: Fringed with fine hairs.
Complete: A flower with all four flower parts (carpel(s), stamens, petals, and sepals).

Cilia

Compound: A leaf that is divided into distinct leaflets.
  Trifoliate/Ternate – A compound leaf with three leaflets
  Palmate – Radiately compound, the axes of the individual segments originating at a common point or nearly so.
  Pinnate – Once-divided with leaflets arranged on both sides of a central axis.
  Bipinnate/Two times divided – A pinnately compound leaf where each primary leaflet is further divided into secondary leaflets.
  Tripinnate/Three times divided – A bipinnately compound leaf where each secondary leaflet is further divided into tertiary leaflets.

Trifoliate   Palmate   Pinnate   Bipinnate   Tripinnate

Cordate: Heart-shaped.
Coriaceous: Thick, leathery leaves.
Corolla: The collective group of petals, found between the calyx and androecium in a flower.
Corymb: A flat-topped inflorescence where the outer flowers open first.
Costa: The midrib of a moss leaf; not found in liverworts.
Cotyledon: A seed leaf.
Crenate: Rounded teeth on leaf margins.
Crustose: A lichen growth form closely attached to the substrate, crust-like, no lower cortex (see Lichen chapter, Fig. 5, 430).
Cryptogam: an organism that reproduces from spores (see also **Phanerogam**).
Cuneate: Narrowly triangular.
Cupule: A cup-like structure of bracts surrounding a fruit; characteristic of the family Fagaceae (e.g. acorns, chesnuts, and beech nuts).
Cyme: A flat-topped inflorescence with the inner flowers opening first.
Deciduous: falling off; e.g. leaves, flowers, moss capsules.
Decompound: More than once-compound, e.g. bipinnate, biternate etc.
Decurrent: Basal leaf margins extending down the stem past the leaf insertion as ridges or narrow wings.
Dehiscent: A fruit that opens at maturity to release the seeds. Opposite of **Indehiscent**.
Deltoid: Broadly triangular.
Dicot/Dicotyledon: A plant with two cotyledons; usually has flower parts in multiples of four or five and net-veined leaves.
Dimorphic: Having two morphologies or forms.
Dioecious: Pertaining to plants, individuals that bear either staminate or pistillate flowers but not both.
Diploid: Possessing two complete chromosome sets.
Dorsal: Relating to the back or outer surface of an organ; opposite of ventral. In moss leaves the abaxial, back, or lower surface; however in bryophyte stems or thalli it refers to the upper surface, away from the substrate.
Drupe: A single locule fleshy or pulpy fruit with a hard or stony center.
Elater: A cell or part of a cell which assists in dispersing spores. The elaters change shape as they lose or acquire water, and they will then push against surrounding spores.
Elliptic: Ellipse- or oval-shaped.
Emetic: Causes vomiting.
Emmenagogue: A plant that stimulates menstruation.
Endemic: A species that only occurs in a particular region.
Entire: Margins without crenation, serration, or dentition; even though the margin may be variously ciliate or pubescent.
Epigynous: A flower in which the other flower parts (stamens, petals, and sepals) attach above the ovary. Same as **Inferior**; see also **Hypogynous/Superior** and **Perigynous**.
Epipetalous: Attached to the petals; usually refers to stamens.

Cordate leaf

Epiphyte: A plant that grows upon another plant.
Equitant: Leaves that are overlapping and clasping at the base, as in the genus *Iris*.
Falcate: Strongly curved and turned to one side.
Fascicle: In reference to conifers, a cluster of needles.
Filament: The stalk of a stamen attached to the anther.
Foliose: A lichen growth form, leaf-like, typically has a distinct upper and lower surface (see Lichen chapter, Fig. 3, 430).
Follicle: A dry fruit that opens along one side.
Forb: See **Herb**.
Frond: the leaf of a fern, including the stipe, rachis, and blade.
Fruticose: A lichen growth form, branch-like or bushy, typically without a distinct upper and lower surface (see Lichen chapter, Fig.4, 430).
Gametophyte: The sexual stage in a plant's life cycle where sperm are produced in antheridia and eggs in archegonia; develops from spores produced by the sporophyte.
Gemma [pl. gemmae]: An asexual cluster of cells capable of developing into a new individual.
Genome: The basic chromosome/DNA complement of an organism.
Glabrous: smooth; hairless.
Gland: A protuberance, appendage, or depression on the surface of a leaf, petal or stem, which produces a sticky or greasy viscous substance.
Glandular: Having glands.
Glaucous: Often used to describe a light waxy layer on a plant, or blue-gray hue.
Globose: Globe-shaped.
Guard cells: Pair of cells that surround a stoma and regulate its size by altering their shape.
Gymnosperm: A group of plants that produces seeds that are not enclosed in ovaries or fruits; includes the conifers.
Gynoecium: Collective term for all of the female reproductive organs (carpels/pistils) in a flower.
Gynostemium: The central reproductive stalk of an orchid, which consists of a stamen and pistil fused together.
Haploid: Possessing one complete chromosome set.
Head: A very dense cluster of flowers, with individual flowers lacking stalks.
Hepatic: A liverwort (or something pertaining to liverworts).
Herb: Generally any plant that is not woody or grass-like.
Herbaceous: Pertaining to plants that lack above-ground woody tissue, thus dying back to the ground every year.
Heterosporangiate: Producing two different kinds of sporangia, specifically microsporangia and megasporangia.
Heterosporous: Producing two different sizes or kinds of spores. These may come from the same or different sporangia, and may produce similar or different gametophytes.
Hexaploid: Possessing six complete chromosome sets.
Homomallous: Pointing the same way.

Homosporous: Producing only one size or kind of spore.
Hybrid: The offspring of a cross between two taxa (e.g. two species).
Hybridization: Crossbreeding between two taxa (e.g. two species) to produce **Hybids**.
Hypanthium: a fusion of the calyx, corolla, and androecium forming a cup or tube-like structure that surrounds the gynoecium.
Hypogynous: A flower in which the other flower parts (stamens, petals, and sepals) attach below the ovary. Same as **Superior**; see also **Epigynous/Inferior** and **Perigynous**.
Imperfect: Unisexual flowers; individual flowers lack either stamens or carpels (see also **Perfect**).
Incomplete: A flower lacking one or more whorls (carpel(s), stamens, petals, and sepals); note that an incomplete flower can still be perfect (i.e. is only lacking petals and/or sepals) but an imperfect flower is always incomplete.
Incubous: Leaves overlapping from the base of the stem upwards. Opposite of **Succubous**.
Indehiscent: A fruit that does not split open at maturity.
Indusium [pl. indusia]: The tissue covering a sorus of a fern.
Inferior: An ovary that is positioned below the point where the stamens, petals, and sepals are attached. Same as **Epigynous**; see also **Hypogynous/Superior** and **Perigynous**.
Inflorescence: A cluster of flowers.
Internode: The region of a stem between two nodes, when there is no branching of the vascular tissue.
Irregular: A zygomorphic/bilaterally symmetric flower that has only one plane of symmetry.
Isidia: Lichen vegetative propagules attached to the thallus surface or the margins of apothecia, with a cortex, contains both the mycobiont and photobiont, cylindrical or globular (see Lichen chapter, Fig. 2, 429).
Julaceous: Worm-like, smoothly cylindric, like a catkin, referring to stems or branches with strongly imbricate leaves.
Keel: Part of a plant or animal that resembles a ship's keel (e.g. a pronounced ridge running along the centre line of the bottom part of the hull); in the bean family (Fabaceae), the two united boat-like petals, usually at the bottom of the flower.
Keeled: Part of a plant or animal (e.g. a leaf or a bird's tail) with a keel-like structure.
Lacerate: appearing deeply and irregularly slashed or torn.
Lamina: the flattened, generally unistratose and green part of the leaf blade excluding the midrib (costa in mosses) and border.
Laminal: The upper surface of the lichen thallus, but does not include the area near the margins.
Lanceolate: Lance-shaped; widest below the middle and tapering to a point.
Leaflet: A smaller division of a leaf.
Lecanorine: See **Apothecia**.
Lecideine: See **Apothecia**.

Laminal Isidia

Liana: A woody vine.

Linear: Long and narrow; used to describe leaves.

Lirellae: See **Apothecia**.

Lobed: A leaf blade or petal indented along its margins from at least ¼ of the way to its base or midrib.

Locule: A separate cavity or space within an ovary, fruit, or anther.

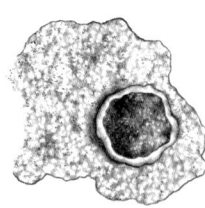

Maculae

Maculae: Pale to white spotting on the upper surface of the lichen thallus, typically slightly raised.

Maculate: Lichens with maculae.

Marginal: On the edge or margin of lobes.

Medial: Occurring near the middle.

Medulla: Layer within lichen that is below the upper cortex and composed of hyphae and the photobiont (see Lichen chapter, Fig. 2, 429).

Megaspore: In plants that are heterosporous, the larger kind of spore is called a megaspore; it usually germinates into a female (egg-producing) gametophyte.

Membranous: Pertaining to leaves, meaning thin and somewhat transparent.

Merous: A term describing the number of parts of a flower, i.e. 3-merous means a flower with parts in multiples of 3.

Microspore: In plants that are heterosporous, the smaller kind of spore is called a microspore; it usually germinates into a male (sperm-producing) gametophyte.

Midrib: See **Midvein**.

Midnerve: See **Midvein**.

Midvein: The central or principal vein of a leaf, bract, sepal or petal; in mosses this is called a **Costa**.

Monocarpous: A gynoecium with a single carpel. See also **Apocarpous** and **Syncarpous**.

Monocot/Monocotyledon: A plant with one cotyledon; usually has flower parts in multiples of three and parallel-veined leaves.

Monoecious: Pertaining to plants, individuals of which bear both staminate and pistillate flowers but not perfect flowers.

Mottled: Irregularly blotched.

Mucilaginous: Pertaining to being sticky, wet, or secreting mucous.

Obcordate Oblanceolate

Mycobiont: The fungal partner of a lichen.

Node: The region of a stem between two internodes, where there is branching of the vascular tissue into leaves or other appendages.

Obcordate: Heart-shaped with the widest end opposite the stem or petiole.

Oblanceolate: Lance-shaped; similar to lanceolate except inverted and tapering to a point near the petiole, not at the leaf tip.

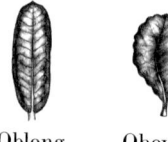

Oblong Obovate

Oblong: Longer than wide, with parallel sides.

Obovate: Egg-shaped with the widest end opposite to the stem or petiole.

Operculum: In mosses, the lid of the sporophyte capsule, which pops off when mature.

Opposite: Leaves occurring in pairs at each node, one on either side of the stem. See also **Alternate** and **Whorled**.

Oval: Round.

Ovary: In angiosperms, the structure that surrounds the ovules/seeds and develops into the fruit.

Ovate: Egg-shaped with the widest end attached to the stem or petiole.

Ovoid: see **Ovate**.

Ovule: In seed plants, the structure which gives rise to the seed.

Opposite

Oval Ovate

Palmate: Radiately lobed or divided, the axes of the individual segments originating at a common point or nearly so. See **Compound**.

Panicle: A much branched inflorescence.

Papilla [pl. papillae]: Minute round bumps on leaves or cells.

Papillose: Covered with short, rounded, blunt bumps.

Paraphyses: In ferns, sterile filaments amongst sporangia.

Pedicel: The stalk of a single flower in an inflorescence.

Peduncle: The stalk of a whole inflorescence OR of a solitary flower.

Peltate: Attaching in the middle; peltate stem/leaf resembles an umbrella.

Pendulous: Hanging, pendent; e.g. stems and branches that hang; hanging fern fronds; moss capsules drooping and inclined beyond horizontal.

Perennial: A plant which continues to grow after it has reproduced, usually meaning that it lives for several years.

Perfect : Bisexual flowers with both male and female parts in the same flower (see also **Imperfect**).

Perianth: Collective term for the calyx and corolla together.

Pericarp: The wall/fleshy part of a fruit.

Perigynous: A flower where the stamens, petals, and sepals are attached to a hypanthium, an open cup-like structure that surrounds the gynoecium. See also **Epigynous/Inferior** and **Hypogynous/Superior**.

Peristome: A set of cells or cell parts that surround the opening of a moss sporangium. In many mosses, they are sensitive to humidity, and will alter their shape to aid in spore dispersal.

Petal: A single part of the corolla.

Petaloid: Parts of a plant that looks like petals.

Petiole: The stalk of a leaf.

Phanerogam: A plant that reproduces from seeds (see also **Cryptogam**).

Photobiont: The photosynthetic partner (algae, cyanobacteria or both) within a lichen.

Pinna [pl. pinnae]: The primary division or leaflet of a pinnately compound fern frond.

Pinnate: Once-divided with leaflets arranged on both sides of a central axis. See **Compound**.

Pinnatifid: Similar to pinnate, except with deeply cut lobes instead of distinct leaflets.

Pinnule: The secondary division or leaflet of a frond; the division of a pinnae.
Pistil: The central part of a flower; can be composed of a single carpel or multiple fused carpels. See also **Carpel**.
Pistillate: An imperfect flower with only female parts.
Pith: A central region of parenchyma tissue within a plant stem.
Plagiotropic: Branches growing at an almost horizontal angle.
Pleurocarpous: A moss gametophyte that produces sporophytes on short lateral branches. Acrocarpous mosses generally grow in mats (rather than tufts) and are freely branched (opposed to **Acrocarpous**).
Plicate: Leaves with longitudinal furrows or pleats (plica).
Plumose: Closely and regularly pinnate; feathery.
Podetium (pl. podetia): A lichen stalk, typically hollow, with cups or fruiting bodies at the tips.
Polarilocular: Lichen spores with a thick septum dividing two cell cavities connected by a narrow channel.
Polyploidy: The occurrence of more than two chromosome sets in an organism.
Pome: A fleshy fruit (e.g. apple), formed from a several loculed inferior ovary.

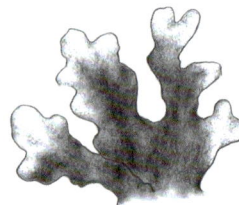

Pruinose Lobe Tips

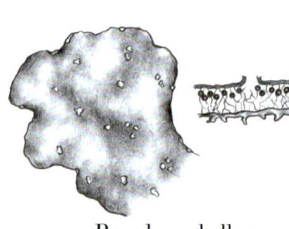 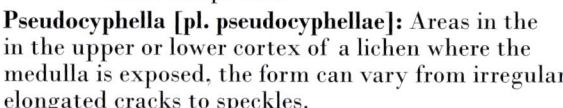

Pseudocyphellae

Prickle: A sharp outgrowth from the epidermis or bark of a plant, e.g. as in roses. See also **Spine** and **Thorn**.
Prostrate: Growing flat along a substrate or horizontal to ground.
Pruina: Lichen chemical deposits that appear frost-like, typically white, can occur on the upper surface of the thallus and on the disks of the apothecia.
Pruinose: Lichen with pruina.
Pseudocyphella [pl. pseudocyphellae]: Areas in the in the upper or lower cortex of a lichen where the medulla is exposed, the form can vary from irregular elongated cracks to speckles.
Pteridophyte: A vascular plant that reproduces via spores; includes ferns, horsetails, and club-mosses.
Pubescent: Hairy.
Purgative: Causes evacuation of the bowels.
Raceme: An elongate inflorescence that has one flower per node, with each flower on a stalk/pedicel.

Rachis: The central axis of a fern blade.
Radial: An actinomorphic/regular flower with numerous planes of summetry and petals similar in size and shape.
Receptacle: The expanded base of a flower where the flower whorls (calyx, corolla, androecium, and gynoecium) are attached.
Recurved: Curving backwards.
Reflexed: Bending backwards.
Regular: An actinomorphic/radially symmetric flower with numerous planes of summetry and petals similar in size and shape.
Reniform: Kidney-shaped.
Resinous: Of or pertaining to resin, a translucent substance of plant origin.

Reticulate: Interconnecting, like a network.

Rhizines: Root-like growths, typically on the lower surface, anchors lichen to substrates (see Lichen chapter, Fig. 2, 429).

Rhizoid or Radicle: hair-like structure that function in absorption and anchorage; in liverworts and hornworts 1-celled and usually hyaline; in mosses usually brown to reddish, simple or branched, multicellular filaments, generally with oblique end-walls.

Rhizome: An underground stem, typically horizontal.

Rosette: Leaves originating from a central point often lying flat on the ground.

Rotate: Saucer shaped, flat and circular in outline.

Scale: Any small, reduced, leaf-like structure.

Secund: turned to one side; e.g. leaves on a stem.

Sepal: A single part of the calyx.

Septate: With a septum.

Septum [pl. septa]: A division between cells in a lichen spore.

Serrate: With pointed teeth.

Sessile: Without a distinct stalk; often used to describe leaves and flowers.

Seta [pl. setae]: In mosses, the stalk of the sporophyte capsule.

Sheath: generally, refers to leaves or part of leaves that surround the stem; in horsetails, refers to the partially-fused leaves that surround the stem at each node.

Simple: A term usually applied to leaves with a single blade, i.e. not divided into separate leaflet. See also **Compound**.

Sinus: The cleft or space between two lobes.

Solitary: An infloresence composed of a single flower.

Soredia: Vegetative propagules arising from the medulla through the upper or lower cortex, contains both the lichen mycobiont and photobiont, without a cortex, appears powdery or granular (see Lichen chapter, Fig. 2, 429).

Sorus [pl. sori]: a cluster of spore-bearing sporangia; often on the underside of a fern leaf.

Spadix: An inflorescence spike typified by a very fleshy axis; usually with a spathe.

Marginal soredia

Spathe: A foliaceous bract-like or sheathiform structure enclosing or partly enclosing an inflorescence; usually with a spadix.

Spike: A long, narrow inflorescence with sessile flowers.

Spine: A sharp outgrowth from a plant that developed from a leaf or stipule; found at nodes. See aslo **Prickle** and **Thorn**.

Sporangium [pl. sporangia]: A structure that produces and contains spores.

Sporophyll: A leaf that bears sporangia.

Sporophyte: The vegetative stage in a plant's life cycle; produces spores via meiosis; develops following fertilization of an egg during the gametophyte stage.

Sporulation: The development and release of spores.

Spur: A hollow appendage of the calyx or corolla.

Spur shoot: A short, compact shoot where the internodes do not elongate.

Squarrose: Spreading at right angles.

Stalked: A vegetative or reproductive structure on a distinct stalk; see also **Apothecia.**

Stamen: The male section of a flower made up of a filament and pollen-producing anther.

Staminate: An imperfect flower with only male parts.

Stellate: Star-shaped or radiantly branched; usually pertaining to hairs.

Stigma: The part of a pistil that is receptive to pollen, directing it to the ovary.

Stipe: The stalk of a fern leaf up to the beginning of the blade; continuous with the rachis.

Stipules: Paired appendages found at the base of the leaves, but not flowers (see also **Bract**), of many flowering plants.

Stolon: A horizontal, prostrate, running branch or stem, often tending to root at the nodes.

Stoma [pl. stomata]: Openings in the epidermis of a stem or leaf of a plant that permit gas exchange with the air. In general, all plants except liverworts have stomata in their sporophyte stage. In mosses, stoma also refers to the mouth or opening of the sporophyte capsule.

Strobilus [pl. strobili]: A tightly clustered group of sporophylls arranged on a central stalk; common in club-mosses, horsetails, and conifers; the term 'cone' is commonly used for conifers.

Style: The narrowed part of a pistil between the stigma and ovary.

Substrate: A substance that a plant or lichen grows upon.

Succubous: Leaves overlapping from the tip of the stem downward, like the shingles on a roof. Opposite of **Incubous.**

Sulcate: Grooved or furrowed lengthwise.

Superior: An ovary that is positioned above the point where the stamens, petals, and sepals are attached. Same as **Hypogynous**; see also **Epigynous/ Inferior** and **Perigynous.**

Syn.: Abbreviation for synonym. An illegitimate name for a plant or animal.

Syncarpous: A gynoecium with multiple carpels fused into a single structure. See also **Apocarpous** and **Monocarpous.**

Terminal

Tendril: A thread-like structure used for plant support, often found on climbing plants.

Tepal: If the sepals and petals of a flower are indistinguishable, they are referred to as tepals.

Terete: Rounded or circular.

Terminal: At the end of a stem; often used to describe the position of an inflorescence. See also **Axillary.**

Ternate: A compound leaf divided into three leaflets. See **Compound.**

Tetraploid: Possessing four complete chromosome sets.

Thalloid: Plants which have no roots, stems, or leaves are called thalloid, such as liverworts and hornworts.

Thallose: See **Thalloid.**

Thallus: The primary vegetative body of a lichen.

Thorn: A sharp outgrowth from a plant that develped from a modified twig or branch; found at the end of twigs or where a branch would be. See also **Spine** and **Prickle**.

Tomentose: Covered with tangled or matted woolly hairs.

Trifoliate: A compound leaf with three leaflets. See **Compound**.

Tripinnate: Three times pinnate; a bipinnately compound leaf where each secondary leaflet is further divided into tertiary leaflets. See **Compound**.

Triploid: Possessing three complete chromosome sets.

Tuber: An underground stem which has been modified for storage of nutrients, such as a potato.

Umbel: A flat-topped inflorescence where all of the individual flower stalks are attached at the same point; resembles an umbrella.

Unarmed: Lacking thorns, spines, and bristles.

Unicarpellate: A gynoecium with a single carpel. Same as **Monocarpous**; see also **Apocarpous** and **Syncarpous**.

Uncinate: Hooked; tip bent in the form of a hook.

Undulate: Wavy.

Unisexual: Individual flowers of one sex, either staminate or pistillate only.

Urn: The spore bearing portion of a moss capsule.

Ventral: Pertaining to the inner or anterior face of an organ; opposite of dorsal. In Mosses it refers to both the leaf upper surface that is adaxial or the top; however for moss stems or thalli the lower surface is next to the substrate; (opposed to dorsal).

Vestigial: Undeveloped or primitive.

Vivipary: Germinating while still on the plant, as certain bulbs and transformations of floral tissues.

Whorled: When three or more leaves are attached at each node. See also **Alternate** and **Opposite**.

Winged: Pertaining to a stem or leafstalk, meaning with flaps of tissue attached along each side, resembling wings.

Xeromorphic: Adapted for dryness.

Zygomorphic: Divisible into two equal parts only through a single plane; bilaterally symmetrical.

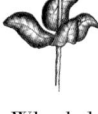

Whorled

Bibliography

Amaral, S., Lurdes M., Nogueira, J.M.F Pereira da Silva, A. and M. Helena Florêncio. 2009. Bioorganic & Medicinal Chemistry. Volume 17(5): 1876-1883.

American Violet Society. http://americanvioletsociety.org/Violets_In_America/Ethnobotanical.htm.

Angiosperm Phylogeny Group. 2009. An update of the Angiosperm Phylogeny Group classification for the orders and families of flowering plants: APG III. Botanical Journal of the Linnean Society, 161(2): 105-121.

Bieryzchek, P. 1982. The demography of jack-in-the-pulpit, a forest perrenial that changes sex. Ecological Monographs, 52(4), pp. 335-351.

Barriault et al. 2009. Flowering period, thermogenesis, and pattern of visiting insects in *Arisaema triphyllum* (Araceae) in Quebec. Botany, 87(3): 324-329.

Blackburn, B. 1952. Trees and shrubs in eastern North America. Oxford University Press. New York, U.S.A. 358 pp.

Boon, H. and M. Smith. 1999. The Botanical Pharmacy. Quarry Press Inc. Kingston, Ontario, Canada. 320 pp.

Bowers, N., R. Bowers and K. Kaufman. 2004. Kaufman Focus Guide: Mammals of North America. Houghton Mifflin, New York. 351 pp.

Brodo, I.M., S.D. Sharnoff and S. Sharnoff. 2001. Lichens of North America. Yale University Press, New Haven, USA. 795 pp.

Burt, W. and R. Grossenheider. 1976. Peterson Field Guides: Mammals. Houghton Mifflin. 289 pp.

Cavers, P. B., M. I. Heagy and R. F. Kokron, 1979. The Biology of Canadian Weeds. 35. *Alliaria petiolata* (M.Bieb.) Cavara & grande. Can. J. Plant Science. 59:217-229.

Cody, W. J. and Britton, D. M. 1989. Ferns and Fern Allies of Canada. Agriculture Canada Research Branch, Ottawa, ON. 430 pp.

Chambers, B., K. Legasy and K. V. Bently. 1996. Forest Plants of Central Ontario. Lone Pine Publishing, Edmonton, AB. 448 pp.

Crum, H. A. and L. E. Anderson. 1981. Mosses of Eastern North America. 2 vols. New York, NY.

Crum, H. A. 1991. Liverworts and Hornworts of Southern Michigan. Ann Arbor, MI. 233 pp.

Dickinson, T., D. Metsger, J. Bull and R. Dickinson. 2004. The ROM Field Guide to Wildflowers of Ontario. Royal Ontario Museum, Toronto, ON. 416 pp.

Duke, J. A. 1986. Handbook of Medicinal Herbs. CRC Press. Boca Raton, Florida. 677 pp.

Duke, J. A. 1997. The Green Pharmacy. Rodale Press. Emmaus, Pennsylvania. 507 pp.

Earley, C. G. 2003. Sparrows and Finches of the Great Lakes Region and Eastern North America. Firefly Books, Toronto. 128 pp.

Earley, C. G. 2003. Warblers of the Great Lakes Region and Eastern North America. Firefly Books, Toronto. 128 pp.

Earley, C. G. 2012. Hawks and Owls of the Great Lakes Region and Eastern North America. Second Edition. Firefly Books, Toronto. 144 pp.

Earley, C. G. 2005. Waterfowl of Eastern North America. Firefly Books, Toronto. 160 pp.

Elbroch, M. 2003. Mammal Tracks and Signs. Stackpole Books. Mechanicsburg, Pennsylvania. 282 pp.

Elias, T. S. 1980. The Complete Trees of North America: Field Guide and Natural History. Van Nostrand Reinhold Co., New York. 948 pp.

Ewan, Joseph. 1952. Pursh and His Botanical Associates. Proceedings, American Philosophical Society of Philadelphia. 96(5). 124 pp.

Farrar, J. 1995. Trees in Canada. Fitzhenry and Whiteside Ltd., Toronto. 540 pp.

Flora of North America Editorial Committee, eds. 1993+. Flora of North America North of Mexico. 16+ vols. New York and Oxford.

Gleason, H. A. and A. Cronquist. 1991. Manual of Vascular Plants of Northeastern United States and Adjacent Canada. Ed. 2. New York Botanical Garden, Bronx, NY. 910 pp.

Goward, T. 1994. The Lichens of British Columbia, Illustrated Keys. Part 1 – Foliose species, Ministry of Forests, Province of British Columbia. Victoria, B.C. 181 pp.

Goward, T. 1999. The Lichens of British Columbia, Illustrated Keys. Part 2 – Fruticose species. Ministry of Forests, Province of British Columbia. Victoria, B.C. 319 pp.

Hausen, B.M., A. Shoji, and O. Jarchow. 1984. Orchid Allergy. Archives of Dermatology. 120(9):1206-1208.

Herrick, J. W. and D. R. Snow. 1995. Iroquois Medical Botany. Syracuse University Press, Syracuse, N.Y. 278 pp.

Heywood, V.H., R.K. Brummitt, A. Culham, and O. Seberg. 2007. Flowering Plant Families of the World. Firefly Books, Ontario, Canada. 424 pp.

Hinds, J.W. and P.L. Hinds. 2007. The Macrolichens of New England. The New York Botanical Garden Press, Bronx, New York. 584 pp.

Hosie, R.C. 1979. Native Trees of Canada, 8th ed. Fitzhenry & Whiteside. Markham, Ontario, Canada. 380 pp.

Integrated Taxonomic Information System (ITIS) online database. http://www.itis.gov.

Kingsbury, J. M. 1964. Poisonous Plants of the United States and Canada. Prentice-Hall Inc., Englewood Cliffs, N.J. pp.626

Lata, H., R.M. Moraes, B. Bertoni and A. M. S. Pereira. 2010. In vitro germplasm conservation of *Podophyllum peltatum* L. under slow growth conditions. Vitro Cellular & Developmental Biology-Plant. 46(1): 22–27.

Lewis, W.H. and M. Elvin-Lewis. 1977. Medical Botany: Plants Affecting Man's Health. John Wiley & Sons, U.S.A. 515 pp.

MaCaulay, J. C. 1987. Orchid allergy. Contact Dermatitis. 17: 112–113. American Contact Dermatitis Society.

Metzgar, J.S., J.E. Skog, E.A. Zimmer, and K.M. Pryer. 2008. The paraphyly of *Osmunda* is confirmed by phylogenetic analyses of seven plastid loci. Systematic Botany, 33(1): 31-36.

Mitchell, J. C. and A. Rook. 1979. Botanical Dermatology. Greenglass Ltd, Vancouver, B.C., Canada. 787 pp.

Moerman, D. E. 1998. Native American Ethnobotany. Timber Press, Oregon, U.S.A. 799 pp.

Montgomery, F.H. 1970. Trees of Canada and the Northern United States. Ryerson Press, Toronto, ON.

Moraes, R. M. , E. Bedir, H. Barrett, C. Burandt Jr, C. Canel and I. A. Khan. 2002. Evaluation of *Podophyllum peltatum* Accessions for Podophyllotoxin Production. Planta Med 68(4): 341–344

Murie, O. 1974. Peterson Field Guides: Animal Tracks, 2nd ed. Houghton Mifflin, New York. 375 pp.

Nash III, T.H. 2008. Lichen Biology 2nd ed. Cambridge University Press, New York, U.S.A. 486 pp.

Native American Ethnobotany. University of Mighigan - Deerborn. [http://herb.umd.umich.edu/]

Newmaster, S.G. and S. Ragupathy. 2010. Flora Ontario Integrated Botanical Information System (FOIBIS). OAC Herbarium, Biodiversity Institute of Ontario, Guelph, Ontario. [http://www.uoguelph.ca/foibis].

Newmaster, S.G., A.G. Harris and L.J. Kershaw. 1998. Wetland Plants of Ontario. Lone Pine Press, Edmonton, Alberta. 241 pp.

Newcomb, L. 1977. Newcomb's Wildflower Guide. Little, Brown and Company (Canada) Limited, Toronto, 490 pp.

Peterson, R. T. 1980. Peterson Field Guides: Eastern Birds, 4th ed. Boston: Houghton Mifflin, Harcourt. 257 pp.

Plants for a Future. Edible, medicinal and useful plants for a healthier world. [http://www.pfaf.org/index.php].

Policansky, D. 1981. Sex choice and size advantage model in jack-in-the-pulpit (*Arisaema triphyllum*). Proc. of Natl. Acad. Sci. USA. 78(2):1306–1308.

Pryer, K.M., H. Schneider, A.R. Smith, R. Cranfill, P.G. Wolf, J.S. Hunt, and S.D. Sipes. 2001. Horsetails and ferns are a monophyletic group and the closest living relatives to seed plants. Nature 409: 618-622.

Pryer, K.M., E. Schuettpelz, P.G. Wolf, H. Schneider, A.R. Smith and R. Cranfill. 2004. Phylogeny and evolution of ferns (monilophytes) with a focus on the early leptosporangiate divergences. American Journal of Botany, 91(10): 1582–1598.

Rasmussen, J. E. 1986. Contact dermatitis from orchids. Clinics in Dermatology. 4(2): 31–35.

Reddoch, A. H. and J. M. Reddoch. 1984. Warning: Lady's-slippers can be hazardous to your health. Plant Press. 2(1): 10.

Rehder, A. 1940. Manual of Cultivated Trees and Shrubs.MacMillan Company, New York, U.S.A. 996 pp.

Rezendes, P. 1992. Tracking and the Art of Seeing. Camden House Publishing Inc., Charlotte, Vermont. 320 pp.

Reznicek, A.A., E. G. Voss, and B. S. Walters. 2011. Michigan Flora Online. University of Michigan, Ann Arbor, U.S.A. http://www.michiganflora.net/home.aspx.

Rothfels, C.J., M. A. Sundue, L.-Y. Kuo, A. Larsson, M. Kato, E. Schuettpelz, and K.M. Pryer. 2012. A revised family-level classification for eupolypod II ferns (Polypodiidae: Polypodiales). Taxon, 61(3): 515-533.

Schuettpelz, E. and K.M. Pryer. 2007. Fern phylogeny inferred from 400 leptosporangiate species and three plastid genes. Taxon 56(4): 1037-1050.

Schuster, R. M. 1953. Boreal Hepaticae, a manual of the liverworts of Minnesota and adjacent regions. American Midland Naturalist 49: 257-684.

Scott, S.L. 1987. Field Guide to the Birds of North America, 2nd ed.. National Geographic. Washington D.C. 464 pp.

Sibley, D. 2000. National Audobon Society, The Sibley Guide to Birds. Alfred A. Knopf Inc., New York, N.Y. 544 pp.

Smith, A.R., K.M. Pryer, E. Schuettpelz, P. Korall, H. Schneider, and P.G. Wolf. 2006. A classification for extant ferns. Taxon, 55(3): 705-731.

Smith, C.W., A. Aptroot, B.J. Coppins, A. Fletcher, O.L. Gilbert, P.W. James and P.A.Wolseley (eds.). 2009. Lichens of Great Britain and Ireland. Richmond Publishing, Slough, UK. 1046 pp.

Smith Jr., J. P. 1977. Vascular Plant Families. Mad River Press Inc., Eureka California. pp. 320

Soper, J.H. and M.L. Heimburger. 1982. Shrubs of Ontario. The Royal Ontario Museum. Toronto, Canada. 494 pp.

Stokes, D.W. and L.Q. Stokes. 1986. Stokes Guide to Animal Tracking and Behaviour. Little, Brown, and Company, Boston, Massachusetts. 418 pp.

Wildflowers of the southeastern U.S. http://www.2bnthewild.com/plants/H53.htm.

Wong, P. Y., and I. M. Brodo. 1992. The lichens of southern Ontario, Canada. Syllogeus, 69, Canadian Museum of Nature, Ottawa. 79 pp.

Xu, S. and G. W. Patterson. 1990. Sterol composition of the Phtolaccaceae and closely related families. Lipids. 25(4): 230-234.

Zennie, T.M. and C.D. Ogzewalla. 1977. Ascorbic acid and vitamin A contents of edible wild plants of Ohio and Kentucky. Economic Botany. 31: 76-79.

Photo Credits

All photos taken by Sean Fox, Thomas Henry, Brian Lacey, Carole Ann Lacroix, Jose Maloles, Troy McMullin, Sean Rapai, Lyndsay Schram, Brittany Shum, Lisa Steele, and Royce Steeves, except where noted below.

Baird-White, Linda: 228g
Barrett, G.: 479e
Brierley, J., www.naturesbestcreations.com: 476, 478c, 478d, 483a, 486a, 486b, 486c, 484a, 484b, 484c, 484d, 484e, 485a, 485b, 487a, 489a, 489b, 490a, 491a, 491b, 492a, 492b, 492c, 492d, 493a, 493b, 494a, 495a, 496a, 496b, 497b, 498a, 499a, 500a, 500b, 501a, 501b, 501c, 502a, 502b, 502c, 502d, 503a, 503b, 504a, 504b, 504c, 504d, 505d, 506a, 506b, 506c, 506d, 507a, 507b, 507c, 508a, 508b, 508c, 508d, 509a, 509b, 510c, 511b, 511c, 512b, 512c, 529a, 532a, 533a, 538a
Brock, Mason: 214a
Canne, Judy: 393b, 398b, 422b
Caverhill, Brennan: 578d
Chen, Peter – College of Dupage: 309f, 312f, 312g, 322e, 322f, 322g
Earley, Chris: 375b, 376b, 386c, 387a, 387b, 394a, 402a, 404a, 404b, 409c, 415b, 505b, 514, 515a, 515b, 516a, 516b, 516c, 516d, 517a, 517b, 517c, 517d, 517e, 517f, 518b, 518c, 518d, 519b, 519c, 519d, 520a, 520b, 520c, 520d, 521b, 521c, 522b, 523a, 523b, 523c, 524c, 524d, 524e, 524f, 525b, 525c, 526a, 526b, 526c, 526d, 527a, 527b, 527c, 527d, 528a, 528b, 528c, 529b, 530b, 530c, 531b, 531c, 532c, 532d, 532e, 533b, 533c, 534a, 534b, 534c, 535b, 535c, 536a, 536b, 536c, 536d, 537a, 537b, 537c, 537d, 538b, 538c, 538d, 538e, 539b, 539c, 539d, 540b, 540c, 541b, 541c, 541d
Egressy, K., www.kegressy.com: 474, 477a, 477b, 478a, 478b, 479a, 479b, 479c, 479d, 480a, 480b, 480c, 481a, 481b, 482a, 482b, 483b, 485c, 487b, 488a, 488b, 488c, 488d, 489c, 490b, 494b, 495b, 497a, 498b, 498c, 498d, 499b, 505a, 505c, 507a, 507e, 509c, 509d, 510a, 510b, 510d, 511a, 511d, 512a, 531a, 535a
Fitzgerald, Z.: 394c
Fraser, Lyndsay: 532b
Gillies, Marg: 375a, 375c, 376a, 376c, 377a, 378a, 378b, 379b, 380a, 380b, 381a, 381b, 382a, 382b, 383a, 383b, 385a, 385b, 385c, 385d, 385e, 388b, 389a, 393a, 395a, 396a, 397a, 397b, 400a, 400b, 402a, 406a, 406c, 408a, 411a, 411b, 412b, 413c, 414a, 415a, 416a, 417a, 417b, 420a, 420b, 421a, 421b, 423a, 423b
Glover, J. – Atlanta, Georgia: 530a
Gonthier, Gilles: 519a, 524b
Hille, Rob: 215f
Lockwood, Tracy: 521a
Lohr, Rhiannon: 541a
McCaw, Robert, robertmccaw.com: 535a
O'Brien, Tom: 518a
Patrikeev, Michael, www.wildnatureimages.org: 522a
Reaume, J.: 540a
Riemer, Joan: 8, 24, 36, 50, 100, 108, 370, 374, 386b
Watson, A.: 539a

Photos Licensed Under Creative Commons

Barra, A. (CC BY-SA 3.0): 368f
Berger, Joseph – Bugwood.org (CC BY 3.0 US): 197b
Bonner, Franklin – USFS (ret.), Bugwood.org (CC BY 3.0 US): 173e, 248f, 252g, 329c
Bryson, Charles T. – USDA Agricultural Research Service, Bugwood.org (CC BY 3.0 US): 344h, 346c
Caie, Stuart (CC BY 2.0): 260f
Cook, Bill – Michigan State University, Bugwood.org (CC BY 3.0 US): 215e, 220j, 221c, 265c, 333f, 333g
Cranshaw, Whitney – Colorado State University, Bugwood.org (CC BY 3.0 US): 345e
Cutler, Wendy (CC BY 2.0): 158e, 158g, 333c,
Dumat, Maja (CC BY 2.0): 252f
Enos, Jacob (CC BY 2.0): 319f
Fischer, Christian (CC BY-SA 3.0): 54c, 57a, 57b, 91d
Flogaus-Faust, Robert (CC BY 3.0): 327c
Guanandi, Mauro (CC BY 2.0): 244k, 245g
Grandmont, Jean-Pol (CC BY-SA 3.0): 174h
Hollinger, Jason (CC BY 2.0): 68e, 77a, 78a, 80e
Jasiutowicz, Krzysztof P. (CC BY-SA 3.0): 325c
Johnson, Steven G. (CC BY-SA 3.0): 214d
Kanoti, Keith – Maine Forest Service, Bugwood.org (CC BY 3.0 US): 148b, 207g, 207h, 208e, 211a, 332c, 333a, 344a, 345a, 345c
Katovich, Steven – USDA Forest Service, Bugwood.org (CC BY 3.0 US): 221b
Lavin, Matt – Montana State University (CC BY-SA 2.0): 196j, 197f, 209b, 265d, 284f, 327d, 350h, 352d
Loarie, Scott – Stanford University (CC BY 2.0): 208f, 211c
Lomas, Alex (CC BY 2.0): 91c
Luc, Valérie (CC BY-SA 3.0): 412a
MacInnes, Bob (CC BY 2.0): 174f
Manners, Malcolm (CC BY 2.0): 309b, 309d, 312c
Mayer, Joshua (CC BY-SA 2.0): 319g, 419
Mayfield, Frank (CC BY-SA 2.0): 187b, 311g, 312e
Marin, Phil (CC BY-SA 3.0): 283d, 288e
de Martigny, Charles (CC BY 2.0): 361d, 364e
Mehrhoff, Leslie J. – University of Connecticut, Bugwood.org (CC BY 3.0 US): 149f, 196e, 196c, 236g, 237d, 237f, 237g, 237h, 237j, 240b, 240c, 240d, 240e, 240f, 240g, 241a, 241b, 241d, 241h, 304f
Merridew, Jack (CC BY-SA 3.0): 326e
Miller, James H. – USDA Forest Service, Bugwood.org (CC BY 3.0 US): 308d
Miller, James H. and Bodner, Ted – Southern Weed Science Society, Bugwood.org (CC BY 3.0 US): 317a, 320c, 328e, 329b, 345b
O'Brien, Joseph – USDA Forest Service Bugwood.org (CC BY 3.0 US): 259f, 317b, 321a
Opiola, Jerzy (CC BY-SA 3.0): 360e, 362g
Payne, Jerry A. – USDA Agricultural Research Service, Bugwood.org (CC BY 3.0 US): 151b
Pennsylvania Department of Conservation and Natural Resources – Forestry Archive, Bugwood.org (CC BY 3.0 US): 326g
Porse, Sten (CC BY-SA 3.0): 158f, 177f, 177g, 251e, 257e, 309a, 311f
Powell, Dave – USDA Forest Service, Bugwood.org (CC BY 3.0 US): 270i, 272b, 344f, 345d, 345f
Price, Homer Edward (CC BY 2.0): 91b, 92a, 92b, 98a, 148f, 272c, 316d, 318e, 364f

Richardson, Jane Shelby – Duke University (CC BY 3.0): 366b
Richmond, Jamie (CC BY 2.0): 171k, 173d
Robertson, D. Gordon E. (CC BY-SA 3.0): 311e
Routledge, Rob - Sault College, Bugwood.org (CC BY 3.0 US): 29, 35a, 56b, 61, 81a, 148f, 165e, 175f, 180h, 181g, 187g, 208i, 213c, 214b, 217e, 220l, 221g, 226a, 226b, 226c, 236c, 236d, 236e, 236h, 236j, 236k, 236m, 237b, 237c, 238a, 238d, 238e, 238g, 238h, 239a, 239b, 239c, 239d, 239e, 240a, 241c, 241g, 247a, 247b, 247c, 253f, 270b, 271g, 290i, 291e, 291f, 299e, 299g, 300b, 301c, 302d, 302e,302b, 303f, 303h, 310b, 316e, 316f, 316g, 318f, 319b, 319c, 327a, 335e, 336a, 340b, 350c, 350d, 350e, 350f, 351e, 351f, 352b, 360a, 362a, 366c
Samanek, Jan – State Phytosanitary Administration, Bugwood.org (CC BY 3.0 US): 251f, 251g
Schlossberg, Jane (CC BY-SA 2.0): 342e
Setaro, Sabrina (CC BY 2.0): 214c
Siegmund, Walter (CC BY-SA 3.0): 300c, 303g, 312c, 319a, 325f
Slickers, Georg (CC BY-SA 2.0): 274f
Sturner, Jason (CC BY 2.0): 281i, 297b
Superior National Forest (CC BY 2.0): 229d, 236o, 238b, 238c, 239f, 332b, 333b, 363c, 363g
Trekell, Larry – Bugwood.org (CC BY 3.0 US): 174g
United States Department of Agriculture (CC BY 2.0): 271c
Vicol, Emilian Robert (CC BY 2.0): 244i, 245c
Waylett, Doug (CC BY 2.0): 265b
Xaver, Franz (CC BY-SA 3.0): 325b

License Types
CC BY 2.0: Creative Commons Attribution 2.0 Generic
   http://creativecommons.org/licenses/by/2.0/
CC BY 3.0: Creative Commons Attribution 3.0 Unported
   http://creativecommons.org/licenses/by/3.0/
CC BY 3.0 US: Creative Commons Attribution 3.0 United States
   http://creativecommons.org/licenses/by/3.0/us/
CC BY-SA 2.0: Creative Commons Attribution-ShareAlike 2.0 Generic
   http://creativecommons.org/licenses/by-sa/2.0/
CC BY-SA 3.0: Creative Commons Attribution-ShareAlike 3.0 Unported
   http://creativecommons.org/licenses/by-sa/3.0/

Index

Note: **Bolded** entries indicate genus or species profiles.

A

Abies **140**
  balsamea 126, **140**
Accipiter
  cooperii **478**
  striatus **478**
Acer 115, 116, 128, **141–151**
  × *freemanii* 141
  negundo 127, 143, **147**
  nigrum 142, **150**
  pensylvanicum 142, **148**
  platanoides 143, **149**
  pseudoplatanus 144, **149**
  rubrum 144, **151**
  saccharinum 136, 144, **151**
  saccharum 136, 142, **150**
  spicatum 139, 141, **148**
Actaea
  pachypoda **374, 375,** 376
  rubra **374, 376**
Adiantum
  pedatum 69, **73**
Adoxaceae 324–325, 356–366
Aesculus 116, 127, **152–153**
  glabra 152, **153**
  hippocastanum 138, 152, **153**
Agelaius
  phoeniceus **511**
Aix
  sponsa **480**
Alder **154–155**
  black 154
  green **155**
  speckled 154, **155**
Alliaria
  petiolata 372, **377**
Alnus 118, 121, 132, **154–155**
  glutinosa 154
  incana ssp. *rugosa* **155**
  viridis ssp. *crispa* **155**
Amelanchier 120, 122, 134, **156**
Anacardiaceae 293–297, 343–346
Anas
  platyrhynchos **480**
Aneuraceae 32
Annonaceae 159

Apple **243**
  domesticated 243
Aquifoliaceae 218–219
Aquilegia
  canadensis 374, **378**
Araceae 379
Arbutus
  trailing **199**
Archilochus
  colubris **489**
Arctostaphylos **157**
  uva-ursi 112, 134, **157**
Arisaema
  triphyllum 373, **379**
Aristolochiaceae 380
Aronia **158**
  melanocarpa 122, 133, **158**
Arrow-wood
  downy 358, **363**
  southern 359, **364**
Arthonia
  caesia 433, **438,** 450
Asarum
  canadense 373, **380**
Ash **204–211**
  black 205, **210**
  blue 204, **208**
  European 204, **210**
  pumpkin 206, **209**
  red 206, **209**
  white 137, 206, **211**
Asimina **159**
  triloba 118, 135, **159**
Asparagaceae 395–397, 401–402, 407
Aspen **262–266**
  largetooth 262, **266**
  trembling 138, 262, **266**
Aspleniaceae 74–75
Asplenium 70, **74–75**
  platyneuron **75**
  rhizophyllum 74, **75**
  ruta-muraria **75**
  scolopendrium **75**
  trichomanes 74, **75**
    ssp. *quadrivalens* **75**
    ssp. *trichomanes* **75**
  trichomanes-ramosum 74, **75**
Asteraceae 408, 414
Athyriaceae 76, 82–83
Athyrium
  filix-femina 70, **76**
Atrichum 37, **38**
  angustatum **38**

undulatum 38
Autumn-olive 196, **198**
Avens
 white **392**
 yellow **391**

B

Baltimore oriole **510**
Baneberry
 red **376**
 white **375**
Barberry **160–161**
 European 160, **161**
 Japanese 160, **161**
Basswood **342**
 American 137, **342**
Bat
 eastern red **521**
 little brown **521**
Bayberry **246–247**
 northern 246, **247**
Bazzania
 trilobata **35**
Bearberry **157**
 common **157**
Beaver
 American **531**
Beech **203**
 American 136, **203**, 347, 449
 blue **168**
Bellwort
 large-flowered **415**
 perfoliate **416**
Berberidaceae 160–161, 384, 399
Berberis 113, 127, **160–161**
 thunbergii 160, **161**
 vulgaris 160, **161**
Betula 119, 120, 121, 132. **162–166**
 alleghaniensis 162, **165**
 lenta 162, **165**
 papyrifera 138, 163, **166**
 populifolia 163, **166**
Betulaceae 154–155, 162–166, 168, 189–190, 249
Bignoniaceae 167, 175
Birch **162–166**, 168
 cherry 162, **165**
 grey 163, **166**
 white 138, 163, **166**
 yellow 162, **165**, 203, 347
Bittersweet **176**

Blackberry
 common 315, **322**
 smooth 313, **320**
Blackbird
 red-winged **511**
 rusty **511**
Black gum **248**
Bladdernut **338**
Blarina
 brevicauda **519**
Bloodroot 12, **406**
Blue beech **168**
Blueberry 103, 109, **354–355**
 lowbush **355**
 velvet-leaf **355**
Bluebird
 eastern **496**
Blue cohosh **384**
Blue jay **491**
Bombycilla
 cedrorum **497**
 garrulus **497**
Bonasa
 umbellus **482**
Boraginaceae 394
Botrychium
 virginianum 68, **77**
Brachytheciaceae 48
Brachythecium 37, 47, **48**
Branta
 canadensis **481**
Brassicaceae 377, 382–383
British soldiers **444**
Brown bat
 little **521**
Brown thrasher **499**
Bryaceae 39
Bryum **39**
Bubo
 virginianus **479**
Buckeye
 Ohio 152, **153**
Buckthorn 110, **289–292**
 alder-leaved 289, **291**
 common 11, 12, 13, 290, **291**
 glossy 289, **292**
Buffalo berry **327**
Bunchberry **386**
Bunting
 indigo **501**
Burning bush 139, 200, **202**

563

Bush honeysuckle 193
  northern 193
Buteo
  jamaicensis 477
Butternut 137, 220, **221**
Buttonbush 178

C

Callicladium
  haldanianum 47
Caloplaca
  cerina 433, **439**
  flavorubescens 440
  flavovirescens 433, **440**
Caltha
  palustris 372, **381**
Campsis 167
  radicans 114, 127, **167**
Candelaria
  concolor 432, 434, **441**, 468
Candelariella
  aurella 433, **442**
  vitellina 442
Canis
  familiaris 537
  latrans 536
Cannabaceae 177
Caprifoliaceae 193, 233–241
Cardamine
  concatenata 372, **382**, 383
  diphylla 372, 382, **383**
Cardinal
  northern 502
Cardinalis
  cardinalis 502
Carduelis
  flammea 507
  pinus 506
  tristis 507
Carpinus 168
  caroliniana 120, 133, **168**, 249
Carpodacus
  mexicanus 508
  purpureus 508
Carya 118, 130, **169–173**
  cordiformis 169, **173**
  glabra 169, **173**
  laciniosa 170, **172**
  ovata 137, 170, **172**
Castanea 174
  dentata 117, 120, 132, 138, **174**

Castor
  canadensis 531
Cat
  domestic 537
Catalpa 175
  northern 175
  speciosa 114, 116, 129, **175**
Catbird
  grey 499
Catharus
  guttatus 495
Caulophyllum
  thalictroides 373, 375, 376, 378, **384**, 409
Cedar 341
  eastern red 224
  eastern white 341
Celandine
  lesser 404
Celastraceae 176, 200–202
Celastrus 176, 200
  scandens 112, 126, **176**
Celtis 177
  occidentalis 116, 119, 132, **177**
Cephalanthus 178
  occidentalis 114, 115, 129, **178**
Cercis 179
  canadensis 119, 120, 135, 139, **179**
Certhia
  americana 493
Chamaedaphne 180
  calyculata 112, 134, **180**
Cherry 268–273
  black 269, **273**
  choke 269, **272**
  pin 138, 268, **272**
Chesnut 174
  American 138, **174**
Chickadee
  black-capped 491
Chipmunk
  eastern 528
Chokeberry 158
  black 158
Cinquefoil 267
  shrubby 267
Cladonia
  chlorophaea 445
  coniocraea 434, **443**
  cristatella 434, **444**
  fimbriata 434, **445**
  ochrochlora 443

rei 434, **446**
Claytonia
  caroliniana 372, **385**
  virginiana 385
Clematis 114, 127, **181**
  occidentalis 181
  virginiana 181
Cliff-brake
  slender **78**
Climaciaceae 44
Climacium
  dendroides 44
Club-moss 53–59
  bristly **59**
  common **59**
  northern bog **57**
  tree 59
    Hickey's 59
  prickly **59**
Coccothraustes
  vespertinus **509**
Coccyzus
  erythropthalmus **483**
Coffee tree
  Kentucky **216**
Colaptes
  auratus **486**
Colchicaceae 415–416
Coltsfoot 12, 414
Columbine **378**
Comptonia **182**
  peregrina 116, 121, 131, **182**
Condylura
  cristata **522**
Conocephalaceae 31
Conocephalum
  conicum **31**
Contopus
  virens **487**
Cornaceae 183–188, 386
Cornus 115, 128, **183–188**, 289
  alternifolia 120, 122, 134, 139, 183, **186**
  canadensis 183, 372, **386**
  florida 115, 183, **186**
  racemosa 115, 184, **187**
  rugosa 184, **187**
  sericea 184, **188**
Corvus
  brachyrhynchos **493**
Corylus 117, 121, 133, **189–190**
  americana 189, **190**
  cornuta 189, **190**

Cottontail
  eastern **533**
Cottonwood
  eastern 263, **265**
Cowbird
  brown-headed **512**
Coyote 535, **536**
Crabapple **243**
  wild 243
Cranberry
  highbush 358, **363**
Cranberry bush
  European 358, **362**
Crataegus 113, 127, 131, 139, **191**
Creeper
  brown **493**
  thicket 250, **251**
  trumpet **167**
  Virginia 250, **251**
Crow
  American **493**
Crowfoot
  hooked **405**
  small-flowered **403**
Cryptogramma
  stelleri 69, **78**
Cuckoo
  black-billed **483**
Cucumber tree **242**
Cupressaceae 222–224, 341
Currant
  black
    American 298, **301**
    swamp **303**
  skunk 298, **302**
  wild red 299, **302**
Cyanocitta
  cristata **491**
Cypripedium
  calceolus 373, **387**
Cystopteridaceae 79–81, 88
Cystopteris 72, **79–81**
  bulbifera 79, **80**
  fragilis **80**
  laurentiana 79, 80, **81**
  montana 81
  protrusa 81
  tenuis 80, **81**

D

Dandelion 408
Daphne 192
   mezereum 117, 135, 192
Deer
   white-tailed 11, 532
Deer mouse
   North American 523
Dendroica
   petechia 500
   pinus 500
Dennstaedtiaceae 96
Dentaria
   diphylla — see Cardamine diphylla
   laciniata — see Cardamine concatenata
Deparia
   acrostichoides 70, 82
Dewberry
   northern 320
   swamp 321
Dicranaceae 40
Dicranella 40
   heteromalla 40
Dicranum 40
   flagellare 40
   montanum 40
   polysetum 40
Didelphis
   virginiana 518
Diervilla 193
   lonicera 114, 128, 193
Diphasiastrum
   digitatum 54, 55
Diplazium
   pycnocarpon 70, 83
Dirca 194
   palustris 118, 135, 194
Dog
   domestic 537
Dogwood 109, 183–188
   alternate-leaved 139, 183, 186
   eastern flowering 183, 186
   grey 184, 187
   red-osier 184, 188
   round-leaved 184, 187
Doll's eyes 375
Dove
   mourning 483
Dryocopus
   pileatus 485
Dryopteridaceae 84–87, 95

Dryopteris 72, 84–87
   carthusiana 85, 87
   clintoniana 85, 86
   cristata 85, 86
   expansa 87
   filix-mas 87
   fragrans 87, 99
   goldiana 85, 86
   intermedia 85, 87
   marginalis 84, 86
   × triploidea 87
Duck
   wood 480
Dumetella
   carolinensis 499

E

Elaeagnaceae 195–198, 327
Elaeagnus 112, 135, 195–198, 327
   angustifolia 195, 197
   commutata 195, 197
   umbellata 196, 198
Elder 324
Elderberry 324–325
   American 324, 325
   eastern red 324, 325
Elm 348–353
   English 349, 351
   rock 137, 348, 353
   Scotch 349, 351
   slippery 137, 349, 352
   white 137, 348, 352
Epigaea 199
   repens 112, 134, 199
Equisetaceae 60–63
Equisetum 60–63
   arvense 61, 62
   fluviatile 62
   hyemale ssp. affine 61, 63
   laevigatum 63
   palustre 62
   pratense 61, 62
   scirpoides 61, 63
   sylvaticum 60, 62
   variegatum ssp. variegatum 61, 63
Erethizon
   dorsatum 530
Ericaceae 157, 180, 199, 212–214, 225–226, 229, 354–355
Ermine 540, 541
Erythronium
   americanum 388

Euonymus 115, 128, **200–202**
  atropurpureus 139, 200, **202**
  europaeus 200, **202**
  obovatus 200, **201**
Euphagus
  carolinus **511**
Evernia
  mesomorpha 434, **447**
  prunastri 447

F

Fabaceae 179, 215–216, 304
Fagaceae 174, 203, 275–288
Fagus 203
  grandifolia 117, 120, 132, 136, **203**, 449
Felis
  catus **537**
Fern 65–99
  beech
    northern **93**
  bladder **79–81**
    bulblet **80**
    mountain 81
    southern 81
  bracken **96**
  Christmas **95**
  cinnamon **92**
  fragile **80**
    Laurentian **81**
    Mackay's **81**
  glade
    narrow-leaved **83**
    silvery **82**
  hart's-tongue 75
  holly 95
    Braun's 95
  interrupted **91**
  lady **76**
  maidenhair
    northern **73**
  male 87
  Marsh **98**
  New York **98**
  oak **88**
    limestone 88
    Nahanni 88
  Ophioglossoid 51, 65, 77
  ostrich **89**
  rattlesnake **77**
  royal **91**
  sensitive **90**
  walking **75**
  wood **84–87**
    Clinton's **86**
    crested **86**
    evergreen **87**
    fragrant 87
    Goldie's **86**
    marginal **86**
    northern 87
    spinulose **87**
Finch
  house **508**, 509
  purple **508**, 509
Fir **140**
Fir-moss
  shining **56**
Fissidens **41**
Fissidentaceae 41
Flavoparmelia
  caperata 436, **448**
Flicker
  northern **486**
Flycatcher
  great crested **488**
Foamflower **410**
Fox
  red **535**
Fragaria
  virginiana 372, **389**
Frangula
  alnus – see *Rhamnus frangula*
Fraxinus 116, 127, **204–211**
  americana 137, 206, **211**
  excelsior 204, **210**
  nigra 205, **210**
  pennsylvanica 206, **209**
    var. *subintegerrima* 206. **211**
  profunda 206, **209**
  quadrangulata 204, **208**
Frullania **34**

G

Garlic mustard 11, 12, 13, **377**
Gaultheria 112, **212–213**
  hispidula 134, **213**
  procumbens 134, **213**
Gaylussacia **214**
  baccata 117, 135, **214**
Gaywings **400**
Geothlypis
  trichas **501**
Geraniaceae 390

Geranium
　robertianum 372, **390**
Geum
　aleppicum 372, **391**
　canadense 372, **392**
Ginger
　wild **380**
Gleditsia **215**
　triacanthos 113, 119, 129, 138, **215**
Goldfinch
　American **507**
Goose
　Canada **481**
Gooseberry
　prickly **303**
Grackle
　common **512**
Grape **367**
Graphis
　scripta 433, **449**
Greenbrier **328–329**
　prickly 328, **329**
　round-leaved **329**
Grey squirrel
　eastern **526**
Grosbeak
　evening **509**
　pine **510**
　rose-breasted **502**
Grossulariaceae **298–303**
Groundhog **529**
Grouse
　ruffed **482**
Gymnocarpium
　dryopteris 71, **88**
　jessoense ssp. parvalum **88**
　robertianum **88**
Gymnocladus **216**
　dioicus 118, 119, 129, **216**

H

Hackberry **177**
Hamamelidaceae **217**
Hamamelis **217**
　virginiana 121, 132, **217**
Hawk
　Cooper's 477, **478**
　red-tailed **477**
　sharp-shinned 477, **478**
Hawthorn 110, 139, **191**
Hazel **189–190**
　American 189, **190**

　beaked 189, **190**
Hemlock 203, **347**
　eastern 136, 140, **347**
Hepatica 12, **393**
　acutiloba 372, 374, **393**
Herb-robert **390**
Hickory **169–173**
　bitternut 169, **173**
　red 169, **173**
　shagbark 137, 170, **172**
　shellbark 170, **172**
Hobblebush **364**
Holly **218–219**
　English **218**
　mountain **219**
Honey locust 138, **215**
Honeysuckle 139, 193, **233–241**
　Amur 235, **240**
　Bell's 234, **241**
　fly **239**
　glaucous 233, **238**
　hairy **238**
　Morrow's 235, **240**
　mountain fly **239**
　northern bush **193**
　Tatarian 234, **241**
Hop-tree **274**
　common **274**
Hornbeam **168**
　American **168**
Horsechesnut 138, **153**
Horsetail **60–63**
　field **62**
　marsh 62
　meadow **62**
　water 62
　wood **62**
Huckleberry **214**
　black **214**
Hummingbird
　ruby-throated **489**
Huperzia
　lucidula 54, **56**, 59
Hydrangeaceae **252**
Hydrophyllum
　virginianum 372, **394**
Hylocichla
　mustelina **495**
Hylocomiaceae **45**
Hypnaceae **46–47**
Hypnum **46**

I

Icterus
   glabula **510**
Ilex 117, **218–219**
   aquifolium 218
   mucronata 135, **219**
   verticillata 134, 218, **219**
Indigo bunting **501**
Ironwood 138, **249**

J

Jack-in-the-pulpit **379**
Jay
   blue **491**
Jubulaceae 34
Juglandaceae 169–173, **220–221**
Juglans 118, 120, 130, **220–221**
   cinerea 137, 220, **221**
   nigra 220, **221**
Jumping mouse
   woodland **522**
Junco
   dark-eyed **506**
   hyemalis **506**
Jungermanniaceae 33
Juniper **222–224**
   common **223**
   creeping **223**
Juniperus 126, **222–224**
   communis 126, **223**
   horizontalis **223**
   virginiana **224**

K

Kalmia 112, 128, **225–226**
   angustifolia **225**
   polifolia **226**
Kentucky coffee tree **216**
Kingbird
   eastern **488**

L

Labrador tea **229**
Lady's slipper **387**
Lanius
   excubitor **498**
Larch **227–228**
   European 227, **228**

Larix 116, 126, **227–228**
   decidua 227, **228**
   laricina 227, **228**
Lasiurus
   borealis **521**
Lauraceae 231, 326
Laurel **225–226**
   bog **226**
   sheep **225**
Leatherleaf **180**
Leatherwood **194**
   eastern **194**
Lecanora
   thysanophora 434, 438, **450**, 451
Ledum **229**
   groenlandicum 112, 134, **229**
Lepidoziaceae 35
Lepraria
   finkii 433, 450, **451**
Leucobryaceae 42
Leucobryum
   glaucum **42**
Lichen
   abraded camouflage **453**
   boreal oakmoss **447**
   bottlebrush frost **464**
   candleflame **441**
   common greenshield **448**
   common script **449**
   fluffy dust **451**
   frosted comma **438**
   frosted grain-spored **467**
   grey-rimmed firedot **439**
   hairy shadow **457**
   hammered shield **454**
   hidden goldspeck **442**
   hooded rosette **461**
   hooded sunburst **468, 469**
   lesser powderhorn **443**
   Maritime sunburst **471**
   mealy rosette **462**
   mealy shadow **458**
   orange-cored shadow **460**
   pincushion orange **472**
   pompon-tipped shadow **459**
   powder-tipped shadow **456**
   rough speckled shield **466**
   star rosette **463**
   sulphur-firedot **440**
   trumpet **445**
   wand **446**
   yellow-edged frost **465**

Ligustrum **230**
  vulgare 115, 129, **230**
Lilac **339**
Liliaceae **388**
Lily
  trout 12, **388**
Lindera **231**
  benzoin 121, 135, **231**
Linnaea
  borealis 212
Liriodendron **232**
  tulipifera 117, 131, **232**
Little brown bat **521**
Liverwort 29–35
  alligator **31**
  umbrella **30**
Locust **304**
  black 138, **304**
  honey 138, **215**
Lonicera 114, 115, 127, 129, **233–241**
  × *bella* 234, 237, **241**
  canadensis **239**
  dioica 233, **238**
  hirsuta **238**
  maackii 235, **240**
  morrowii 235, 237, **240**
  tatarica 234, 237, **241**
  villosa **239**
Lophozia 11, 29, **33**
Lycopodiaceae 53–59
Lycopodiella
  inundata 54, **57**
Lycopodium 54, **58–59**
  annotinum 56, 58, **59**
  clavatum 58, **59**
  dendroideum 58, **59**
  hickeyi 59
  lagopus 59
  obscurum 59

M

Magnolia **242**
  acuminata 117, 118, 135, **242**
Magnoliaceae 232, 242
Maianthemum
  canadense 372, **395**
  racemosum 373, **396**
  stellatum 373, **397**
Mallard **480**
Malus 113, 120, 127, 131, 132, 134, **243**
  coronaria 243
  domestica 243

Malvaceae 342
Maple **141–151**
  black 142, **150**
  Manitoba 143, **147**
  mountain 139, 141, **148**
  Norway 13, 143, **149**
  red 144, **151**
  silver 136, 144, **151**
  striped 142, **148**
  sugar 136, 142, **150**, 203
  sycamore 144, **149**
Mapledust **450**
Marchandiomyces
  corallinus 433, **452**
Marchantia
  polymorpha **30**
Marchantiaceae 30
Marigold
  marsh **381**
Marmota
  monax **529**
Matteuccia
  struthiopteris 69, **89**
Mayapple **399**
Mayflower
  Canada **395**
Meadow-rue
  early **409**
Meadowsweet
  broad-leaved 334, **336**
  narrow-leaved 335, **337**
Meadow vole **524**
Megascops
  asio **479**
Melanelixia
  fuliginosa 453
  subargentifera 453
  subaurifera 434, 435, **453**
Melanerpes
  carolinus **486**
Melanthiaceae 412–413
Meleagris
  gallopavo **482**
Melospiza
  melodia **503**
Mephitis
  mephitis **538**
Microtus
  pennsylvanicus **524**
Mink
  American **539**
Mitella
  diphylla 373

Mniaceae 43
Mnium 43
Mock-orange **252**
Mole
   star-nosed **522**
Molothrus
   ater **512**
Montiaceae 385
Moraceae 244–245
Morus 117, 131, **244–245**
   alba 244, **245**
   rubra 244, **245**
Moss 37–49
   beautiful **47**
   broom **40**
   electrified cat tail **45**
   fern **49**
   pigtail **46**
   pin cushion **42**
   sidewalk **39**
   tree **44**
Mountain-ash **331–333**
   American 331, **333**
   European 331, **332**
   showy 331, **333**
Mourning dove **483**
Mouse
   North American deer **523**
   white-footed **523**
   woodland jumping **522**
Mulberry **244–245**
   red 244, **245**
   white 244, **245**
Mustela
   erminea 540, **541**
   frenata **540**, 541
   vison **539**
Myiarchus
   crinitus **488**
Myotis
   lucifuga **521**
Myrica 133, **246–247**
   gale 121, 246, **247**
   pensylvanica 122, 135, 246, **247**
Myricaceae 182, 246–247

N

Nannyberry 357, **365**
Nightshade **330**
   bittersweet **330**
Ninebark **253**

Nuthatch
   red-breasted **492**
   white-breasted **492**
Nyssa **248**
   sylvatica 120, 135, **248**
Nyssaceae 248

O

Oak **275–288**
   black 276, **286**
   bur 278, **284**
   chinquapin 278, **285**
   dwarf chinquapin 278, **285**
   northern pin 277, **288**
   pin 277, **288**
   red 137, 276, **286**
   scarlet 276, **287**
   Shumard 275, **287**
   Swamp white 279, **284**
   white 279, **283**
Odocoileus
   virginianus **532**
Ohio buckeye 152, **153**
Oleaceae 204–211, 230, 339
Oleaster **195–198**
Onoclea
   sensibilis 68, **90**
Onocleaceae 89–90
Ophioglossaceae 77
Opossum
   Virginia **518**
Orchid
   Lady's slipper **387**
Orchidaceae 387
Oriole
   Baltimore **510**
Osmunda 68, **91**, 92
   cinnamomea – see *Osmundastrum cinnamomeum*
   claytoniana **91**, 92
   regalis **91**
Osmundaceae 91–92, 92
Osmundastrum
   cinnamomeum 68, **92**
Ostrya 121, **249**
   virginiana 133, 138, 168, **249**
Owl
   eastern screech **479**
   great horned **479**

P

Papaveraceae 406
Parmelia
  saxatilis 454
  squarrosa 454
  sulcata 432, 436, **454**
Parthenocissus 113, 130, **250–251**
  quinquefolia 250, **251**
  vitacea 250, **251**
Passerella
  iliaca 503
Passerina
  cyanea 501
Pawpaw **159**
Peromyscus
  leucopus 523
  maniculatus 523
Phaeocalicium
  polyporaeum 433, **455**
Phaeophyscia
  adiastola 435, 436, **456**, 458, 459, 460
  hirsuta 435, 436, **457**
  hirtella **457**
  hispidula 456, 458
  kairamoi **457**
  orbicularis 435, 436, 456, **458**, 459
  pusilloides 435, 436, 456, **459**, 460
  rubropulchra 435, 436, 456, 459, **460**
Phegopteris
  connectilis 71, **93**
Pheucticus
  ludovicianus **502**
Philadelphus 114, 128, **252**
Phlox
  divaricata 372, **398**
  wild blue **398**
Phoebe
  eastern **487**
Physcia
  adscendens 432, 435, **461**
  aipolia 454, 461, 463
  dubia 463
  millegrana 432, 436, **462**
  stellaris 432, 435, 452, 454, **463**
  subtilis 462
Physciella
  melanchra 458
Physconia
  detersa 435, 436, **464**, 465
  enteroxantha 435, 436, **464**, **465**
  grisea 464, 465
  leucoleiptes 464, 465

Physocarpus 253
  opulifolius 118, 121, 131, **253**
Picea 126, **254–257**
  abies 256
  glauca 257
  mariana 136, 256
  pungens 257
  rubens 255
Picoides
  pubescens **484**
  villosus **484**
Pinaceae 140, 227–228, 254–260, 347
Pine **258–260**
  jack 136, **260**
  red 136, **259**
  Scots 136, **260**
  white 136, **259**, 347
Pine siskin **506**
Pinicola
  enucleator **510**
Pinus 126, **258–260**
  banksiana 136, **260**
  resinosa 136, **259**
  strobus 136, **259**
  sylvestris 136, **260**
Plagiomnium **43**
  cuspidatum 43
Platanaceae 261
Platanus **261**
  occidentalis 118, 131, **261**
Plum **268–273**
  American 268, **271**
  Canada 269, **271**
Podophyllum
  peltatum 373, **399**
Poecile
  atricapilla **491**
Poison ivy **343–346**
Poison sumac **345**
Polemoniaceae 398
Polygala
  paucifolia 374, **400**
Polygalaceae 400
Polygonatum
  biflorum 373, **401**
  pubescens 373, 401, **402**
Polypodiaceae 94
Polypodium
  virginianum 71, **94**
Polypody
  rock **94**

Polystichum
   acrostichoides 72, 83, **95**
   braunii 95
   lonchitis 95
Polytrichaceae 38
Polytrichum 38
Poplar **262–266**
   balsam 263, **265**
Populus 119, 134, **262–266**
   balsamifera 263, **265**
   deltoides ssp. *deltoides* 263, **265**
   grandidentata 262, **266**
   tremuloides 138, 262, **266**, 439
Porcupine
   North American **530**
Porpidia
   albocaerulescens 467
   cinereoatra 467
Potentilla **267**
   fruticosa 116, 130, **267**
Prickly-ash **368**
Primulaceae 411
Privet **230**
   common **230**
Procyon
   lotor **534**
Prunus 113, 120, 122, 127, 133, **268–273**
   americana 268, **271**
   nigra 269, **271**
   pensylvanica 138, 268, **272**
   serotina 269, **273**
   virginiana 269, **272**
Ptelea **274**
   trifoliata 119, 121, 130, **274**
Pteridaceae 73, 78
Pteridium
   aquilinum 69, **96**
Punctelia
   rudecta 435, **466**

Q

Quercus 117, 131, 132, **275–288**
   alba 279, **283**
   bicolor 279, **284**
   coccinea 276, **287**
   ellipsoidalis 277, **288**
   macrocarpa 278, **284**
   muehlenbergii 278, **285**
   palustris 277, **288**
   prinoides 278, **285**
   rubra 137, 276, **286**
   shumardii 275, **287**

   velutina 276, **286**
Quiscalus
   quiscula **512**

R

Raccoon
   northern **534**
Ranunculaceae 181, 375–376, 378, 381, 393, 403–405, 409
Ranunculus
   abortivus 372, **403**
   ficaria 372, **404**
   recurvatus 372, **405**
Raspberry
   black 315, **322**
   dwarf **319**
   purple-flowering 313, **318**
   wild red 314, **321**
Rat
   Norway **525**
Rattus
   norvegicus **525**
Red bat
   eastern **521**
Redbud 139, **179**
Red fox **535**
Redpoll
   common **507**
Red squirrel **527**
Rhamnaceae 289–292
Rhamnus **289–292**
   alnifolia 121, 132, 289, **291**
   cathartica 114, 128, 290, **291**
   frangula 121, 134, 135, 289, **292**
Rhizomnium 43
Rhus 118, 130, **293–297**, 343
   aromatica 117, 130, 293, **296**
   copallina 131, 294, **296**
   glabra 293, **297**
   typhina 294, **297**
Rhytidiadelphus
   triquetrus **45**
Ribes 113, 121, 131, **298–303**
   americanum 298, **301**
   cynosbati **303**
   glandulosum 298, **302**
   lacustre **303**
   triste 299, **302**
Riccardia
   latifrons **32**
   multifida 32
   palmata 32

Riccia **33**
Ricciaceae **33**
Robin
   American **496**
Robinia **304**
   pseudoacacia 113, 129, 138, **304**
Rock-shield
   Cumberland **470**
Rosa 113, 130, **305–312**
   acicularis 306, **312**
   blanda 122, 307, **312**
   carolina **311**
   eglanteria 307, **310**
   multiflora 305, **310**
   palustris 306, **311**
   setigera 305, **309**
Rosaceae 156, 158, 191, 243, 253, 267–273, 305–322, 331–337, 389, 391–392, 425
Rose **305–312**
   multiflora 305, **310**
   pasture **311**
   prairie 305, **309**
   swamp 306, **311**
   wild
      prickly 306, **312**
      smooth 307, **312**
Rubiaceae 178
Rubus 113, 116, 130, 131, **313–322**
   allegheniensis 315, **322**
   canadensis 313, **320**
   flagellaris **320**
   hispidus **321**
   idaeus 314, **321**
   occidentalis 315, **322**
   odoratus 313, **318**
   parviflorus **319**
   pubescens **319**
Running-pine
   southern **55**
Russian olive 195, **197**
Rutaceae 274, 368

S

Salicaceae 262–266, 323
Salix 114, 119, 121, 128, 133, 135, 138, 139, 323
Sambucus 114, 127, **324–325**
   canadensis 324, **325**
   pubens 324, **325**
Sanguinaria
   canadensis 374, **406**

Sapindaceae 141–153
Sarcogyne
   regularis 433, **467**
Sassafras **326**
   albidum 116, 131, **326**
Saxifragaceae 410
Sayornis
   phoebe **487**
Scilla
   siberica 373, **407**
Sciurus
   carolinensis **526**
Scolopax
   minor **481**
Scouring rush 60–63
   common **63**
   dwarf **63**
   smooth **63**
   variegated **63**
Screech-owl
   eastern **479**
Serviceberry **156**
Shadbush **156**
Shepherdia **327**
   canadensis 115, 128, **327**
Shrew
   masked **520**
   northern short-tailed **519, 524**
Shrike
   northern 477, **498**
Sialia
   sialis **496**
Silverberry 195, **197**
Siskin
   pine **506**
Sitta
   canadensis **492**
   carolinensis **492**
Skunk
   striped **538**
Smilacaceae 328–329
Smilacina
   racemosa – see *Maianthemum racemosum*
   stellata – see *Maianthemum stellatum*
Smilax 112, 126, **328–329**
   rotundifolia **329**
   tamnoides 328, **329**
Snowberry
   creeping **213**
Solanaceae 330
Solanum **330**
   dulcamara 112, 126, **330**

Solomon's seal
    false **396**
    hairy **402**
    smooth **401**
    starry false **397**
Sorbus 118, 130, **331–333**
    americana 331, **333**
    aucuparia 331, **332**
    decora 331, **333**
Sorex
    cinereus **520**
Sparrow
    American tree **505**
    chipping **505**
    fox **503**
    song **503**
    white-crowned **504**
    white-throated **504**
Spicebush **231**
Spindle-tree **200–202**
    European 200, **202**
Spiraea 117, 133, **334–337**
    alba 335, **337**
    alba var. *latifolia* 334, **336**
    tomentosa 334, **336**
Spizella
    arborea **505**
    passerina **505**
Spleenwort **74–75**
    ebony **75**
    green **75**
    maidenhair **75**
Spring beauty **385**
Spruce **254–257**
    black 136, **256**
    Colorado **257**
    Norway **256**
    red **255**
    white **257**, 347
Squill
    Siberian **407**
Squirrel
    eastern grey **526**
    red **527**
Staff-tree **176**
Staphylea **338**
    trifolia 114, 115, 127, **338**
Staphyleaceae **338**
Starflower **411**
Starling
    European **498**
Star-nosed mole **522**

Steeple-bush 334, **336**
Strawberry
    barren **425**
    wild **389**
Strawberry-bush
    running 200, **201**
Stubble
    polypore **455**
Sturnus
    vulgaris **498**
Sumac **293–297**
    fragrant 293, **296**
    poison 139, **345**
    shining 294, **296**
    smooth 293, **297**
    staghorn 294, **297**
Swallow
    tree **489**
Sweetbrier 307, **310**
Sweet-fern **182**
Sweet gale 246, **247**
Sycamore **261**
Sylvilagus
    floridanus **533**
Syringa 115, 129, 230, **339**
    vulgaris 339

T

Tachycineta
    bicolor **489**
Tamarack 227, **228**
Tamias
    striatus **528**
Tamiasciurus
    hudsonicus **527**
Taraxacum
    officinale 374, **408**
Taxaceae **340**
Taxus 222, **340**
    canadensis 126, **340**
Thalictrum
    dioicum 374, 375, 376, 378, **409**
Thelypteridaceae **97**
Thelypteris 71, **97**
    noveboracensis 97, **98**
    palustris 97, **98**
Thicket creeper 250, **251**
Thimbleberry **319**
Thrasher
    brown **499**
Thrush
    hermit **495**

wood **495**
Thuidiaceae 49
Thuidium **49**
Thuja 222, **341**
  occidentalis 126, **341**
Thymelaeaceae 192, 194
Tiarella
  cordifolia 372, **410**
Tilia **342**
  americana 117, 119, 132, 137, **342**
Toothwort
  broadleaf **383**
  five-parted **382**
Toxicodendron 118, 293, **343–346**
  radicans 130, 343
    ssp. *radicans* 112, 130, 344, **346**
    ssp. *rydbergii* 118, 130, 343, **345**
  vernix 117, 119, 131, 139, **345**
Toxostoma
  rufum **499**
Tree sparrow
  American **505**
Trientalis
  borealis 372, **411**
Trillium
  erectum 373, **412**
  grandiflorum 373, **413**
  red **412**
  white **413**
Troglodytes
  aedon **494**
  troglodytes **494**
Trumpet creeper **167**
Tsuga **347**
  canadensis 126, 136, 140, **347**
Tulip tree **232**
Tupelo **248**
Turdus
  migratorius **496**
Turkey
  wild **482**
Tussilago
  farfara 374, **414**
Twinflower 212
Tyrannus
  tyrannus **488**

U

Ulmaceae 348–353
Ulmus 119, 120, 132, **348–353**
  americana 137, 348, **352**
  glabra 349, **351**

procera 349, **351**
rubra 137, 349, **352**
thomasii 137, 348, **353**
Uvularia
  grandiflora 373, **415**
  perfoliata 373, **416**

V

Vaccinium 117, **354–355**
  angustifolium 133, **355**
  myrtilloides 135, 214, **355**
Viburnum 114, 115, 128, **356–366**
  acerifolium 359, **362**
  lantana 356, **365**
  lantanoides **364**
  lentago 357, **365**
  maple-leaved 359, **362**
  nudum var. *cassinoides* 129, 357, **366**
  opulus 114, 358, **362**
  rafinesquianum 358, **363**
  recognitum 359, **364**
  trilobum 358, **363**
Viola
  canadensis 374, **417**, 420
  conspersa 374, **418**, 422
  cucullata 373, **419**, 424
  macloskeyi 373, **420**
  odorata 373
  pubescens 374, **421**
  rostrata 374, 418, **422**
  selkirkii 373, **423**
  septentrionalis 373, 419, **424**
Violaceae 417–424
Violet
  Canada **417**
  dog **418**
  downy yellow **421**
  great-spurred **423**
  long-spurred **422**
  marsh blue **419**
  northern blue **424**
  northern white **420**
Vireo
  gilvus **490**
  olivaceus **490**
  red-eyed **490**
  warbling **490**
Virginia creeper 250, **251**
Virgin's-bower **181**
Vitaceae 250–251, 367
Vitis 112, 126, **367**
  vinifera 367

Vole
    meadow **524**
Vulpes
    vulpes **535**

W

Waldsteinia
    fragarioides 372, **425**
Wall-rue 75
Walnut **220–221**
    black 220, **221**
Warbler
    pine **500**
    yellow **500**
Waterleaf **394**
Waxwing
    Bohemian **497**
    cedar **497**
Wayfaring tree 356, **365**
Weasel
    long-tailed **540**, 541
    short-tailed 541
Wild ginger **380**
Wild raisin 357, **366**
Wild turkey **482**
Willow 138, 139, 262, **323**
Winterberry
    common 218, **219**
Wintergreen 212, **213**
Witch-hazel 12, 217
Woodchuck **529**
Woodcock
    American **481**
Wood duck **480**
Woodland jumping mouse **522**
Woodpecker
    downy **484**, 485
    hairy **484**, 485
    pileated **485**
    red-bellied **486**
Wood-pewee
    eastern **487**
Woodsia
    ilvensis 72, **99**
    rusty 99
Woodsiaceae 99
Wood thrush **495**
Wren
    house **494**
    winter **494**

X

Xanthomendoza
    fallax 432, 434, 441, **468**, 469
    hasseana 472
    ulophyllodes 434, 468, **469**
Xanthoparmelia
    cumberlandia 435, **470**
    viriduloumbrina 470
Xanthoria
    parietina 12, 434, 468, 469, **471**
    polycarpa 434, 468, **472**

Y

Yellowthroat
    common **501**
Yew **340**
    Canada **340**

Z

Zanthoxylum **368**
    americanum 113, 130, **368**
Zenaida
    macroura **483**
Zonotrichia
    albicollis **504**
    leucophrys **504**

The Authors

Dr Steven Newmaster
Herbarium Director
Biodiversity Institute of Ontario (BIO)
Associate Professor,
Department of Integrative Biology
University of Guelph

Chris Earley
Interpretive Biologist &
Education Coordinator,
The Arboretum
University of Guelph

Dr Aron Fazekas
Forest Diversity Research Associate
(BIO) Herbarium
University of Guelph

Dr Troy McMullin
Lichenologist
(BIO) Herbarium
University of Guelph

Carole Ann Lacroix M.Sc.
Botanist, Curator
(BIO) Herbarium
University of Guelph

Brian Lacey
Botanical Technician
(BIO) Herbarium
University of Guelph

Jose Maloles
Botanical Technician
(BIO) Herbarium
University of Guelph

Dr S. Ragupathy (Ragu)
Plant Barcoding Lead
(BIO) Herbarium
Center for Biodiversity Genomics
University of Guelph

Peter Williams M.Sc., R.P.F.
Forest Biologist
President, Williams & Associates,
Forestry Consulting Ltd.

Thomas Henry
Botanical Technician
(BIO) Herbarium
University of Guelph

Kevan Berg
Ethnoecologist
(BIO) Herbarium
University of Guelph

Notes